INVASIVE PLANTS *of*
CALIFORNIA'S WILDLANDS

GEOGRAPHIC SUBDIVISIONS OF CALIFORNIA

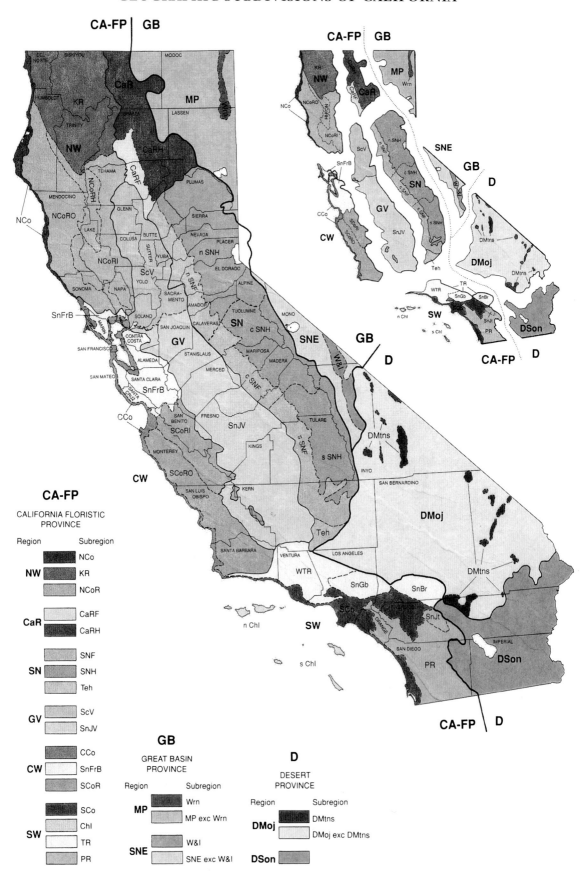

CA-FP

CALIFORNIA FLORISTIC
PROVINCE

Region	Subregion
NW	NCo
	KR
	NCoR
CaR	CaRF
	CaRH
SN	SNF
	SNH
	Teh
GV	ScV
	SnJV
CW	CCo
	SnFrB
	SCoR
SW	SCo
	Chl
	TR
	PR

GB

GREAT BASIN
PROVINCE

Region	Subregion
MP	Wrn
	MP exc Wrn
SNE	W&I
	SNE exc W&I

D

DESERT
PROVINCE

Region	Subregion
DMoj	DMtns
	DMoj exc DMtns
DSon	

INVASIVE PLANTS *of* CALIFORNIA'S WILDLANDS

EDITED BY

Carla C. Bossard,

John M. Randall, and Marc C. Hoshovsky

A Phyllis M. Faber Book

UNIVERSITY OF CALIFORNIA PRESS

Berkeley • Los Angeles • London

DEDICATION

We dedicate this book to

Oren Pollak

JUNE 1959–NOVEMBER 1998

our friend, colleague, and dedicated
steward of California's wildlands;

and to

Professor Marcel Rejmánek

who sparked and encouraged our interest in invasive plants,
continues to share his knowledge of plant ecology,
and endeavors to extend understanding of invasive plants.

◆ ◆ ◆

University of California Press
Berkeley and Los Angeles, California

University of California Press, Ltd.
London, England

Library of Congress Cataloging-in-Publication-Data

Invasive plants of California's wildlands / Carla C. Bossard, John M. Randall, and Marc C.
Hoshovsky, editors.
    p. cm.
  Includes bibliographical references.
  ISBN 0-520-22546-5 (cloth : alk. paper) — ISBN 0-520-22547-3 (paper : alk. paper)
    1. Invasive plants—Control—California. 2. Invasive plants—Ecology—California. I.
Bossard, Carla C., 1945- II. Randall, John M., 1960- III. Hoshovsky, Marc C., 1957-

SB612.C2 158 2000
639.9'9'09794—dc21

                                    00-024607

First Edition

Book produced by Phyllis M. Faber, Pickleweed Press,
for the University of California Press

Cover and book design and typesetting by Beth Hansen-Winter
Printed in Hong Kong through Global Interprint, Santa Rosa, CA

9 8 7 6 5 4 3 2 1

THE EDITORS WISH TO ACKNOWLEDGE

United States Fish and Wildlife Service
for financial support of this project.

California Exotic Pest Plant Council
*for assistance in gathering illustrations, providing financial support,
and sharing their slide collection and insights.*

Nelroy Jackson, Dave Cudney, and Joe DiTomaso
for valuable assistance in reviewing sections of the book pertaining to chemical control.

Carri Benefield, Victor Lee, Jenny Drewitz, Steven Bossard
for doing many tasks necessary for completing this book when no one else would do them.

Fred Hrusa, California Department of Food and Agriculture Herbarium
for providing several photographs.

University of California, Davis, Herbarium
for access to herbarium specimens.

Lesley Randall
for being our illustrator par excellence.

Beth Hansen-Winter
for diligence as our designer with an eye for clarity and beauty.

Phyllis Faber
for guiding and cheering us on.

Nora Harlow
for persevering through many rounds of proofreading.

Barry A. Meyers-Rice
for efforts above and beyond.

Earl Bossard
for scanning and computer consultation.

Our families
*for enduring many nights and weekends alone while we worked on the book
and encouraging us anyway.*

◆　◆　◆

The publisher wishes to acknowledge with gratitude
Elizabeth Durein
David B. Gold Foundation
Moore Family Foundation
for their generous contribution to this book.

CONTENTS

FOREWORD

The heart of this book is the species accounts, which provide detailed information about the biology and control of seventy-eight non-native plant species that are listed as Exotic Invasive Plants of Greatest Ecological Concern as of 1996 by the California Exotic Pest Plant Council (CalEPPC). We decided to cover only the species on this list because it is the best effort to date[1] to determine which of the non-native plants already growing wild in California cause or have the potential to cause serious damage in the state's parks, preserves, and other wildlands. We are convinced that non-native invasive plants pose one of worst threats, perhaps the worst of all, to the state's remaining populations and communities of native species. We hope the information on the pages that follow will be used to help promote the survival and growth of native plants and animals threatened by these invaders.

CalEPPC was established in 1992 in response to growing concern about invasive non-native plants in the state's wildlands. In 1994 CalEPPC canvassed its members and other land managers and researchers around the state for information about non-native plants that invade California's preserves, parks, and other wildlands. This information was used to develop a list of Exotic Invasive Plants of Greatest Ecological Concern in California. The species were grouped into several categories to indicate how severe and/or widespread they are. List A-1 includes the most invasive and damaging species that are widespread in the state. List A-2 includes highly damaging species that are invasive in fewer than five of the geographic subdivisions designated in *The Jepson Manual: Higher Plants of California*. List B includes less invasive species that move into and degrade wildlands. The Red Alert List includes species whose ranges in California currently are small but that are believed to have the potential to spread explosively and become major pests. Species for which there was insufficient information to determine their ability to invade and degrade natural areas were placed on a Need More Information list and only a few for whom strong evidence is mounting are included in this book. As the list was being compiled and categorized it was reviewed, re-reviewed, and finally approved by a group of respected researchers. In 1996 the CalEPPC list was updated based on new information and expanded to include a total of seventy-eight species.

We begin this book with a brief overview of the impacts of invasive plants and what we know about the characteristics of plant species most likely to invade and the habitats and communities most likely to be invaded. This is followed by a discussion of strategies and methods appropriate for the control of invasive plants in parks, preserves, and other wildlands. The remainder of the book consists of species accounts for seventy-eight invasive non-native species. Each account helps readers to identify the species and understand important aspects of its biology and lists specific control methods that are regarded as relatively effective, as well as some found to be ineffective.

---

[1] We acknowledge that several non-native invaders that have caused severe damage to wildlands in California are not on the 1996 edition of the list, as does CalEPPC. As we write this CalEPPC is preparing an updated version of the list, but it will not be ready in time for us to include newly listed species.

*Invasive plants invade native habitat areas and vastly alter the ecological landscape by outcompeting and exclud-
ing native plants and animals; altering nutrient cycles, hydrology, and wildfire frequencies; and hybridizing
with native species.* Bromus madritensis rubens (TOP) *has vastly altered a creosote bush scrub habitat, subject-
ing it to the possibility of frequent fires which, in time, eliminates creosote scrub. Photo by Matt Brooks. In the
great dune system at Humboldt Bay, the invasive, non-native* Lupinus arboreus (BOTTOM) *is being removed
from the dunes in order to restore the native dune flora. Photo by Andrea Pickart.*

CALIFORNIA'S WILDLAND INVASIVE PLANTS

John M. Randall and Marc C. Hoshovsky

The focus of this book is non-native plants that invade parks, preserves, and other wildlands in California, but our real concern is the survival and growth of the native plants and animals these invaders threaten. Unfortunately, some non-native invasive plant species inflict so much damage that, unless they are controlled, it will be impossible to preserve viable populations of many native species or many of the state's natural communities and ecosystems.

The good news is that many plant invasions can be halted or slowed, and, in certain situations, even badly infested areas can be restored to relatively healthy communities dominated by native species. Weed control and restoration are now widely regarded as necessary in many wildlands across the state and around the world. We hope this book will help land managers, volunteer stewards, and others to recognize some of California's most damaging wildland invaders, to better understand their impacts, and to minimize the damage they do to native biological diversity.

Invasive species are now widely recognized worldwide as posing threats to biological diversity second only to direct habitat loss and fragmentation (Pimm and Gilpin 1989, Scott and Wilcove 1998). In fact, when biological invasion by all types of organisms is considered as a single phenomenon, it is clear that to date it has had greater impacts on the world's biota than have more notorious aspects of global environmental change such as rising CO_2 concentrations, climate change, and decreasing stratospheric ozone levels (Vitousek *et al.* 1996). Compared to other threats to biological diversity, invasive non-native plants present a complex problem that is difficult to manage and has long-lasting effects. Even when exotics are no longer actively introduced, these plants continue to spread and invade new areas. Effective control will require awareness and active participation of the public as well as natural resource managers and specialists.

California's invasive plant problems are widespread and severe. The state's varied topography, geology, and climates have helped to give rise to the state's extraordinary native biological diversity and high levels of endemism. However, these varied conditions also provide suitable habitat for a wide variety of non-native plant species, many of which have readily established and rapidly spread in the state. Fewer than ten percent of the 1,045 non-native plant species that have established in California are recognized as serious threats (Randall *et al.* 1998), but these have dramatically changed California's ecological landscape. They alter ecosystem functions such as nutrient cycles, hydrology, and wildfire frequency, outcompete and exclude native plants and animals, harbor dangerous animal invaders, and hybridize with native species. Some spread into national parks, preserves, and other wildlands and reduce or eliminate the species and communities these sites were set aside to protect.

Rare species appear to be particularly vulnerable to the changes wrought by non-native invaders. For example, the California Natural Diversity Database indicates that 181 of the state's rare plant species are experiencing threats from invasive weeds (California Department of Fish and Game, Natural Heritage Division). Habitats for rare animals such as the San Clemente sage sparrow and the Palos Verde blue butterfly are also being invaded. Even more common species could be driven to rarity or near extinction by particularly disruptive invaders, as evidenced by the fate of the American chestnut (*Castanea dentata*) in the eastern hardwood forest following introduction of chestnut blight, *Cryphonectria parasitica* (National Academy of Science 1975).

IMPACTS OF INVASIVE PLANTS ON WILDLANDS

Non-native plant invasions can have a variety of effects on wildlands, including alteration of ecosystem processes; displacement of native species; support of non-native animals, fungi, or microbes; and alteration of gene pools through hybridization with native species.

Ecosystem Effects

The invasive species that cause the greatest damage are those that alter ecosystem processes such as nutrient cycling, intensity and frequency of fire, hydrological cycles, sediment deposition, and erosion (D'Antonio and Vitousek 1992, Vitousek 1986, Vitousek and Walker 1989, Vitousek *et al.* 1987, Whisenant 1990). These invaders change the rules of the game of survival and growth, placing many native species at a severe disadvantage (Vitousek *et al.* 1996). Cheat grass (*Bromus tectorum*) is a well studied example of an invader that has altered ecosystem processes. This annual grass has invaded millions of acres of rangeland in the Great Basin, leading to widespread increases in fire frequency from once every sixty to 110 years to once every three to five years (Billings 1990, Whisenant 1990). Native shrubs do not recover well from more frequent fires and have been eliminated or reduced to minor components in many of these areas (Mack 1981).

Some invaders alter soil chemistry, making it difficult for native species to survive and reproduce. For example, iceplant (*Mesembryanthemum crystallinum*) accumulates large quantities of salt, which is released after the plant dies. The increased salinity prevents native vegetation from reestablishing (Vivrette and Muller 1977, Kloot 1983). Scotch broom (*Cytisus scoparius*) and gorse (*Ulex europaea*) can increase the content of nitrogen in soil. Although this increases soil fertility and overall plant growth, it gives a competitive advantage to non-native species that thrive in nitrogen-rich soil. Researchers have found that the nitrogen-fixing firetree (*Myrica faya*) increases soil fertility and consequently alters succession in Hawaii (Vitousek and Walker 1989).

Wetland and riparian invaders can alter hydrology and sedimentation rates. Tamarisks (*Tamarix chinensis, T. ramosissima, T. pentandra, T. parviflora*) invade wetland and riparian areas in southern and central California and throughout the Southwest, and are believed to be responsible for lowering water tables at some sites. This may reduce or eliminate surface water habitats that native plants and animals need to survive (Brotherson and Field 1987, Neill 1983). For example, tamarisk invaded Eagle Borax Spring in Death Valley in the 1930s or 1940s. By the late 1960s the large marsh had dried up, with no visible surface water. When managers removed tamarisk from the site, surface water reappeared, and the spring and its associated plants and animals recovered (Neill 1983). Tamarisk infestations also can trap more sediment than stands of native vegetation and thus alter the shape, carrying capacity, and flooding cycle of rivers, streams, and washes (Blackburn *et al.* 1982). Interestingly, the only species of *Tamarix* established in California that is not generally regarded as invasive (athel, or *T. aphylla*) is regarded as a major riparian invader in arid central Australia.

Other wetland and riparian invaders and a variety of beach and dune invaders dramatically alter rates of sedimentation and erosion. One example is saltmarsh cordgrass (*Spartina alterniflora*), native to the Atlantic and Gulf coasts and introduced to the Pacific Coast, where it invades intertidal habitats. Sedimentation rates may increase dramatically in infested areas, while nearby mudflats deprived of sediment erode and become areas of open water (Sayce 1988). The net result is a sharp reduction in open intertidal areas where many migrant and resident waterfowl feed.

Coastal dunes along the Pacific Coast from central California to British Columbia have been invaded and altered by European beachgrass (*Ammophila arenaria*). Dunes in infested areas are generally steeper and oriented roughly parallel to the coast rather than nearly perpendicular to it as they are in areas dominated by *Leymus mollis*, *L. pacificus*, and other natives (Barbour and Johnson 1988). European beachgrass eliminates habitats for rare native species such as Antioch Dunes evening-primrose (*Oenothera deltoides* ssp. *howellii*) and Menzies' wallflower (*Erysimum menziesii* ssp. *menziesii*). Species richness on foredunes dominated by European beachgrass may be half that on adjacent dunes dominated by *Leymus* species (Barbour *et al.* 1976). Changes in the shape and orientation of the dunes also alter the hydrology and microclimate of the swales and other habitats behind the dunes, affecting species in these areas.

Some upland invaders also alter erosion rates. For example, runoff and sediment yield under simulated rainfall were fifty-six percent and 192 percent higher on plots in western Montana dominated by spotted knapweed (*Centaurea maculosa*) than on plots dominated by native bunchgrasses (Lacey *et al.* 1989). This species is already established in northern California and the southern Peninsular Range and recently was found on an inholding within Yosemite National Park (Hrusa pers. comm.).

Some invasive plants completely alter the structure of the vegetation they invade. For example, the punk tree (*Melaleuca quinquenervia*) invades marshes in southern Florida's Everglades that are dominated by sedges, grasses, and other herbaceous species, rapidly converting them to swamp forest with little or no herbaceous understory (LaRoche 1994, Schmitz *et al.* 1997). Such wholesale changes in community structure may be expected to be followed by changes in ecosystem function.

Habitat Dominance and Displacement of Native Species

Invaders that move into and dominate habitats without obviously altering ecosystem properties can nevertheless cause grave damage. They may outcompete native species, suppress native species recruitment, alter community structure, degrade or eliminate habitat for native animals, and provide food and cover for undesirable non-native animals. For example, edible fig (*Ficus carica*) is invading riparian forests in the Central Valley and surrounding foothills and can become a canopy dominant. Invasive vines are troublesome in forested areas across the continent. In California, cape ivy (*Delairea odorata*) blankets riparian forests along the coast from San Diego north to the Oregon border (Elliott 1994).

Non-native sub-canopy trees and shrubs invade forest understories, particularly in the Sierra Nevada and Coast Ranges. Scotch broom (*Cytisus scoparius*), French broom (*Genista monspessulana*), and gorse (*Ulex europaea*) are especially troublesome invaders of forests and adjacent openings and of coastal grasslands (Bossard 1991a, Mountjoy 1979). Herbaceous species can colonize and dominate grasslands or the ground layer in forests. Eupatory (*Ageratina adenophora*) invades and dominates riparian forest understories along California's southern and central coast. Impacts of these ground-layer invaders have not been well studied, but it is suspected that they displace native herbs and perhaps suppress recruitment of trees.

Annual grasses and forbs native to the Mediterranean region have replaced most of California's native grasslands. Invasion by these species was so rapid and complete that we do not know what the dominant native species were on vast areas of bunchgrasses in the Central Valley and other valleys and foothills around the state. The invasion continues today as medusahead (*Taeniatherum caput-medusae*) and yellow starthistle (*Centaurea solstitialis*) spread to sites already dominated by other non-natives. Yellow starthistle is an annual that produces large numbers of seeds and grows rapidly as a seedling. It is favored by soil disturbance, but invades areas that show no sign of being disturbed by humans or livestock for years and has colonized several relatively pristine preserves in California, Oregon, and Idaho (Randall 1996b).

In some situations invasive, non-native weeds can prevent reestablishment of native species following natural or human-caused disturbance, altering natural succession. Ryegrass (*Lolium multiflorum*), which is used to reseed burned areas in southern California, interferes with herb establishment (Keeley *et al.* 1981) and, at least in the short term, with chaparral recovery (Schultz *et al.* 1955, Gautier 1982, Zedler *et al.* 1983).

Hybridization with Native Species

Some non-native plants hybridize with natives and could, in time, effectively eliminate native genotypes. The non-native *Spartina alterniflora* hybridizes with the native *S. foliosa* where they occur together. In some *Spartina* populations in salt marshes around south San Francisco Bay, all individual plants tested had non-native genes (Ayres *et al.* in press).

Promotion of Non-Native Animals

Many non-native plants facilitate invasions by non-native animals and vice versa. *Myrica faya* invasions of volcanic soils in Hawaii promote populations of non-native earthworms, which increase rates of nitrogen burial and accentuate the impacts these nitrogen-fixing trees have on soil nutrient cycles (Aplet 1990). *M. faya* is aided by the non-native bird, Japanese white-eye (*Zosterops japonica*), perhaps the most active of the many native and non-native species that consume its fruits and disperse its seeds to intact forest (Vitousek and Walker 1989).

EARLY INVASIONS BY NON-NATIVE PLANTS

The first recorded visit by European explorers to the territory now called California occurred in 1524, but people of Old World ancestry did not begin to settle here until 1769. Available evidence indicates that the vast majority of non-native plants now established in California were introduced after this time. There is compelling evidence that red-stem filaree (*Erodium cicutarium*), and perhaps a few other species, may have established even earlier, perhaps after being carried to the territory by roaming animals or by way of trading networks that connected Indian communities to Spanish settlements in Mexico (Hendry 1931, Hendry and Kelley 1925, Mensing and Byrne 1998). Once settlers began to arrive, they brought non-native plants accidentally in ship ballast and as contaminants of grain shipments and intentionally for food, fiber, medicine, and ornamental uses (Frenkel 1970, Gerlach 1998).

The number of non-native species established in California rose from sixteen during the period of Spanish colonization (1769-1824) to seventy-nine during the period of Mexican occupation (1825-1848) to 134 by 1860 following American pioneer settlement (Frenkel 1970). Jepson's *A Manual of the Flowering Plants of California* (1925), the first comprehensive flora covering the entire state, recognized 292 established non-native species. Rejmánek and Randall (1994) accounted for taxonomic inconsistencies between the 1993 *Jepson Manual* and earlier floras and found that Munz and Keck's 1959 *A Flora of California* included 725 non-native plants species and their 1968 *A California Flora and Supplement* included 975. The 1993 *Jepson Manual* recorded 1,023 non-natives, and subseqent reports in the literature have brought the number up to 1,045 (Randall *et al.* 1998). Rejmánek and Randall (1994) remarked that, although non-native species continue to establish in California, the rate of increase in their number appears to be slowing after roughly 150 years of rapid growth.

Most non-native plants introduced to California in earlier times first established at coastal sites near ports and around missions and other settlements. In recent times, first reports of new non-native species have come from every major geographic subdivision of the state (Rejmánek and Randall 1994). Apparently, the great speed and reach of modern transportation systems and the increasing global trade in plants and other commodities have enabled non-natives to spread to sites throughout the state. A variety of human activities continue to introduce new species to California and to spread those that have established populations in only a few areas. For example, land managers still introduce non-native species to control erosion or provide forage for livestock. New ornamental plants and seeds are imported and sold. Movement of bulk commodities such as gravel, roadfill, feed grain, straw, and mulch transport invasive plant propagules from infested to uninfested areas (OTA 1993). The rate of spread is often alarming. For example, within California, yellow starthistle has expanded its range at an exponential rate since the late 1950s, increasing from 1.2 to 7.9 million acres by 1991 (Maddox *et al.* 1996, Thomsen *et al.* 1993).

Problems caused by invasive plants in California were recognized by Frederick Law Olmsted in 1865 in a report he filed on the newly set-aside Yosemite Valley, noting that, unless actions were taken, its vegetation likely would be diminished by common weeds from Europe. The report pointed out that this had already happened "in large districts of the Atlantic States." Botanists and other students of natural history noted the establishment of non-native species in the state in published papers, and by the 1930s natural area managers in Yosemite and scattered parks and preserves around the state began controlling invading non-native species that were recognized as agricultural pests (Randall 1991). The issue was brought into mainstream ecology in the late 1950s with the publication of Charles Elton's book, *The Ecology of Invasions by Animals and Plants* (1958). Concern and interest among both land managers and researchers have grown since that time, particularly since the mid-1980s.

SPECIES MOST LIKELY TO BE INVASIVE

Many people have wondered if certain traits distinguish species that become invasive. Despite a great deal of study, no single answer presents itself, and researchers have been surprised by the success of some species and the failure of others. Studies conducted in 1980 in central California on Peruvian pepper (*Schinus molle*) and its close relative Brazilian pepper (*Schinus terebinthifolius*) failed to determine why the former was spreading in California (Nilsen and Muller 1980a, 1980b). Instead the studies suggested Brazilian pepper was the more invasive species. Recently, Brazilian pepper has been found to be invasive in southern California, so perhaps studies of this type do have some predictive power.

Despite these puzzling cases, recent work has pointed to several factors that may help to predict which species are likely to be invasive. In two studies the best predictor was whether a species was invasive elsewhere (Panetta 1993, Reichard and Hamilton 1997). For example, if a species native to Spain is invasive in Western Australia, it is likely to be invasive in California and South Africa as well. Rejmánek and Richardson (1996) analyzed characteristics of twenty species of pines and found that the invasive species were those that produce many small seeds and that begin reproducing within their first few years. When they extended the analysis to a group of flowering trees, these same characters usually discriminated between invasive and non-invasive species. This study and several others also found plants with animal-dispersed seeds, such as bush honeysuckles or ligustrums, are much more likely to be invasive in forested communities (Reichard 1997, Reichard and Hamilton 1997). It has also been suggested that species capable of reproducing both by seed and by vegetative growth have a better chance of spreading in a new land (Reichard 1997).

Self-compatible species, with individuals that can fertilize themselves, have been thought more likely to invade, since a single plant of this type could initiate an invasion (Baker 1965). However, many self-incompatible species are successful invaders, including some with male and female flowers on separate plants. It is also thought that plants dependent on one or a few other species for pollination, fruit dispersal, or the uptake of nutrients from the soil are less likely to invade new areas unless these organisms are introduced at the same time. As a group, figs may be relatively poor invaders because, with few exceptions, each species is pollinated by a distinctive species of wasp that is in turn dependent on that species of fig. However, the edible fig's pollinator was introduced to promote fruit production, and now the species is invasive in parts of California. Other plant invasions may be promoted by introduced animals as well. For example, honeybees boost seed production of invaders whose flowers they favor (Barthell pers. comm.). In Hawaii feral pigs promote the spread of banana poka (*Passiflora mollissima*) and other species by feeding voraciously on their fruits and distributing them in their scat, often in soil they have disturbed while rooting for food.

It has also been suggested that species with relatively low DNA contents in their cell nuclei are more likely to be invasive in disturbed habitats (Rejmánek 1996). Under certain conditions, cells with low DNA contents can divide and multiply more quickly, and consequently these plants grow more rapidly than species with higher cellular DNA content. Plants that germinate and grow rapidly can quickly occupy such areas and exclude other plants following disturbance.

It is generally agreed that a species is most likely to invade an area with a climate similar to that of its native range, but some non-native species now thrive in novel conditions. An analysis of the distribution of non-native herbs of the sunflower and grass families in North America indicated that species with a larger native range in Europe and Asia are more likely to become established and to have a larger range here than species with small native ranges (Rejmánek 1995). It is thought that species with large native ranges are adapted to a variety of climate and soil conditions and are more likely to find suitable habitat in a new area. This ability to cope with different conditions can be attributed in part to genetic plasticity (genetic differences among individuals of a species) or to phenotypic plasticity (the ability of any given individual of some species to cope with a variety of conditions). Another factor that may help to determine whether a plant will invade a site is whether it is closely related to a native species (e.g., in the same genus). Plants without close relatives appear more likely to become established (Rejmánek 1996).

A species may be more likely to become established if many individuals are introduced at once or if they are introduced repeatedly. Introductions of many individuals may help to ensure that they will mate and produce offspring and that there will be sufficient genetic variability in the population for the species to cope with a wider variety of conditions. In addition, if sites where the species can successfully germinate and grow are limited in number, the chance that at least one seed scattered at random will land on an appropriate site increases with the number of seeds dispersed. Chance may be important in other ways. For example, species that happen to be introduced at the beginning of a drought may be doomed to fail, although they might easily establish following a return to normal rainfall. An early introduction may by chance include no individuals with the genetic makeup to thrive in an area, while a later introduction may include several.

There is often a time lag of many decades between the first introduction of a plant and its rapid spread. In fact, some species that rarely spread today may turn out to be troublesome forty, fifty, or more years from now. This makes it all the more urgent that we find some way of determining which species are most likely to become invasive so that we can control them while their populations are still small.

HABITATS AND COMMUNITIES MOST LIKELY TO BE INVADED

Another question that has long intrigued ecologists is why some areas appear more prone to invasion than others. Again, many hypotheses have been advanced, but we have few solid answers. There is even some question about which areas have suffered the highest numbers of invasions, since this may differ depending on the type of organism considered and which species are regarded as firmly established. A given area may be highly susceptible to invasion by one type of organism and highly resistant to another, while the situation might be reversed in other areas.

It is generally agreed that areas where the vegetation and soil have been disturbed by humans or domestic animals are more susceptible to invasion. In North America disturbed sites are commonly invaded by species native to the Mediterranean region and the fertile crescent of the Old World where the plants had millennia to adapt to agricultural disturbances. Changes in stream flows, the frequency of wildfires, or other environmental factors caused by dam building, firefighting, and other human activities may also hinder survival of native plants and promote invasion by non-natives. Nonetheless, reserves and protected areas are not safe from exotic species. In a 1996 poll, sixty-one percent of National Park Service supervisors throughout the United States reported that non-native plant invasions are moderate to major problems within their parks. In more than half (fifty-nine percent) of The Nature Conservancy's 1,500 preserves exotic plants are considered one of the most important management problems (TNC 1996a, 1997).

It is also safe to say that remote islands in temperate and tropical areas appear to be highly susceptible to invasions by non-native plants and animals. For example, nearly half (forty-nine percent) of the flowering plant species found in the wild in Hawaii are non-native as are twenty-five percent of plants on California's Santa Cruz Island (Junak *et al.* 1995). Most remote islands had no large native herbivores, so pigs, cattle, sheep, and other grazers introduced by humans found the native plants unprotected by spines or foul-tasting chemicals. Introduced grazers often denuded large areas of native vegetation, leaving them open to colonization by introduced species adapted to grazing. There is also speculation that islands, peninsulas such as southern Florida, and other areas with low numbers of native species or without any representative or distinctive groups are more prone to invasion. For example, there are no rapidly growing woody vines native to the Hawaiian Islands, where several introduced vines have become pests. Some researchers theorize that where such gaps exist, certain resources are used inefficiently if at all. Such "open niches" are vulnerable to invasion by non-native species capable of exploiting these resources. Other researchers reject this concept, maintaining that "open niches" are impossible to identify in advance and that when new species move in they do not slip into unoccupied slots but instead use resources that would have been used by organisms already present.

History likely also plays a large role in determining the susceptibility of a site to invasion. Busy seaports, railroad terminals, and military supply depots are exposed to multiple introductions. People from some cultures are more likely to introduce plants from their homelands when they migrate to new regions. In fact, colonization of much of the Americas, Australia, and other areas of the world by western Europeans and the plants and animals from their homelands may go hand in hand, the success of one species promoting the success of others. European colonists were followed, sometimes preceded, by animals and plants with which they were familiar and that they knew how to exploit. The plants and animals benefited in turn when these people cleared native vegetation and plowed the soil.

DEFINITIONS OF TERMS USED IN THIS BOOK

Native plants are those growing within their natural range and dispersal potential. They are species or subspecies that are within the range they could occupy without direct or indirect introduction and/or care by humans. Most species can be easily classed as either native or non-native using this definition, but there are some gray areas. Natural ranges should not be confused with political or administrative boundaries. Bush lupine (*Lupinus arboreus*), for example, may be thought of as a California native, but its native range is only along the central and southern coasts of the state. It is not native along the north coast, where it was intentionally planted outside its natural range (Miller 1988, Pickart this volume). All hybrids between introduced or domesticated species and native species are also non-native.

Non-native plants are those species growing beyond their natural range or natural zone of potential dispersal, including all domesticated and feral species and all hybrids involving at least one non-native parent species. Other terms that are often used as synonyms for non-native include alien, exotic, introduced, adventive, non-indigenous, non-aboriginal, and naturalized. With rare exceptions, conservation programs are dedicated to the preservation of native species and communities. The addition of non-native species rarely contributes positively to this unless these plants alter the environment in ways that favor native species as do some grazers and biological control agents.

Natural areas are lands and waters set aside specifically to protect and preserve undomesticated organisms, biological communities, and/or ecosystems. Examples include most national parks, state and federally designated wilderness areas, and preserves held by private organizations such as The Nature Conservancy and the National Audubon Society.

Wildlands include natural areas and other lands managed at least in part to promote game and/or non-game animals or populations of native plants and other organisms. Examples include federal wildlife refuges, some national and state forests, portions of Bureau of Land Management holdings, including some areas used for grazing, and some lands held by private landowners.

Pest plant and **weed** are used interchangeably in this book to refer to species, populations, and individual plants that are unwanted because they interfere with management goals and objectives. Plants regarded as pests in some wildlands may not be troublesome elsewhere. For example, the empress tree (*Paulownia tomentosa*) is a pest in deciduous forests of the eastern United States, particularly in the southern Appalachians, but it is not known to escape from cultivation in California, where it is used as an ornamental landscape tree. Some species that are troublesome in agricultural or urban areas rarely, if ever, become wildland weeds. The term "environmental weeds" is used by many Australians (Groves 1991, Humphries *et al.* 1991b) to refer to wildland weeds, but few North American land managers or researchers use this term.

Invasive species are those that spread into areas where they are not native, according to Rejmánek (1995), while other authors define as invasives only species that displace natives or bring about changes in species composition, community structure, or ecosystem function (Cronk and Fuller 1995, White *et al.* 1993). Most wildland weeds are both invasive and non-native, but not all non-native plants are invasive. In fact, only a small minority of the thousands of species introduced to California have escaped cultivation, and a minority of those that have escaped spread into wildlands.

MANAGEMENT OF INVASIVE PLANT SPECIES

Marc C. Hoshovsky and John M. Randall

Before embarking on a weed management program, it is important to develop a straightforward rationale for the actions you plan to take. We believe this is best accomplished using an adaptive management approach as follows: (1) establish management goals and objectives for the site; (2) determine which plant species or populations, if any, block or have potential to block attainment of management goals and objectives; (3) determine which methods are available to control the weed(s); (4) develop and implement a management plan designed to move conditions toward management goals and objectives; (5) monitor and assess the impacts of management actions in terms of effectiveness in moving toward goals and objectives; and (6) reevaluate, modify, and start the cycle again (Figure 1). Note that control activities are not begun until the first three steps have been taken.

It is vital to establish management goals before embarking on any management activities. What is it you want to protect or manage? Is your objective to protect or enhance a certain species or community, preserve a vignette of pre-Columbian America, preserve certain ecosystem attributes, or preserve a functioning ecosystem? A weed control program is best viewed as part of an overall restoration program, so focus on what you want in place of the weed, rather than simply eliminating the weed. Keep in mind that the ultimate purpose of a weed control program is to further the goal of preserving a species, community, or functioning ecosystem.

In many cases it will be easy to identify species that degrade the site or threaten to do so. If impacts of a species are not clear, you may need to monitor its abundance and effects on the natural community. Set priorities to minimize your total, long-term workload. This often means assigning highest priority to preventing new invasions and to quickly detecting and eliminating any new invasions that occur. High priority should also be assigned to the species with the most damaging impacts, to infestations that are expanding rapidly, and to infestations that affect highly valued areas of the site. Also consider the difficulty of control. It is of little use to spend time and resources to attack an infestation you have little hope of controlling.

Consider all control options available: manual, mechanical, encouraging competition from native plants, grazing, biocontrol, herbicides, prescribed fire, solarization, flooding, and other, more novel techniques. Each of these methods has advantages and disadvantages, and often the best approach is to use a combination of methods. Frequently, one or more methods will not be appropriate for a given situation because they do not work well, their use is objectionable to people in the area, or they are too costly. Herbicides may kill important non-target plants. Mechanical methods often disturb soil and destroy vegetation, providing ideal conditions for establishment of weedy species. It will often be best to employ two or more methods. For example,

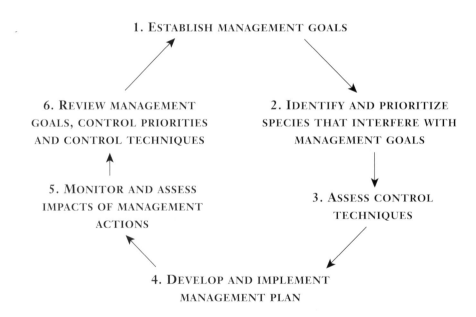

Figure 1. Flow Chart for Adaptive Management of Weeds describing management actions and decisions confronting wildland managers (from Randall 1997, based on a diagram by Oren Pollak).

cutting and herbicides or prescribed fire and herbicides have been used successfully in combination in many weed control programs.

Biological control can be an extremely selective control tool, but there is some risk that control agents may attack desirable species. The best known example of a biocontrol agent attacking desirable species is that of *Rhinocyllus conicus*, a beetle first released to control non-native thistles in North America in the 1960s that was recently found attacking native thistles and reducing their populations at some sites (Louda *et al.* 1997).

Some native animals use invasive non-native species for food and cover and may have difficulty finding replacements if infestations are removed and not replaced with non-invasive native or introduced species. For example, huge numbers of monarch butterflies (*Danaus plexippus*) roost in some groves of *Eucalyptus globulus* in coastal California. In addition, elimination of plants in a natural area can be alarming to some people, particularly when herbicides are used, so it is important to explain the threats posed by the pest and the reasons why you chose the methods you did.

There is much room for improvement in control methods for many of the species described in this book. Readers may want to experiment with methods that may more effectively and efficiently control these invaders and promote native species.

The Bradley Method is a sensible approach to weed management (Bradley 1988, Fuller and Barbe 1985). In this approach, weed control is begun in portions of the site with the best stands of desirable native vegetation (those with few weeds) and proceeds slowly to areas with progressively worse weed infestations. This is similar to Moody and Mack's (1988) advice to attack outlying satellite weed populations first rather than larger, denser source populations. They based this advice on modeling work that indicated that the rate of spread of small satellite poplations is generally significantly higher than that of older, larger populations and that containing or eliminating the outliers saves time and effort in the long run. The Bradley Method dictates that the area under control should expand at a rate that allows previously treated areas to be monitored and kept in satisfactory condition. It also advocates the use of techniques that minimize damage to native plants and disturbance to the soil so that the natives can thrive and defend against reinvasion. This approach is particularly promising for small preserves or sites with access to large pools of volunteer labor. More detailed information on the Bradley Method is contained in Fuller and Barbe (1985).

PREVENTION

The most effective and efficient weed control strategies are preventing invasions by new plants species and quickly detecting invasions that occur so weeds can be eradicated or contained before they spread. The California Department of Food and Agriculture (CDFA) has long recognized this, and the state's Noxious Weed List gives highest priority to species that either are not yet established in the state or whose populations are not yet widespread. The state's native species will be better protected if new invaders are detected quickly and word of their discovery is communicated to those who can take action to prevent their spread, such as the staff at the CDFA Control and Eradication Branch or Plant Pest Diagnostics Branch.

There are already at least 1,045 non-native plant species established in California (Randall *et al.*1998), and more continue to arrive and become established. If allowed to spread, some of these new species could impact native species and communities as severely as yellow starthistle and tamarisk do now. Preventing or stopping just one new invasive weed would be of greater conservation benefit in the long run than far more costly and difficult efforts to control an already widespread pest.

Taking precautions in normal resource management activities can halt or slow the establishment and spread of weeds in a given area. Wise precautions include: removing seed sources from roads, trails, rights-of-way, watercourses, and other dispersal routes; closing unnecessary roads and trails where possible; planning work projects to minimize soil disturbance and reestablish vegetation as quickly as possible where disturbance does occur; limiting the use of construction materials such as gravel, fill, mulch, straw, and seed mixes that may carry weeds or buying from suppliers who guarantee their products are weed-free; washing vehicles and equipment to remove weed seeds and other propagules before they are used in another area; follow-up monitoring of work sites to detect new weed populations while they are still small and easily controlled; and public education and outreach regarding the importance of weed detection and prevention of invasion.

ERADICATION

Eradication is the complete elimination of a species from a given area. The great appeal of eradicating a weed is, of course, that once the project achieves success no more work is required and the species cannot spread unless it is re-introduced. Unfortunately, it is rarely possible to eradicate an established weed from a large area. In fact, the history of CDFA's eradication projects indicates that there is little likelihood of eradicating a species from California once it has spread to a few tens of acres in the state.

It may be possible to eradicate a weed from a given area, such as a preserve or national park if it has not yet become widespread there, but it is likely to re-invade from adjacent lands unless there is some barrier that will prevent it from doing so. Eradication is most likely when the species has just begun to establish in a new area, which underscores the importance of efforts to detect new invaders at national, state, and local levels.

PHYSICAL CONTROL

Physical methods of weed control generally are labor intensive and often are used for small populations or where other control methods are inappropriate, such as near sensitive water supplies. Nonetheless, physical methods have been used successfully by volunteer groups and paid workers to control weed infestations on several large sites in California (e.g., Pickart and Sawyer 1998). Physical methods can be highly selective, targeting only the pest species, but they can also disturb the soil or damage nearby vegetation, thereby promoting germination and establishment of weedy species. Physical control methods may also produce large amounts of debris, disposal of which is sometimes difficult.

Physical control methods range from manual hand pulling of weeds to the use of hand and power tools to uproot, girdle, or cut plants. Two companies produce tools specifically for pulling shrubs such as scotch broom, tamarisk, and Russian olive. The Weed Wrench (see Resources section) and the Root Jack (see Resources section) are lever arms with a pincher or clamp at the bottom that grips the plant stem. Once the stem is secured, the user leans back, tightening the clamp in the process. After a little rocking, the entire plant comes up, roots included (Hanson 1996). Other tools for weed control, including girdling knives, axes, machetes, loppers, clippers, chainsaws, and brush cutters, are available from hardware stores and gardening and forestry supply companies. Various attachments are available for bulldozers and tractors to clear and uproot woody plants. Brush rakes or blades may be mounted on the front of the bulldozer, and brushland disks or root plows may be pulled behind. Mowing can prevent seed formation on tall annual and perennial weeds and deplete food reserves of shoots and roots. Unfortunately, repeated mowing can favor low-growing weeds or damage desirable native species (Ashton and Monaco 1991).

Prescribed Fire

Fire can be an effective means of reducing weed infestations, particularly for shrubby weeds and in native communities that evolved with fire. Fire may sometimes be the only element necessary to give native species a chance to recover. Fire may also be used to eliminate old vegetation and litter in areas infested with perennial herbs such as fennel (*Foeniculum vulgare*) or leafy spurge (*Euphorbia esula*) prior to treating the area with herbicide. This allows more herbicide to reach the living leaves and stems of target plants, potentially enhancing its effectiveness. Fire can also be used to induce seeds of some species to germinate so the seedbank can be flushed and the resulting seedlings can then be killed with another fire or some other method (e.g., Bossard 1993).

Conducting a prescribed burn is not a simple or risk-free operation. Managers considering prescribed burning should be trained and certified and should work closely with the local office of the California Department of Forestry and Fire Protection to ensure safe, effective, and legal burns. Good logistical planning, coordination of work teams, careful timing with respect to weather (winds, moisture conditions), coordination with air quality agencies, and attention to other details are required to carry out an effective and safe burn. In most parts of California it is necessary to address air quality concerns and to obtain permission from the regional air quality board. Escaped fires are costly and can be disastrous.

Prescribed fires may promote certain invasive, non-native species, and so should be used with caution. Non-native annual and biennial species, such as cheat grass (*Bromus tectorum*) and bull thistle (*Cirsium vulgare*), are most likely to be favored in the years immediately following a burn and in repeatedly burned areas. Hot fires can also sterilize the soil, volatilizing important nutrients and killing microorganisms on which native plants rely. Removal of vegetation by fire can also increase soil erosion and stream sedimentation. Construction of firebreaks and associated soil disturbance can increase erosion and provide a seedbed for invasive weeds.

Blowtorches and flamethrowers can also be used to burn individual plants or small areas. This method has been used with some success on thistles in several areas. Flamethrowers have also been used to heat-girdle the lower stems of shrubs such as scotch broom (*Cytisus scoparius*). This technique has the advantages of being less costly than basal and stem herbicide treatments and suitable for use during wet weather. On the other hand, it is time-consuming and not viable in areas where wildfire is a danger.

Flooding and Draining

Prolonged flooding can kill plants that infest impoundments, irrigated pastures, or other areas where water levels can be controlled. This method may be even more effective if plants are mowed or burned before flooding. Spotted knapweed (*Centaurea maculosa*) is sensitive to flooding, and its populations can be reduced by flood irrigation in pastures. Flooding may also help to control non-natives by promoting the growth and competitive ability of certain native species in some situations. Unfortunately, flooding will not kill the seeds of many target species.

Draining water from ponds and irrigation canals may control aquatic weeds such as reed canary grass (*Phalaris arundinacea*) (Schlesselman *et al.* 1989). Drainage can be conducted in different ways, including seasonal, intermittent (within-season), or partial draw downs (McNabb and Anderson 1989).

Mulching

Mulching excludes light from weeds and prevents them from photosynthesizing. Commonly used mulches are hay, manure, grass clippings, straw, sawdust, wood chips, rice hulls, black paper, and black plastic film. The most effective mulches are black paper or plastic because of their uniform coverage. Particle mulches cannot prevent all weeds from breaking through (Schlesselman *et al.* 1989). Mulch materials and application can be expensive and may be suitable only for small infestations. Particle mulches should be weed-free to avoid introduction of other weeds.

Soil Solarization

Soil solarization is a technique for killing weed seeds that have not yet germinated. A clear polyethylene plastic sheet is placed over moist soil and kept in place for a month or more. The incoming solar radiation creates a greenhouse effect under the plastic, increasing soil temperatures. High temperatures kill some seeds outright and weaken others, making them more susceptible to attack by pathogens (Schlesselman *et al.* 1989).

BIOLOGICAL CONTROL

Biological control, or biocontrol, involves the use of animals, fungi, or other microbes that prey upon, consume, or parasitize a target species. Target species are frequently non-natives whose success in new environments may be due in part to the absence of their natural predators and pathogens.

Classical biological control involves careful selection and introduction of one or more natural enemies to the target species' new habitat to reduce target populations. Successful control programs of this kind result in permanent establishment of the control agent or agents and permanent reduction in target species populations. Such programs are not designed to eliminate the target species completely, and it may take repeated releases to ensure the establishment of an agent. It may take years or decades before their effects are obvious. Some of the greatest strengths of classical biological control are that once an agent is established it will last indefinitely and it may spread on its own to cover most or all of the area infested by the weed, generally without additional costs. On the other hand, these strengths can become liabilities if the agent begins to attack desirable species as well as the pest it was introduced to control. Biocontrol researchers take great pains to locate and use agents that are highly specific to the targeted weed. This contributes to the high cost and long time required for development and approval of new biological control agents. Several of the species covered in this book are the subjects of ongoing classical biological control programs.

As opposed to classical biocontrol, inundative or augmentative biocontrol involves mass releases of pathogens whose effects on the target are normally limited by their inability to reproduce and spread. Inundative biocontrol agents that are non-native and/or not target-specific may be sterilized or otherwise rendered incapable of establishing permanent populations before they are released. Because they do not become established, they must be reared and released again each time weed populations erupt. There have, however, been instances in which mistakes or back mutations allowed some of these species to establish permanent wild populations.

The USDA must approve biocontrol agents for use. Approved biological control agents have been studied, and their host specificity determined. Accidentally introduced species have unknown host species, are not permitted for distribution, and should not be redistributed. If you have questions about any potential biocontrol agents, contact the CDFA Biological Control Program (see Resources section).

Competition and Restoration

The use of native plants to outcompete alien weeds is a frequently overlooked but potentially powerful technique. Sometimes the natives must be planted into the habitat and given some care until they are well established. This may be appropriate where a native forest community is to be reestablished in an old field currently occupied by a thick cover of alien grasses and forbs. Reseeding with native species also works well in some grasslands. In other cases all that may be required is time; the native community may reestablish itself once human-caused disturbance ceases. Even in these cases, it may be important to locate and remove certain weeds capable of hindering succession. You can also enhance other weed control methods by encouraging competition from native species.

Ideally, seeds or cuttings used in restoration should be collected on the site or from adjacent properties. Unfortunately, in many cases the only available or affordable seeds and plants are from distant or unidentified populations. Potential impacts of using seeds and plants collected at distant sites include project failure if genotypes used are unable to survive conditions on the site, introduction of diseases, and loss of genetic diversity through overwhelming or contaminating locally adapted genotypes.

Grazing

Grazing animals may be used to selectively control or suppress weeds, but grazing is also known to promote certain invaders in some circumstances. Cattle, sheep, goats, geese, chickens, and grass carp have been used to graze undesirable species at sites around the nation. Often grazing must be continued until the weed's seedbank is gone, as the suppressed plants may otherwise quickly regain dominance. Another drawback to using grazing animals is that they sometimes spread weed seeds in their droppings.

CHEMICAL CONTROL

Herbicides are chemicals that kill or inhibit plant growth. They can be extremely effective tools when used to eliminate certain species. They can also be dangerous and should be used only after careful consideration of other options and only with extreme care. Each species treatment in this book provides specific information on the herbicides, rates, and times that have been found most effective against that species. However, the effectiveness of a given treatment may vary with climate and environmental conditions, and some populations of a given species may be more tolerant of, or even resistant to, a particular herbicide than other populations of the same species. It may be necessary to conduct trials to identify the most effective techniques for controlling a particular problem species.

The most important safety rule for herbicide use is to read the label and follow the directions. Applicators must wear all protective gear required on the label of the herbicide they are using. It is also important to adopt or develop protocols for storing, mixing, transporting, cleaning up, and disposing of herbicides and for dealing with medical emergencies and spills.

California's programs to regulate pesticides and pesticide applicators are regarded as the most stringent in the nation and as such are the standard against which many other states measure their programs. California's Department of Pesticide Regulation reviews health effects of pesticides independently of the federal Environmental Protection Agency and has more stringent registration requirements. California also has the most stringent pesticide use reporting requirements. Agricultural pesticide use is broadly defined and includes applications made in nature preserves, parks, golf courses, and cemeteries and along roadsides. Such applications are regulated by the CDFA, and county agricultural commissioners' offices enforce the regulations. Pest control businesses, agricultural pest control advisors, and pest control aircraft pilots must register in each county where they operate. Anyone who wants to buy a restricted pesticide must have a permit from the commissioner's office. All agricultural pesticide use must be reported monthly to the commissioner's office. Home-use pesticides (those purchased over-the-counter in small volumes) are exempt. There are also more detailed requirements for applicator training and protective gear. Inspectors from county commissioners' offices conduct thousands of compliance inspections every year and have the authority to halt pesticide applications if they believe an applicator's safety is in danger or the pesticide is likely to drift off-site. Contact your county agriculture commissioner's office for details on training and other regulations before purchasing or applying herbicides. County agricultural agents can answer questions about both wildland and agricultural uses of herbicides, as can certified herbicide applicators.

Environmental risks posed by herbicide use include drift, volatilization, persistence in the environment, groundwater contamination, and harmful effects on animals. Drift and resulting death or damage to non-target plants may occur when herbicides are applied as a spray; chances of drift increase with decreasing size of spray droplets and increase with increasing wind speeds. Volatilization and subsequent condensation on non-target plants resulting in their death or damage is another risk of herbicide use. Some herbicides are much more likely to volatilize than others, and likelihood of volatilization increases with increasing temperature. Some herbicides are more persistent in the environment and thus have a greater opportunity for harmful effects. Most herbicides will decompose more rapidly with increasing temperature and soil moisture, and some are decomposed by ultra-violet light. Chances of groundwater contamination generally increase with increasing solubility and persistence of the herbicide, increasing porosity of the soil, and decreasing depth to the water table. Herbicides with potential to cause direct harm to animals (e.g., diquat) are rarely used in natural areas. Animals may, however suffer from indirect impacts if, for example, their food plants are killed.

In order to minimize these environmental risks, look for compounds that can be used selectively (to kill one or a few species); that degrade rapidly under conditions found at the site; that are immobilized on soil particles and unlikely to reach groundwater; that are non-toxic to animals; and that are not easily volatilized.

Also choose an application method that minimizes risks of harming non-target plants and environmental damage. Possible application methods include: spraying on intact, green leaves (foliar spray); spot application (usually from backpack or handheld sprayer); wick application; boom application (from a boom mounted on a vehicle or aircraft); single spot or around the circumference of the trunk on intact bark (basal bark); cuts in the stem (frill or hack and squirt); injected into the inner bark; cut stems and stumps (cut stump); spread in pellet form at the plant's base; and sprayed on the soil before seeds germinate and emerge (pre-emergent).

Mix a dye with the herbicide so applicators can see which plants have been treated and if they have gotten any on themselves or their equipment. Some pre-mixed herbicides include a dye (e.g., Pathfinder II® includes the active ingredient triclopyr, a surfactant and a dye). Ester-based herbicides such as Garlon4® require oil-soluble dyes such as colorfast purple, colorfast red, and basoil red (for use in basal bark treatments), which are sold by agricultural chemical and forestry supply companies. Clothing dyes such as those produced by Rit® will work in water-soluble herbicides such as Garlon3A®, and they are inexpensive and available at most supermarkets and drugstores.

Detailed information on herbicides is available in the Weed Science Society of America's *Herbicide Handbook* (Ahrens 1994) and *Supplement* (Hatzios 1998). This publication gives information on nomenclature, chemical and physical properties, uses and modes of action, precautions, physiological and biochemical behavior, behavior in or on soils, and toxicological properties for several hundred chemicals (see Resources section). Critical reviews of several common herbicides are available at a small charge from the Northwest Coalition for Alternatives to Pesticides (see Resources section).

Beyond this book, additional information and training on weeds and their control can be found by contacting local universities, extension agents, county weed and pest supervisors, and the California Department of Food and Agriculture. The California Exotic Pest Plant Council can direct readers to other local experts on weeds. The Bureau of Land Management offers an Integrated Pest Management and Pesticide Certification course in Denver, Colorado, and the Western Society of Weed Science offers a Noxious Weed Management Short Course in Bozeman, Montana.

SPECIES ACCOUNTS

The scientific names (binomials) used in the species accounts and throughout the book follow *The Jepson Manual: Higher Plants of California* (Hickman, 1993). The accepted scientific name for each species is given at the beginning of each account and in parentheses after the common name the first time it is mentioned in any chapter. Scientific names used in earlier floras but no longer accepted are listed as synonyms. Common names mentioned in any of the sources used in the preparation of each species account are listed at the beginning of the account. The common name deemed most widely used in California is listed first and is used in the text of the species account when referring to the species. The number of closely related native and non-native species (if any) are also listed at the top of each species account.

Each species rating on the CalEPPC list (A-1, A-2, B, or Red Alert) and its rating on the California Department of Food and Agriculture noxious weed list (A, B, C, or nl = not listed) are also indicated at the top of each account.

Ageratina adenophora (Sprengel) R. King & H. Robinson

Common names: eupatory, Crofton weed, sticky snakeroot, catweed, hemp agrimony, sticky agrimony, sticky eupatorium

Synonymous scientific names: *Eupatorium adenophorum, E. glandulosum, E. pasadense*

Closely related California natives: *Ageratina herbacea, A. occidentalis, A. shastensis*

Closely related California non-natives: 0

Listed: CalEPPC B; CDFA nl

by Richard Lichti and Marc C. Hoshovsky

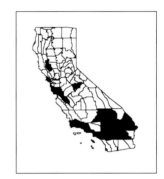

HOW DO I RECOGNIZE IT?

Distinctive Features

Eupatory (*Ageratina adenophora*) is a shrub three to five feet tall with trailing purplish branches that root on contact with the soil, resulting in dense thickets. The base of the plant is woody with a glandular-hairy stem. It seeds abundantly.

Description

Asteraceae. Stem: 1.5-5 ft (0.5-1.7 m) erect, purplish, glandular-hairy. Leaves: opposite, blade deltoid-ovate, serrate, purple below, glandular-puberulent esp. below, leaf blade generally 2-4 in (5-10 cm). Inflorescence: clustered, phyllaries glandular-puberulent, heads approx 0.3 in (0.65 cm). Flowers: 10-60 per head, cylindric, corollas white, pink tinged. Fruit: 5 angled, usually 5 ribbed, pappus 5-40, slender scabrous bristles, often easily detached 0.1-0.2 in (2.5 mm), petiole generally 1-1.3 in (2.5-3.3 cm) (Hickman 1993).

WHERE WOULD I FIND IT?

Eupatory is found in most California coastal counties, but is particularly common from the San Francisco Bay Area to Santa Barbara and in Riverside, Los Angeles, and Orange counties. It also occurs in the Mojave Desert and in the central Sierra Nevada foothills (Mariposa County). In California it generally is found below 1,000 feet elevation (300 m), but in northeast India it has been a problem at nearly 6,600 feet (2,000 m) (Baruah *et al.* 1993). Eupatory occurs mainly in creek beds and forest clearings, in areas with steep (more than 20 percent) slopes, or in disturbed areas. It prefers frost-free areas with abundant rainfall (Parsons 1992). It can be found in redwood forest and coastal canyon habitats, but is most seriously invasive in mild coastal riparian areas, forested areas, or grasslands (Erasmus *et al.* 1991).

WHERE DID IT COME FROM AND HOW DOES IT SPREAD?

Native to Mexico, eupatory is widely naturalized elsewhere. It was introduced to many parts of the world as an ornamental during the nineteenth century and is now an established pest in many tropical and subtropical areas, especially northeastern India, Nigeria, Southeast Asia, the Pacific Islands, South Africa, New Zealand, and Australia (Morris 1989, Parsons 1992). It is unknown how or when it was introduced to California. However, by 1935 it was reported as a "rare escape" in the San Francisco Bay Area and along the south coast (Robbins 1940). In Australia it spread slowly at first. Following a ten-year drought, combined with overgrazing, the infestations expanded rapidly, overrunning large areas of dairy pastures and horticultural land along the border of New South Wales and Queensland. Spread was so fast that in some areas farmers abandoned their holdings (Everist 1959, Dodd 1961).

Eupatory reproduces by prolific asexual seed production and spreads by dispersal of seeds (Muniappan and Viraktamath 1993). Seeds are easily dispersed by wind and water because of their pappus of hairs. Seeds are also spread as an impurity in agricultural produce, in sand and gravel used for road making, in mud sticking to animals, machinery, and other vehicles, and by adhering to footwear or clothing (Parsons 1992).

WHAT PROBLEMS DOES IT CAUSE?

Eupatory is considered a serious weed in agriculture, especially in rangeland, because it often replaces more desirable vegetation or native species (Erasmus *et al.* 1991). It is fatally toxic to horses and most livestock, and apparently is unpalatable to cattle. The toxic disease caused in horses, known as "blowing disease," may take several years to become evident. The symptoms of coughing, difficult breathing, and violent blowing after exertion are the result of acute edema (swelling) of the lungs, leading to hemorrhaging. This plant may reduce growth of nearby vegetation by releasing inhibitors, perhaps allelopathic compounds, into the soil. Eupatory is potentially a problem weed in forestry (Morris 1989).

HOW DOES IT GROW AND REPRODUCE?

Buds appear in late winter and flowering begins in March. Seeds are set without pollination or fertilization, and some 15 to 30 percent of the 7,000 to 10,000 seeds produced by each plant are not viable. Seeds mature and are shed between April and mid-June, the lower leaves of the plant dropping after seed fall. Dense stands can contribute up to 60,000 viable seeds per square meter to the seedbank. Buried seeds lose their viability at a constant rate, averaging 20 percent of all viable seeds per year. Nevertheless, because of high seed production, this high mortality has little effect on the plant's potential for spreading.

Germination occurs between June and March, with peak germination (over 80 percent of viable seeds) in August and September. Light is necessary for seeds to germinate, so unshaded conditions, such as bare soil, are essential for establishment. Eupatory does not invade managed, densely growing pastures. Once germinated, seedlings can withstand a considerable amount of shading, compensating for reduced light intensity by increasing leaf area. Deep shade, however, will kill seedlings. Seedlings grow rapidly and are fully established and able to regenerate from the crown, if damaged, within eight weeks of germination. In second-year and older plants, new growth begins with the first major summer rains, usually in June. Growth rate of seedlings and mature plants remains high during summer but tapers off in the cooler winter months (Parsons 1992).

Eupatory grows rapidly and produces many shoots and branches, which form dense thickets. Plant colonies increase in density and local coverage when bent-over and broken stems take root where they contact soil or when root fragments with attached crown pieces are moved during cultivation. Growth in this latter instance occurs only from buds on the fragment of the crown, not from the roots (Parsons 1992).

HOW CAN I GET RID OF IT?

Physical Control

Mechanical control is difficult because of the species' preference for steep slopes. Where practical, eupatory can be controlled by slashing followed by ripping or plowing and then sowing other plant species to outcompete seedlings. This is best done in spring, using crawler tractors and tandem offset discs (Parsons 1992).

Biological Control

Insects and fungi: No insects or fungi have been approved by the USDA for introduction as biological control agents against eupatory in the United States. However, most control work on this species has focused on biological methods. In one study, the gall fly *Procecidochares utilis* did not significantly reduce seed germination. However, the galls produced by this insect did either temporarily stop growth or completely killed growth of stems beyond the gall. In all cases the galls resulted in the production of underdeveloped capitula and a reduced number of capitula (Van Staden and Bennett 1990). This gall fly has been fairly successful in Hawaii, but its performance has been more variable in Australia due to parasitism by some indigenous hymenopterous insects (Parsons 1992).

A host-specific leaf spot fungus (*Phaeoramularia* sp.) is also being used in South Africa to control eupatory (Morris 1989, Kluge 1991)). The plant is also attacked by the fungus *Cercospora eupatorii*, which may have been introduced accidentally with *P. utilis*, and by a crown-boring cerambycid, *Dihammus argentatus*. None of these organisms offers any real control by itself, but the combined effect has reduced plant vigor in Australia (Parsons 1992).

Chemical Control

Herbicides can be effective against eupatory. Parsons (1992) recommends a label strength, high-volume application of glyphosate, or dicamba + MCPA, or triclopyr in late summer or autumn when the weed is growing actively. Plants should be thoroughly wetted, particularly at the base. In less accessible, steep or rocky areas, treat scattered plants with granular formulations of these herbicides, or use a gas gun and apply low-volume, high-concentration treatments of triclopyr (Parsons 1992). Examine labels for current registered uses of these herbicides in California.

Ailanthus altissima (Miller) Swingle

Common names: ailanthus; tree-of-heaven

Synonymous scientific names: *Ailanthus glandulosa, Toxicodendron altissimum*

Closely related California natives: 0

Closely related California non-natives: 0

Listed: CalEPPC B;
    CDFA nl

by John Hunter

HOW DO I RECOGNIZE IT?

Distinctive Features

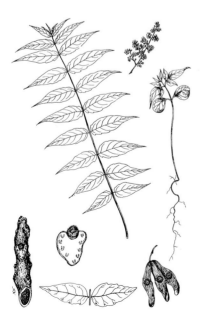

Ailanthus (*Ailanthus altissima*) is a deciduous tree thirty to sixty-five feet high, with gray bark, and generally with root sprouts. Its branches have a large pith and prominent heart-shaped leaf scars. Ailanthus has large compound leaves with several circular glands on the underside of most leaflets. The crushed foliage has an unpleasant odor. Flowers, which are small, greenish, and in large clusters at the branch tips, develop into con-

spicuous and distinctive clusters of fruits. The dry straw-colored or reddish brown fruits are winged with a single seed in the center, their shape resembling an airplane propeller.

Description

Simaroubaceae. Bark: gray and more or less smooth, but develops shallow diamond-shaped fissures with age. Branches have a large pith, alternate leaves, and prominent heart-shaped leaf scars. Leaves: pinnate compound, and 1-3+ ft (30-90+ cm) long with 10-40 lanceolate leaflets. Leaflets have 2-4 rounded teeth near the base, most with a round gland on the lower surface. Inflorescence: large at branch tips. Flowers: small, unisexual, and yellow-green. Flower parts can vary in number, but typically flowers have 5 sepals, 5 petals, 10 functional stamens in male flowers, and 2-5 carpels in female flowers. Female flowers ill-smelling. Each carpel can mature into a winged fruit containing a single central seed. Fruits: dry, indehiscent, 1-2 in (2.5-5 cm) long, propeller-shaped, and straw-colored or reddish brown. (Description based on Bailey 1949, Harlow *et al.* 1979, McClintock 1993).

WHERE WOULD I FIND IT?

Ailanthus is widely but discontinuously distributed in California. It is most abundant along the coast and in the Sierra foothills, primarily in wastelands and disturbed, semi-natural habitats. However, it also occurs in riparian areas and other naturally disturbed habitats throughout California's mid-lower elevations, below 6,600 feet (2000 m). For example, it is found in disturbed woodland and grassland at John Muir National Monument, along a creek bed at the Carrizo Plain, and in the riparian zone of the American and Sacramento rivers. It grows in humid and sub-humid, temperate regions, mainly on clay or loam soils in areas with moderate rainfall. Ailanthus is tolerant of extreme soil conditions (Miller 1990). The tree has been used in revegetating acid mine spoils, tolerating a pH of less than 4.1, soluble salt concentrations up to 0.25 mmhos/cm, and phosphorus levels as low as 1.8 ppm (Plass 1975). It withstands harsh urban environments better than most plants and is used as a street tree in many cities.

WHERE DID IT COME FROM AND HOW DOES IT SPREAD?

A native of eastern China, ailanthus has been introduced throughout the northern hemisphere. In the 1740s it was introduced to Europe in the mistaken belief that it was the source of lacquer used in the production of polished wooden ware (Hu 1979). Because of its tropical look, rapid growth, and legendary tolerance of urban life, ailanthus was planted throughout Europe and the United States during the nineteenth century. In California it was planted widely until the 1890s (McClintock 1981). During the days of the California gold rush, Chinese miners also may have brought ailanthus seeds with them as they settled in California. In this century, its popularity as an ornamental has declined primarily because of its unpleasant fragrance and prolific root sprouting. Today, it is still occasionally planted in California (Perry 1989).

Ailanthus escapes from cultivation and spreads by root sprouts and wind-dispersed seeds. Seeds may also spread by water, birds, and on farm machinery. However, new patches do not occur as frequently as would be expected from the amount of seed produced (Parsons 1992).

WHAT PROBLEMS DOES IT CAUSE?

By producing abundant root sprouts, ailanthus creates thickets of considerable area, displacing native vegetation (Kowarik 1983, 1995). Although it may suffer from root competition by other trees already established, usually it competes successfully with other plants (Cozzo 1972, Hu 1979). In California its most significant displacement of native vegetation is in riparian zones. It also produces allelopathic chemicals that may contribute to displacement of native vegetation (Lawrence 1991). A high degree of shade tolerance gives ailanthus a competitive edge over other plant species (Grime 1965).

HOW DOES IT GROW AND REPRODUCE?

Ailanthus reproduces by seed and vegetatively by root sprouts. Trees generally become reproductive at ten to twenty years, but younger shoots also may produce fruits. Trees are deciduous, leaving clumps of stark, bare stems over the winter. Most trees produce only male or only female flowers (dioecy). Flowering follows leaf expansion in late spring. Female trees may produce several hundred inflorescences per year, and at maturity an inflorescence contains hundreds of seeds (Hunter 1995). A single tree can produce up to a million seeds per year.

Seeds ripen in large, crowded clusters from September to October of the same year and may persist on the tree through the following winter (Little 1974, Hu 1979). Most ailanthus seeds are viable, even those that have overwintered on the tree (Little 1974, Hunter 1995).

Fruits are winged, containing a single light seed. The fruits mature in late summer and are dispersed by wind throughout fall, winter, and even during the following spring. After exposure to cold temperatures of 32 to 42 degrees F for more than forty days, seed will germinate readily in moist soil (Little 1974). Germination ranges from 14 to 75 percent. Because its seeds do not remain dormant for more than a year, ailanthus does not have a persistent soil seedbank.

Despite prolific seed production, seedling establishment of ailanthus is infrequent in California. Most new shoots are root sprouts. Young root sprouts are distinguished by a cluster of leaves with a variable number of leaflets attached to a thick rope-like root, whereas seedlings are thin-stemmed and have round cotyledons followed by trifoliate leaves.

Root sprouts are produced up to fifty feet (15 m) away from the nearest shoot. Their initial growth is rapid, commonly over a meter per year (Miller 1990, Hunter unpubl. data). In favorable settings, rapid growth continues, but for shaded sprouts growth drops to several centimeters per year (Hunter unpubl. data).

Initially, growth is concentrated along a single axis and branching is rare. Even when browsed or cut, most stems continue to grow with a single main trunk. Damage, however, often gives rise to root sprouts (Bailey 1930). Once branching begins, the branches diverge at wide angles and a broad dome-shaped crown develops. Within twelve to twenty-five years, the tree typically reaches heights of thirty-three to sixty-three feet (10-20 m) in California. Ailanthus shoots typically have a life span of thirty to fifty years (Miller 1990), but production of multiple trunks from root sprouts allows genetic individuals to occupy a considerable area, sometimes over an acre (0.4 ha), for a prolonged period of time.

HOW CAN I GET RID OF IT?

Physical Control

If mechanical and/or chemical control is attempted, sites should be monitored several times per growing season. All new root sprouts should be removed, and monitoring should be continued for one year after the last sprout is removed.

Hand pulling: Young seedlings are best pulled after a rain when the soil is loose. This allows removal of the root system, which may resprout if left in the ground. After the tap root has developed, root removal is more difficult. Plants should be pulled as soon as they are large enough to grasp but before they produce seeds.

Hand digging: Removal of rootstocks by hand digging is a slow but sure way of destroying weeds that resprout from roots. The work must be thorough to be effective as every piece of root that breaks off and remains in the soil may produce a new plant. This technique is suitable only for small infestations and around trees and shrubs where other methods are not practical.

Cutting: Manually operated tools such as brush cutters, power saws, axes, machetes, loppers, and clippers can be used to cut ailanthus. This is an important step before many other methods are tried, as it removes the above-ground portion of the plant. For thickly growing, multi-stemmed shrubs and trees, access to the base of the plant may be not only difficult but dangerous where footing is uncertain.

Girdling: This involves manually cutting away bark and cambial tissues around the trunks of undesirable trees. A relatively inexpensive method, girdling is done with an ordinary ax in spring when trees are actively growing. Hardwoods are known to resprout below the girdle unless the cut is treated with herbicides. Although it may be undesirable to leave standing dead trees in some areas, this technique has been shown to reduce stump sprouting in live oaks, and may be a useful technique for controlling ailanthus.

Cutting an ailanthus stem induces prolific root suckering and the production of stump sprouts. After a stem is cut, its stump sprouts may grow over ten feet (3 m) per year and its root sprouts three to seven feet (1-2 m) per year (Pannill 1995). As a consequence, mechanical removal will be ineffective unless all stems are cut at least several times per year (Pannill 1995).

Prescribed burning: This is probably not an effective technique for controlling ailanthus. Fire may kill main stems, but this will result in prolific sprouting.

Biological Control

Insects and fungi: Biological control of ailanthus has not been investigated. The species is not significantly affected by insects or disease (Miller 1990). French (1972) reports that the zonate leafspot fungus (*Cristulariella pyramidalis*) causes defoliation of ailanthus in Florida. In India the insect *Atteva fabricella* is considered an ailanthus defoliator (Misra 1978), and seedlings in Italy, weakened by cold, were weakly parasitized by the fungus *Placosphaeria* sp. (Magnani 1975).

Grazing: Ailanthus may suffer extensive browse herbivory from deer and cattle, particularly the young growth of sprouts, which may aid eradication (Pannill 1995, Hunter 1995).

Plant competition: In most cases ailanthus prevents the establishment of other native plants and must be initially removed. Following removal of mature plants, root crowns must be treated to prevent resprouting. Seedlings of native plant species usually cannot establish fast enough to compete with sprout growth from untreated stumps. Ailanthus is shade tolerant, so presumably will sprout under other plants.

Chemical Control

Herbicide applications should be most effective in spring, just after leaves are fully expanded. Smaller sprouts probably can be controlled by spraying foliage with 4 percent glyphosate (as Roundup®). Young stems usually can be killed by generously applying 15-20 percent triclopyr (sold as Garlon®) to all of the bark from the stem base to twenty inches above the ground. The thicker bark of larger plants interferes with uptake of herbicide, and therefore, to kill larger individuals, the stem needs to be frilled (have an encircling ring of bark removed) before herbicide is applied. In order to damage the root system, concentrated herbicide (such as 15 to 20 percent triclopyr or 15 to 40 percent glyphosate) needs to be applied with brush or wick to the freshly exposed surface immediately after cutting (Pannill 1995). Applying herbicide to freshly cut stumps is probably the most effective technique for controlling ailanthus. Wiping the stump with full strength, 41 percent glyphosate within several minutes of cutting should reduce or even eliminate subsequent root suckering.

Alhagi pseudalhagi (M. Bieb) Desv.

Common names: camelthorn, camels thorn, camel-thorn, Caspian manna, Persian manna

Synonymous scientific names: *Alhagi camelorum, A. maurorum, A. persarum*

Closely related California natives: 0

Closely related California non-natives: 0

Listed: CalEPPC B; CDFA A

by Ross O'Connell and Marc C. Hoshovsky

HOW DO I RECOGNIZE IT?

Distinctive Features

Camelthorn (*Alhagi pseudalhagi*) is a medium-sized, spiny, intricately branched perennial shrub that grows one and a half to three feet tall and two to three feet in diameter. It has many rigid branches armed with numerous sharp yellow-tipped spines (one-half inch to one and a half inches long). Both flowers and seed pods are borne on thorns. The small pea-like flowers, ranging in color from brown to red, maroon, and purple, fade to violet with age. The pea-like seed pods are reddish brown and constricted between seeds to resemble a string of beads, with the end of the pod narrowed to a short beak.

Description

Fabaceae. Stems: rigid, much branched, 1-4 ft (3-12 dm) high, greenish, glabrous, (young branches hairy). Stems have numerous sharp, axillary spines 0.5-1 in (1-2.5 cm) long, with yellow tips, often on short branches. One or more stems produced from each root crown. Leaves:

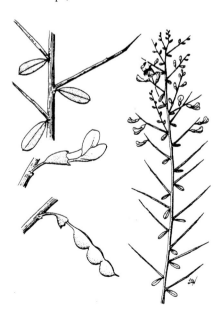

blue-green, alternate, small up to 1 in (2.5 cm); simple, entire, oval to lanceolate, tapering to a short petiole. Leaves finely hairy when young, becoming glabrous with age. Short-stalked, usually produced singly at branch or spine nodes. Leaves tend to fall off in warm weather. Flowers: range in color from brown to red, maroon, and purple, fading to violet with age; pea-like, 0.3- 0.4 in (8-9 mm) long, produced in clusters of 1-8 along short branchlets or spines near the extremities of branches. Fruits: reddish brown indehiscent pods, 0.4-1.2 in (1-3 cm) long, each containing 1-5 (or more) seeds. Pods strongly constricted between seeds to resemble string of beads, with end of pod narrowed to a short beak. Seeds: kidney-shaped and resemble large clover seeds, about 1/8 in (2.5 mm) long, dark brown to yellowish or greenish brown with dark mottling; notched at the scar, com-

monly lighter at notch. Roots: extensive root system consisting of both vertical and horizontal roots extends to 6.6 ft (2 m deep) and 26.5 ft (8 m) laterally. Aerial shoots are formed on horizontal branches (Hickman 1993, Parsons 1992).

WHERE WOULD I FIND IT?

In California camelthorn has been found in the southern Sierra foothills, the Mojave and Colorado deserts, and throughout the Central Valley from Kern County as far north as Tehama County (Barbe 1990). Camelthorn has been reported in sixteen California counties, but eradication efforts since 1923 have eliminated all but four infestations in Tulare, San Bernardino, Inyo, and Imperial counties. These remaining infestations are undergoing active eradication efforts. Large infestations still remain in arid parts of Arizona, Nevada, and Washington (Whitson 1992). In Arizona dense thickets have formed along the Colorado River in the Grand Canyon and along the Little Colorado River.

Camelthorn is able to grow in soils of high carbonate level and in varied soil types, including sand, silt, clay, and even in rock crevices (Kassas 1952a). It grows best in alkaline soils and can grow in pastures, rangeland, and irrigated croplands such as date plantings, alfalfa fields, and citrus groves (Parsons 1992). Possibly the only soil limitation of camelthorn may be an intolerance of high soil salinity (Kassas 1952a).

WHERE DID IT COME FROM AND HOW DOES IT SPREAD?

The native range of camelthorn extends from India and southwest Asia to North Africa. It has spread to many parts of the world, including South Africa and Australia. The present distribution of camelthorn in the United States includes the states of Washington, California, Nevada, Utah, Arizona, New Mexico, and Texas (Kerr 1963). It often occurs first in semi-arid areas and later establishes and thrives under irrigation. The weed was first reported in California in 1915, near Mecca in Riverside County. This weed is believed to have been introduced to California by two means: in camel dung used as packing material around date plants from the Mediterranean region and also as an impurity in Turkestan alfalfa seed. Camelthorn was declared a noxious weed in alfalfa fields in California's Imperial Valley in 1921 (Kerr 1963), and extensive efforts to control the plant were initiated in nine counties by the mid-1920s. However, it was not eradicated at that time (Bottel 1933, Koehler *et al.* 1956). Camelthorn spread from California to Washington in contaminated hay and was declared a primary noxious weed in Washington in 1955 (Kerr 1963). It is now virtually eradicated from California, but large infestations still exist in Arizona, Washington, and Nevada, which could be sources of reinfestation here.

Although not readily propagated by seed, the spread of this species does occur by seed. Camelthorn is spread through livestock, which browse the seed pods in fall after the foliage is dropped or killed by frost, to which it is susceptible in its non-native environment (Richardson 1953, Kerr 1963, Kerr *et al.* 1965). Seedlings are rarely found, however, suggesting that spread by seed may not be important (Parsons 1992).

Once established, the predominant method of spread is through vegetative reproduction. Lateral growth of the extensive root system is an important method of spread, extending the radius of a patch by up to twenty-four feet (7.4 m) per year. Seeds and root pieces can spread by water and high winds, which may blow balls of entangled aerial parts (including seeds) for long distances (Richardson 1953). If roots are cut by cultivation equipment, small pieces can be transported elsewhere to produce new plants.

WHAT PROBLEMS DOES IT CAUSE?

Camelthorn is strongly competitive with other plants. Its rapid and aggressive growth allows it to outcompete both native vegetation and cultivated crops. Because of its rhizomatic growth habit, dense stands may form that are impenetrable because of its spiny stems. It is especially troublesome in cereal and horticultural croplands, where repeated cultivation aids its spread. In the past some California croplands were abandoned because of the high cost of camelthorn control. Specific problems in wildland areas are not well known (Parsons 1992).

HOW DOES IT GROW AND REPRODUCE?

Camelthorn reproduces by seed and vegetatively by rhizomes that send up shoots. Plants usually do not flower until they are at least a year old. Flowers appear in June and fruits in July, but summer rains or irrigation can extend the flowering season later into the summer. Flowering also varies with environmental conditions. In hot, dry, exposed conditions, a single camelthorn plant was reported to produce from 790 to 4,150 flowers. However, in moist or shady conditions camelthorn may not flower at all (Kassas 1952b). Despite its high flowering potential, the percentage of flowers that develop into viable fruits is low (3 to 20 percent in experiments of Kassas

1952b). Seed pods ripen in July. Seeds may remain viable in the soil for many years (Kerr *et al.* 1965) because of the hard, thick seed coat (Bottel 1933).

Germination requires adequate soil moisture and warm (25 to 28 degrees C) temperatures (Kerr *et al.* 1965). Optimal seed depth for germination is 1 cm below the soil surface (Kassas 1952b). Kassas (1952b) reports that exposure to sunlight inhibits germination, yet Kerr *et al.* (1965) report that seedling growth appears quickest under a high light regime.

Seeds in or under cow manure, where soil moisture and temperature are higher, have a higher probability of germination, seedling growth, and survival (van der Walt 1955, Kerr *et al.* 1965). Digestive scarification of the thick, hard seed coat by livestock is important, if not essential, for establishment of this species (Kerr 1963, Kerr *et al.* 1965). Experiments support this possibility, showing that seeds scarified with sulphuric acid have a higher germination rate than untreated seeds (Parsons 1992). In its native range camelthorn is one of the most important food plants of the camel (Carmin 1950) and may have adapted to the plant/animal interactions described above.

Seedling shoot growth is slow. Leaves require up to three months to develop (Kassas 1952b). In contrast, seedling root growth is rapid. Kassas (1952b) reports that in six-month-old plants, the ratio of root to shoot is about 2.5:1.

Vegetative growth is the most important means of spread. Small portions of the rootstock are capable of giving rise to new plants (Richardson 1953, van der Walt 1955). This allows camelthorn to thrive in situations of intensive erosion (such as along riverbanks) by resprouting from underground perennating buds (Kassas 1952b).

Camelthorn has a vigorous root system that is able to tap into a water table up to fifteen meters below the soil surface. This allows camelthorn to thrive in areas of little rainfall and high water table, such as saline meadows, sandbars, playas, riverbanks, irrigation canals, and irrigated croplands (Kerr 1963). The fast-growing rhizomes can enlarge stands by up to twenty-four feet (7.4 m) per year. Many infestations of this plant have been described as nearly circular in form. Camelthorn can send its roots underneath asphalt roads and then produce shoots over twenty feet (6.5 m) from the parent plant. Shoots can also break through asphalt roads to produce new plants (Parsons 1992).

Although the stems are somewhat woody, in cooler environments camelthorn is deciduous and dies back to the ground each winter, remaining dormant until spring (Kassas 1952b).

Camelthorn's growth form may vary with habitat conditions. Plants can be either prostrate (in areas of prevailing winds) or, more typically, erect (Kassas 1952a). This variable morphology may exist within a single clone growing in differing microhabitats. In moist habitats leaves are broader and spines smaller than on plants growing in hot, dry habitats. In hot, dry areas the ratio of spines to leaves increases, slowing water loss. Spines have less transpiring surface than leaves, they are structurally more capable of withstanding water loss, and they have chlorophyllous tissues that can continue photosynthesis in the absence of leaves (Kassas 1952a).

Alhagi species exude a sap (manna) that contains mannitol. This sap, which may be effective as a laxative (Kassas 1952b), is the reason for the common names Caspian manna and Persian manna (Richardson 1953).

HOW CAN I GET RID OF IT?

Camelthorn is a primary noxious weed in California, and all infestations of this plant are under eradication by the California Department of Food and Agriculture. Suspected infestations should be reported to the local county agricultural commisioner or the California Department of Food and Agriculture, Integrated Pest Control Branch, Weed and Vertebrate Program. Recommended preventive measures include promoting the use of certified weed-free hay, not allowing livestock to eat and thus disperse the seeds, and not allowing the spread of camelthorn seed through the use of heavy equipment in infested areas. Historically, eradication efforts have included deep plowing, flooding, arsenical sprays, and fumigation with carbon bisulfide.

Physical Control

Mechanical methods: Cutting, discing, and even deep plowing are not effective in controlling camelthorn because these methods create root fragments that reestablish readily. Roots at depths of six feet (1.85 m) cannot be removed by plowing, and they can regenerate new plants following disturbance. In experiments on camelthorn in Egypt, Kassas (1952b) found that, within three weeks after shearing plants to ground level, 194 new shoots were produced from underground buds up to 60 cm deep. In unsheared plants only eight new shoots were produced from underground buds. Although it seems that mechanical shearing or mowing will aid the spread of camelthorn, Kerr (1963) states that repeated top removal might serve to exhaust the food reserves of the plant and aid its control. It is obvious that camelthorn control should aim at eradicating not only above-ground plant parts but also below-ground parts. Van der Walt (1955) reports that mechanical methods of controlling camelthorn are futile.

Prescribed burning: No control using fire has been reported. This is unlikely to be effective because of the plant's rhizomatous nature. The deep roots would also not be killed by heat at the soil surface.

Flooding: Flooding has been used as a control technique in California, and it can be effective if a depth of five to ten inches of water is maintained over several weeks (Koehler *et al.* 1956, Kerr 1963).

Biological Control

Insects and fungi: No biocontrol agents for camelthorn are available in the United States. Carmin (1950) reports a eurytomous wasp that forms galls on the vascular tissue of camelthorn. There is no mention of this wasp's potential as a biological control agent.

Grazing: Cattle and sheep graze on camelthorn. Plants are sensitive to low temperatures, and they are more readily grazed after being damaged by frost (Parsons 1992).

Chemical Control

Herbicides offer practical control, but the only effective materials are those that act through the soil and thus also affect other plants. Some chemicals that have been effective in the past, such as carbon bisulfide and sodium arsenate, are no longer registered for use in California. Trial work in Australia suggests that glyphosate, fosamine, and clopyralid can be effective (Parsons 1992).

Picloram has been the preferred herbicide for controlling camelthorn by the California Department of Food and Agriculture. However, this chemical currently cannot be purchased for use in California.

Ammophila arenaria (L.) Link

Common name: European beachgrass

Synonymous scientific name: *Arundo arenaria*

Closely related California natives: none

Closely related California non-natives: *Ammophila breviligulata*

Listed: CalEPPC A-1; CDFA nl

by Rachel Aptekar

HOW DO I RECOGNIZE IT?

Distinctive Features

European beachgrass (*Ammophila arenaria*) is a perennial rhizomatous grass occurring in coastal dunes. Stems are clumped, stiff, and upright. Leaves are twelve to forty-four inches long, thick and waxy. The outer surface is smooth and light green; the inner surface has ridges and is covered with a whitish coating. Leaves are often rolled at the edges, covering ridges on the inner side, and leaf tips are pointed and sharp. The inflorescence is cylindrical, six to twelve inches long on stiff, erect stems. Rhizomes are tough. Leaves are narrower, stiffer, and lighter in color than the native beachgrass, *Leymus mollis*.

Description

Poaceae. Stems: 20 to 48 in (50-120 cm). Leaves: involute, 12 to 44 in (40-110 cm) long and 0.1 to 0.25 in (2-5 mm) wide. Ligules 0.4 to 1.2 in (1-3 cm) long. In comparison, the introduced species *A. breviligulata*, which is native to Atlantic and Great Lakes coastal dunes and is only rarely found along the Pacific Coast, has ligules less than 0.25 in (5 mm) long. Inflorescence: in dense spike-like panicles, 6 to 12 in (15-30 cm) long and 0.8 in (2 cm) wide. Flowers: spikelets 0.4 to 0.5 in (10-13 mm) long, subsessile, laterally compressed, strongly keeled; glumes +or- > floret, membranous, obtuse to acuminate, lower generally 1-veined, upper 3-veined; callus hairs 0.1 to 0.2 in (2-4 mm), tufted; floret bisexual, breaking above glumes; lemma membranous, 0.3 to 0.4 in (8-10 mm), 5-7 veined; palea membranous 0.3 to 0.4 in (8-10 mm) (Hickman 1993).

WHERE WOULD I FIND IT?

European beachgrass occurs on sandy coastal dunes from British Columbia to San Diego County, California (Breckon and Barbour 1974, Barbour and Johnson 1977). The plant thrives in unstable dunes where there is continuous sand accretion, but it is also found in stabilized dunes. While European beachgrass appears to spread actively north of San Francisco, it is not as aggressive to

the south (Barbour *et al.* 1976, Barbour and Johnson 1977). However, it is reported as invasive at Guadalupe-Nipomo Dunes in San Luis Obispo County.

WHERE DID IT COME FROM AND HOW DOES IT SPREAD?

European beachgrass is native to the coast of Europe and North Africa, from Scandinavia south to the Mediterranean Sea, and from Great Britain east to Egypt. It was first planted along the Pacific Coast of North

America, around San Francisco's Golden Gate Park, in 1869. It has since been transplanted elsewhere along the coast (Barbour 1970), and there may also have been subsequent introductions from its native range. It was planted extensively to stabilize dunes.

European beachgrass spreads almost exclusively by rhizomes, which can extend laterally over 6.6 feet (2 m) in six months (Aptekar, 1999). Rhizomes are also washed by ocean currents to new sites, where new populations can become established. The rhizomes can survive and sprout new plants after being submerged in seawater for prolonged periods. Bud viability is still 51.2 percent after seven days of submergence, and 8.5 percent after thirteen days submergence (Aptekar, 1999).

WHAT PROBLEMS DOES IT CAUSE?

European beachgrass forms a dense cover that appears to exclude many native taxa. Plant species diversity on beaches dominated by European beachgrass is much less than that of beaches dominated by native beachgrass (Breckon and Barbour 1974, Barbour *et al.* 1976, Pavlik 1983c, Boyd 1992). As European beachgrass cover increases, the cover of native plant species decreases significantly (Aptekar 1999).

Dunes dominated by European beachgrass also have lower arthropod species diversity, and fewer rare arthropod species than dunes dominated by native species (Slobodchikoff and Doyden 1977). The reduction in the amount of open sand areas in dunes dominated by European beachgrass has severely reduced nesting habitat for the federally listed threatened western snowy plover (*Charadrious alexandrinus*) (Pickart and Sawyer 1998).

European beachgrass directly threatens native plant communities, including *Leymus mollis* dominated foredunes (Pitcher and Russo 1988) and dune mat communities (Van Hook 1983, Pickart 1988, Pitcher and Russo 1988, Buell *et al.* 1995, Wiedemann and Pickart 1996, Pickart and Sawyer 1998), as well as the endangered Menzies' wallflower (*Erysimum menziesii*) and other rare dune species (Van Hook 1983, Pickart and Sawyer 1998).

European beachgrass has changed beach topography, creating steep foredunes where none were present prior to its introduction, and altering dune formation to promote dunes that are parallel to the coast, whereas under native grasses, dunes formed roughly perpendicular to the coast (Cooper 1967, Barbour and Johnson 1977, Wiedemann and Pickart 1996).

HOW DOES IT GROW AND REPRODUCE?

European beachgrass primarily reproduces vegetatively through rhizome growth. Plants are initially established by plantings undertaken to stabilize dunes, and by rhizomes washing in on ocean currents from other sites. Once plants are established, they expand their territory through vigorous rhizome growth. European beachgrass rarely becomes established by seed. Seedlings are occasionally found, but most die from desiccation, burial, or erosion. Seedlings are most likely to survive in dune slacks where the sand surface remains damp (Huiskes 1977, Benecke 1990).

European beachgrass is a rhizomatous perennial grass that grows most vigorously under conditions of continued sand accretion, and forms extensive monospecific stands. The plant produces both vertical and horizontal rhizomes. New shoots arise mainly along the vertical rhizomes, forming dense tufts of grass. Horizontal rhizomes are responsible for lateral growth.

European beachgrass is able to withstand up to 3.3 feet (1 m) a year of sand burial (Huiskes 1979), which is greater than what the native beachgrass, *Leymus mollis*, appears to tolerate. Sand burial of European beachgrass promotes leaf and internode elongation, as well as growth of vertical rhizomes from axillary buds on horizontal rhizomes (Gemmell *et al.* 1953). In areas without sand accretion, it is much less robust (Wallen 1980). European beachgrass uses nitrogen more efficiently, and allocates more resources to producing leaves and shoots, than does *L. mollis* (Pavlik 1983). Shoots grow most vigorously in spring. Growth slows during winter, but does not cease entirely (Huiskes 1979).

HOW CAN I GET RID OF IT?

Physical Control

Manual methods: European beachgrass can be removed by intensive repeated digging. Successful manual control at Lanphere-Christensen Dunes Preserve (Pickart and Sawyer 1998) required weekly to monthly digging of European beachgrass from early spring through fall. If sand was "sifted" with rakes to remove rhizome fragments for a depth of 19.5 to thirty-nine inches (0.5 to 1 m) following digging, follow-up treatment was not required the following year. If sand was not sifted following digging, and plots were dug monthly or less frequently, a second year of monthly digging was required. On foredunes, where European beachgrass grows more vigorously than in inland stands, a third year of monthly digging was required. In general, less follow-up digging was required when first-year treatments were more frequent, more thorough, and/or larger or in less dense locations (so that fewer plants re-invaded from surrounding stands).

Mechanical methods: Some attempts have been made at Oregon Dunes National Recreation Area to remove European beachgrass with heavy machinery. In the most ambitious attempt, European beachgrass was removed to a depth of approximately one meter (3.3 feet). The removed material was buried and capped with up to one meter of sand. Moderate resprouting occurred the following spring (Pickart and Sawyer 1998).

Prescribed burning: Burning stimulates European beachgrass to resprout, and so by itself does not appear to be effective in eradicating it (Pickart and Sawyer 1998).

Flooding: Several attempts have been made using salt water or seawater to eliminate European beachgrass (Pickart and Sawyer, 1998). In small test plots in 1983 at Lanphere-Christensen Dunes Preserve, seawater treatments were not effective and rock salt treatments had inconclusive results. A large-scale project using seawater irrigation to control European beachgrass on twenty-six acres (10.5 ha) of the North Spit of Coos Bay, Oregon, in 1996 was not successful (Pickart, pers. comm.).

Biological Control

Insects and fungi: No insects or fungi have have been approved by the USDA for control of European beachgrass in the United States. There is no recorded insect that feeds solely on European beachgrass. *Meromyza pratorum* Meigen (Dipt., Chloropidae), a beetle, feeds only on European beachgrass in western Europe, but it has been found feeding on wheat in Russia and Italy. The larvae destroy vegetative points, killing up to 30 to 40 percent of tillers (Huskies 1979). Many fungi have been recorded as living on European beachgrass, but most of these are non-specific saprophytes or weak parasites on dying parts of the plant.

Grazing: Animals native to California beaches rarely graze on European beachgrass. Even if an animal could be found that would graze on the plant, it would not be an effective control, since European beachgrass resprouts from its numerous below-ground buds and rhizomes.

Chemical Control

In experimental trials conducted in northern California from 1991 to 1994, using a variety of herbicides, the only foliar treatment that consistently reduced live European beachgrass cover by 90 percent or more was glyphosate (as Roundup®), applied at concentrations of 4 percent or

10 percent and mixed with 0.5 percent added surfactant (Citowett® or Silwet L-77® were used) applied at 200 gallons per acre (Aptekar 1999).

Selective application of 33 percent glyphosate (as Roundup®) applied with a wiper or with an herbicide sprayer had mixed results. In some instances it was extremely effective at reducing European beachgrass cover, but in other instances it had practically no effect. Experimental trials conducted by Monsanto from 1992 to 1994 using Rodeo® (glyphosate without surfactants) have resulted in a label recommendation of 8 percent solution of Rodeo® plus 0.5 to 1.5 percent nonionic surfactant on a spray-to-wet basis, applied during active growth. For selective control, application of 33 percent glyphosate and 1 to 2.5 percent non-ionic surfactant, applied with a wiper to avoid non-target plants, is recommended.

The liquid soil fumigant form of metham (as Vapam®) is extremely effective in killing European beachgrass (Aptekar 1999). Applied at label rate, it reduced European beachgrass by 98 to 100 percent, and it was nearly as effective when applied at one-half and one-quarter the label rate. However, there are significant disadvantages to metham. It is difficult to apply and affects all soil organisms. A granular form of the metham, Basimid®, is easy to apply, but is not very effective under dune conditions without sufficient rain or irrigation to move the fumigant to the appropriate soil depth following application, and it also would be detrimental to all soil organisms.

Aptenia cordifolia (L.f.) N.E. Brown

Common names: red apple, baby sun rose, heartleaf iceplant, dew plant

Synonymous scientific name: *Mesembryanthemum cordifolium*

Closely related California natives: 0

Closely related California non-natives: 8

Listed: CalEPPC Red alert; CDFA nl

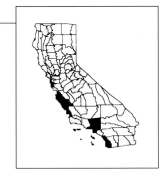

by Jo Kitz

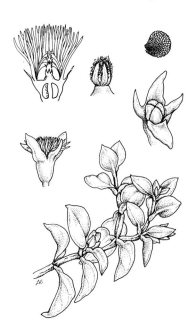

HOW DO I RECOGNIZE IT?

Distinctive Features

Red apple (*Aptenia cordifolia*) is a perennial herb, spreading over ground and neighboring vegetation, with small, heart-shaped, dark green succulent leaves interspersed with small, axillary, many-petaled, bright pink to purple flowers that open only in sun. The hybrid red apple (with *Platythyra haekeliana*) has brighter red flowers.

Description

Aizoaceae. Stems: prostrate, 12-18 in (3-6 dm), nodes widely spaced, base of stems woody. Leaf: 0.4-1.2

in (1-3 cm), petioled, cordate, minutely papillate. Inflorescence: flower solitary, axillary; peduncle 0.25-0.6
in (8-15 mm). Flower: hypanthium 0.2 in (6-7 mm);
sepals, four of unequal size +/-0.2 in (5mm), the largest flat, the others awl-shaped; petals numerous 0.2 in
(3 mm), ovary inferior. Fruit: 13-15 mm four locular
capsule (Hickman 1993).

WHERE WOULD I FIND IT?

In California, red apple is found in disturbed places
and on margins of coastal wetlands, usually less than
100 feet (30 m) elevation. Naturalized in California, it
is reported in central and southern California and on
embankments along irrigation waterways in the Central Valley (Hickman 1993). It has naturalized also in
Oregon, St. Lucie County, Florida, and along the south-

ern coast of Europe. It is marketed as a fire-resistant (unproven), drought-tolerant ground cover
in California. It is planted in parking lot planters, parkways, home gardens, in the urban interface
near parks and preserves, and on brush-cleared hillsides and stream embankments. It can tolerate
some soil salinity and grows well in dryish, frost-free or almost frost free areas in full sun.

WHERE DID IT COME FROM AND HOW DOES IT SPREAD?

Red apple is native to the eastern coastal region of the Cape Province and Kruger National

Park in the Transvaal, both in South Africa, and was brought to California as a horticultural plant (Herre 1971). It is sold widely in plant supply stores. It is known to spread vegetatively by rooting of branches. Seed dispersal capabilty is not known.

WHAT PROBLEMS DOES IT CAUSE?

When watered, red apple overwhelms all neighboring vegetation, climbing over anything in its path. It is listed as a wildland weed red alert as it has only recently become a problem. It has been used in landscaping adjacent to riparian areas within the urban interface, so it can easily spread into and dominate more natural riparian and wetland areas.

The author's first introduction to this plant was at a hilltop residence in San Luis Obispo County, where it had been planted under oak trees and watered daily. It had grown five to six feet (2 m) up the oak trees, cloaked three-foot (1 m) ceanothus shrubs, and formed a thick mat by growing over itself. Distinctive features of the landscape had surrendered to an unbroken cloak of red apple. As testimony to its vigor against other invasive monocultures, it has been seen overgrowing *Vinca major*.

HOW DOES IT GROW AND REPRODUCE?

Red apple grows rampant and leggy with water and/or shade and compactly when not watered. Its morphological characteristics indicate that it can over-summer without water and then grow vigorously during the rainy season, although constantly wet soil can cause it to rot. It spreads vegetatively. Nodes root when they touch the ground (Bailey 1949). It blooms in spring and summer. Seeds grow well in sandy, well drained soil and germinate at 60-65 degrees F. More information is needed on seed viability and potential for seed dispersal.

HOW CAN I GET RID OF IT?

Little is known about control, but red apple should respond to the same methods as the related sea fig (*Carpobrotus edulis*) and New Zealand spinach (*Tetragonia tetragonoides*).

Physical Control

Manual methods: Red apple can be easily removed by hand pulling. However, because of the ability of this plant to grow roots and shoots from any node, all live shoot segments must be removed from contact with the soil to prevent resprouting. If complete removal is not possible, mulching with the removed plant material is adequate to prevent most resprouting, but requires at least one follow-up treatment to remove resprouts.

Chemical Control

The herbicide glyphosate (as Roundup®) has been effectively used to kill related sea fig clones at label-recommended concentrations of 2 percent or higher. The addition of 1 percent surfactant to allow penetration of the cuticle on the leaves should improve effectiveness.

Arctotheca calendula (L.) Levyns

Common names: capeweed, South African capeweed, cape dandelion, cape gold

Synonymous scientific names: none known

Closely related California natives: 0

Closely related California non-natives: 0

Listed: CalEPPC Red Alert; CDFA A

by Maria Alvarez

HOW DO I RECOGNIZE IT?

Distinctive Features

Capeweed (*Arctotheca calendula*) is an annual or perennial evergreen herb that, when young, forms a low-growing rosette of heavily pinnately lobed leaves, with undersides covered by woolly down. With age, it forms an extensive, dense, mat-like groundcover by proliferation of rooting stems (stolons) from rosettes. Leaves are pinnately lobed; fine, dense hairs cause stems and leaves to appear silvery. Flowers are approximately two inches in diameter, lemon yellow, and daisy-like with yellow centers. The plant is conspicuous in late spring and early summer due to its increase in size and the profusion of large yellow daisies. Plants are seldom solitary, and they spread vigorously by creeping stems (Lasca Leaves 1968).

Description

Asteraceae. Stems: creeping or decumbent, originating from an individual rosette; succulent, hairy (tomentose), and ribbed; stems creep along or just below the soil surface, bearing fully formed leaves and reaching lengths of up to nine feet in one growing season. Leaves: pinnately lobed, 2-10 in (5-25 cm) long, upper surface finely hairy (cobwebby), lower surface densely

hairy or silky (white-woolly). The woolly leaf underside is a feature distinguishing capeweed from other herbs in the absence of flowers. Inflorescence: heads 2.4 in (6 cm), peduncles scapose, 6-8 in (15-20 cm) long; outer phyllaries green, woolly tips reflexed. Ray flowers fewer than 20; ligules lemon yellow above. Disk flowers many, bisexual, also yellow. Fruit: sterile, up to 0.2 in (5 mm), covered with hairs (Hickman 1993).

WHERE WOULD I FIND IT?

At present, infestations in California are known only from coastal Marin and Humboldt counties (Barbe 1990), but capeweed can survive in most of this state west of the Sierra Nevada. Hickman (1993) reports a probable range of coastal counties from the Oregon border to Monterey

County. Capeweed grows best in full sun to light shade, can tolerate a wide variety of soils, and needs little water to persist once it is established. It is typically planted by homeowners and is used extensively in San Francisco Bay Area landscaping. It is often planted in or near urban wildlands, where it thrives in seasonally wet meadows. Capeweed also will grow in drier soils, spreading during the wet season, then becoming dormant during periods of low water availability. It is subject to frost damage, but can quickly regenerate from the crowns when the weather warms.

WHERE DID IT COME FROM AND HOW DOES IT SPREAD?

A sterile, vegetatively reproducing race of capeweed was introduced to the United States in 1963 from the Cape of Good Hope in South Africa. Capeweed was propagated by Los Angeles State and County Arboretum, and it was made available to the nursery trade in 1965 (Lasca Leaves 1968). It is still available in nurseries and has been widely used in landscaping. It escapes from cultivation to wildlands. Capeweed spreads vegetatively by rooting stolons.

WHAT PROBLEMS DOES IT CAUSE?

Capeweed grows over and displaces other herbs and in coastal grasslands and riparian zones forms monospecific stands of impenetrable mats up to several thousand square feet (Alvarez unpubl. data). It is a rapidly growing groundcover, and, if planted on one-foot centers, will establish full cover within six months (Sunset 1985). Capeweed is an aggressive competitor for water and space, and it seriously threatens native plant communities by crowding out grasses, herbs, and small shrubs. Once capeweed is established, it is difficult for other plants, particularly perennials, to become established (Frey 1984).

HOW DOES IT GROW AND REPRODUCE?

Propagation or reproduction of capeweed is by vegetative means. Flowering occurs principally from March to June, but this variety does not produce fertile seed. It is a sterile race, extremely successful at spreading by sending out extensive stolons (stems) from one individual crown (rosette). These stolons are capable of rooting at each node and forming a new plant that remains attached to the runner until it is capable of making its own stolons (approximately one season). Above-ground stolons develop from the axil of a leaf at the root crown of the plant. Runners grow horizontally along the ground, and are typically about 6.6 feet (2 m) in length; some up to thirteen feet (3.9 m) long have been extracted from capeweed removal sites in the Golden Gate National Recreation Area. Stolons form principally in late winter and throughout spring until water availability decreases. Another method by which capeweed is known to spread is the mechanical removal of a piece of stem or root tuber from an established patch to a new location. Capeweed infestations are often located along roads and trails, particularly where heavy equipment operators perform routine grading, resurfacing, or fill removal activities.

New growth can grow over other herbs and through shrubs. An old colony often has new stems growing over old stems, forming a thickened mat of mostly capeweed over time. Capeweed also can form small tubers 0.4 inch (1 cm) thick and 1.2 inches (3 cm) long. This plant will grow well in less than favorable conditions, does equally well on slopes, flat areas, or mounds, and performs best in full sun (Lasca Leaves 1973). It is hardy to a few degrees below freezing. If

unchecked, a small plant can cover as much as 200 square feet (18 sq m) in a year or two (Mathias 1982). It has a shallow root system.

HOW CAN I GET RID OF IT?

Physical Control

Manual methods: Hand removal has been the primary method of capeweed removal in the GGNRA since 1987. Many tools can aid in removal, but the best found so far is a lightweight hand pick available at garden or hardware stores. A hand pick 12 to 15 in (45-53 cm) in size, with a 10 in (25 cm) head, consisting of a five-inch (13 cm) pick on one side and a five-inch (13 cm) hoe on the other has been found to be most effective. A spading fork can also be useful for loosening densely rooted capeweed prior to hand pick removal.

Approach the removal of capeweed from the outer perimeter of the infestation, carefully locating the growing tips of the runners and gently prying them up with the hand pick, feeling for resistance where each node may have taken root. Strike the soil around the rooted node, or root crown and gently lift the surrounding soil upward to remove the root intact. If you are removing a well established plant, the root crown will appear semi-woody and small tubers may be left behind. Try to remove the crown and roots as intact as possible to avoid leaving tubers.

Avoid breaking the capeweed stems, since stem pieces with nodes will take root if they are left behind. Dormant crowns can be located by following succulent or shriveled stem runners leading to them. The center of the infestation usually contains older, more established plants with deeper root systems. Capeweed cover in the center is also usually denser, since the runners cross each other and form a mat. In dense infestations, all capeweed crowns are not removed during one control effort, so follow-up will be needed.

Solarization: Application of horticultural grade, polyethylene landscape fabric is a successful alternative to hand removal of most large infestations. Landscape fabric is a supple black fabric that prevents sunlight from reaching the plants but allows water and gases to penetrate. Unable to photosynthesize, capeweed exhausts stored food reserves and eventually dies. A minimum of one and a half years is needed to kill 99 percent of the covered capeweed (GGNRA). Landscape fabric is far superior to black plastic, which photodegrades in the field after several months, turns brittle, and crumbles. Wildlife will also eat plastic. Landscape fabric has been re-used for at least five years. Seeds will stick to the fabric, so be sure to remove them if the fabric is to be taken to another plant community. Not all sites are suitable for the use of landscape fabric, since it must be staked down for a long period. It is difficult to maintain cover on steep, rocky, or extremely windy sites. Jute netting staples typically are used to secure the fabric; in rocky, level sites they are useless, so the edges are weighted down with heavy boards, rocks, or logs. A soil trench can also be excavated to bury the edges of the fabric. If there are still native plants in the capeweed site, holes can be cut through the fabric to preserve large native plants, but it is preferable to relocate desirable plants, if possible, to reduce fabric maintenance. Hand weeding is also critical around the edges when the fabric is first applied or capeweed will spread from the site. Straw mulch can be placed over the fabric to conceal it from vandals.

Mechanical methods: Heavy equipment can be an appropriate means of capeweed removal. A small tractor with a front-end loader was successful in removing most of a dense trailside infestation in the GGNRA by scraping the capeweed off the soil surface into a pile. Capeweed was subsequently bagged, and manual follow-up was conducted for several years. Use of heavy equipment greatly reduced the manual labor needed to eradicate the capeweed.

Prescribed burning: There are no reports of attempts to control capeweed by burning.

Biological Control

Insects and fungi: Capeweed is not known to be eaten by California wildlife or invertebrates and has no known pathogens in the central coast region of California. No effective biological control agents have been reported. Horticultural literature reports that it is occasionally eaten by caterpillars, aphids, mites, slugs, and snails but will usually sustain or recover from such herbivory without any treatment (Lasca Leaves 1973).

Literature exists on the forage value of the related fertile form of capeweed for Australian fauna (Rayner and Langidge 1985). Large established patches may experience some root rot, but plants quickly regenerate from stem nodes. An Ortho book (1977) on groundcovers claimed that capeweed is subject to fungus diseases that damage the foliage but generally not the root system, so the plant will usually recover.

Chemical Control

Herbicide application to large, dense capeweed patches can be successful in reducing the density of the infestation. Repeated application of 3 percent glyphosate may be needed to permanently eliminate capeweed. Glyphosate has been recommended for use on the related fertile capeweed. However, ten years of continual herbicide use on the fertile capeweed in Australia resulted in a herbicide-resistant biotype (Powles *et. al.* 1989).

Arundo donax L.

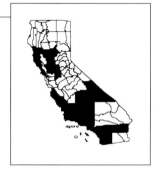

Common names: giant reed, giant cane

Synonymous scientific names: none known

Closely related California natives: 0

Closely related California non-natives: 0

Listed: CalEPPC A-1, CDFG noxious

by Tom L. Dudley

HOW DO I RECOGNIZE IT?

Distinctive Features

Giant reed (*Arundo donax*) is a robust perennial grass nine to thirty feet tall, growing in many-stemmed, cane-like clumps, spreading from horizontal rootstocks below the soil, and often forming large colonies many meters across. Individual stems or culms are tough and hollow, divided by partitions at nodes like bamboo. First-year culms are unbranched, with single or multiple lateral branches from nodes in the second year. The pale green to blue-green leaves, which broadly clasp the stem with a heart-shaped base and taper to the tip, are up to two feet or more in length. Leaves are arranged alternately throughout the culm, distinctly two-ranked (in a single plane). Giant reed produces a tall, plume-like flowerhead at the upper tips of stems, the flowers closely packed in a cream to brown cluster borne from early summer to early fall. Culms may remain green throughout the year, but often fade with semi-dormancy during the winter months or in drought. Giant reed can be confused with cultivated bamboos and corn, and in earlier stages with some large-stature grasses such as *Leymus* (ryegrass), and especially with *Phragmites* (common reed), which is less than ten feet tall and has panicles less than one foot long with long hairs between the florets.

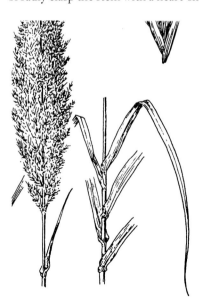

Description

Poaceae. Stems: 30 ft (<9 m) tall, stems erect, hollow, and glabrous 1.6 in (<4 cm) in diameter with somewhat swollen nodes; thick, fleshy rhizomes form creeping rootstocks, yielding dense colonies. Leaves: cauline, sheaths > internodes, ligule thinly membranous and fringed with hairs; blade <3.3 ft (<1 m), 0.8-2.4 in (2-6 cm) wide at base, tapering to a sharp tip, flat or folded, margins scabrous; leaves alternate and conspicuously two-ranked. Inflorescence: as terminal panicle 1-2 ft (30-60 cm) with branches ascending, silver-cream-brown, the numerous spikelets laterally compressed; glumes > florets, membranous and 3-5 veined; florets 4-5, breaking above glumes; lemma 0.3-0.5 in (8-12 mm) and hairy, nerves ending in slender teeth, the middle forming an inconspicuous awn; palea < lemma, 0.12-0.2 in (3-5 mm); anthers 0.1-0.12 in (2.5-3 mm). It does not form viable achenes in North America (Hickman 1993).

WHERE WOULD I FIND IT?

Giant reed occurs in central and southern California and in Baja California, usually below 1,000 feet (350 m) elevation. It has invaded central California river valleys in San Luis Obispo and Monterey counties, the San Francisco Bay Area, and in the Sacramento and San Joaquin River valleys, and is also increasing in the North Coast region (Dudley and Collins 1995). Giant reed has been the most serious problem in coastal river drainages of southern California, especially in the Santa Ana, Santa Margarita, Santa Clara, Tijuana, and other major and minor watersheds, where it sometimes occupies entire river channels from bank to bank (Jackson *et al.* 1994, Bell 1998). Although not currently considered a problem in California deserts, giant reed survives in regularly watered areas of lower-elevation deserts, but does not appear to tolerate high-elevation and continental environments where regular freezing occurs (Sunset 1967).

Giant reed is naturalized and invasive in many regions, including southern Africa, subtropical United States through Mexico, the Caribbean islands and South America, Pacific Islands, Australia, and Southeast Asia (Hafliger and Scholz 1981). In California, the largest colonies occur in riparian areas and floodplains of medium-sized to large streams, from wet sites to dry river banks far from permanent water. Giant reed tends to favor low-gradient (less that 2 percent) riparian areas over steeper and smaller channels, but scattered colonies are found in moist sites or springs on steeper slopes.

Populations also occur in the upper estuaries of coastal streams. It is often found along drainage ditches, where the plant has been used for bank stabilization, and in other moist sites, including residential areas where giant reed is used horticulturally. While it is usually associated with rivers that have been physically disturbed and dammed upstream, giant reed also can colonize within native stands of cottonwoods, willows, and other riparian species, even growing in sites shaded by tree canopy. Plants establish primarily in streamside sites, but expand beyond the margins of riparian vegetation.

Soil preferences are broad, as giant reed is known from coarse sands to gravelly soil to heavy clays and river sediments. It grows best in well drained soil with ample moisture, from freshwater to semi-saline soils at margins of brackish estuaries. In Egypt, Rezk and Edany (1979) found that *Arundo donax* tolerates both higher and lower water table levels than *Phragmites australis*, which is native to California.

WHERE DID IT COME FROM AND HOW DOES IT SPREAD?

Three species of *Arundo* occur worldwide in tropical to warm temperate regions. *A. donax* is

often considered indigenous to the Mediterranean Basin (Hickman 1993) or to warmer regions of the Old World, but apparently it is an ancient introduction into Europe from the Indian sub-continent (Bell 1998). In Eurasia it similarly inhabits low-gradient river courses and may provide useful wildlife habitat in greatly altered river deltas (Granval *et al.* 1993, He 1991).

Giant reed was brought to North America quite early, as it was abundant by 1820 in the Los Angeles River, where it was harvested for roofing material and fodder. This plant has played an important role in the development of music, as the cane was the source of the original Pan pipe or syrinx, and remains the source of reeds for woodwind instruments (Perdue 1958). Commercial plantations exist in California for musical instrument production, and other commercial pos-

sibilities are being explored. Horticultural propagation is widely conducted, and varieties of *Arundo* are available and commonly used in gardens or for erosion control (Sunset 1967). Invasive populations almost certainly resulted from escapes and displacement of plants from managed habitats. It spreads vegetatively either by rhizomes or fragments.

WHAT PROBLEMS DOES IT CAUSE?

Giant reed displaces native plants and associated wildlife species because of the massive stands it forms (Bell 1994, Gaffney and Cushman 1998). Competition with native species has been shown to result from monopolization of soil moisture and by shading (Dudley unpubl. data). It clearly becomes a dominant component of the flora, and was estimated to comprise 68 percent of the riparian vegetation in the Santa Ana River (Douthit 1994). As giant reed replaces riparian vegetation in semi-arid zones, it reduces habitat and food supply, particularly insect populations, for several special status species such as least Bell's vireo, southwestern willow fly-catcher, and yellow-billed cuckoo (Frandsen and Jackson 1994, Dudley and Collins 1995). Un-like native riparian plants, giant reed provides little shading to the in-stream habitat, leading to increased water temperatures and reduced habitat quality for aquatic wildlife. At risk are protected species such as arroyo toad, red-legged frog, western pond turtle, Santa Ana sucker, arroyo chub, unarmored three-spined stickleback, tidewater goby, and southern steelhead trout, among others (Franklin 1996). In the Sacramento-San Joaquin Delta region *Arundo donax* interferes with levee maintenance and wildlife habitat management (Perrine, pers. comm.).

Giant reed is also suspected of altering hydrological regimes and reducing groundwater availability by transpiring large amounts of water from semi-arid aquifers. It alters channel morphology by retaining sediments and constricting flows, and in some cases may reduce stream navigability (Lake, pers. comm., TNC 1996).

Dense growth presents fire hazards, often near urbanized areas, more than doubling the available fuel for wildfires and promoting post-fire regeneration of even greater quantities of giant reed (Scott 1994, Gaffney and Cushman 1998). Uprooted plants also pose clean-up problems when deposited on banks or in downstream estuaries (Douthit 1994) and during floods create hazards when trapped behind bridges and other structures. Although often planted for erosion control, giant reed can promote bank erosion because its shallow root system is easily undercut and bank collapse may follow.

HOW DOES IT GROW AND REPRODUCE?

Plants in North America do not appear to produce viable seed, and seedlings are not seen in the field. Population expansion here occurs through vegetative reproduction, either from underground rhizome extension of a colony or from plant fragments carried downstream, primarily during floods, to become rooted and form new clones. Horticultural propagation is routinely done by planting rhizomes, which readily establish, but stems with no basal material are less likely to root. Fresh stems form roots at nodes under laboratory conditions, but survival is poor (Zimmerman and Bunn unpubl. data), and root formation does occur where an attached culm has fallen over and is in contact with the substrate.

New shoots arise from rhizomes in nearly any season, but are most common in spring. Growth likewise occurs in all seasons, but is highly sensitive to temperature and moisture (Perdue 1958). During warm months with ample water culms are reported to attain growth rates of 2.3 feet (70 cm) per week or about four inches (10 cm) per day, putting it among the fastest growing terrestrial plants. Biomass production has been estimated at 8.3 tons dry weight per acre (Perdue 1958). Young stems rapidly achieve the diameter of mature canes, with subsequent growth involving thickening of the walls (Perdue 1958). Age of individual culms is certainly more than one year, and branching seems to represent stem growth in later years, while rhizomes show indeterminant growth. Branches also form when a stem is cut or laid over. Dieback is infrequently observed, but culms fade or partially brown out during winter, apparently becoming dormant under cold conditions. The outstanding growth trait of this plant is its ability to survive and grow at almost any time under a wide variety of environmental conditions.

HOW CAN I GET RID OF IT?

Studies of giant reed invasion in California are underway, so more data on its biology and management will be available soon. For further information about monitoring and managing infestations, contact Team Arundo in southern California or Team Arundo del Norte in central and northern parts of the state (see Resources section).

Physical Control

Manual methods: Minor infestations can be eradicated by manual methods, especially where sensitive native plants and wildlife may be damaged by other methods. Hand pulling is effective with new plants less than six feet (2 m) in height, but care must be taken that all rhizome material is removed. This may be most effective in loose soils and after rains have made the substrate workable. Plants can be dug up using hand tools (pick-ax, mattock, and shovel), especially in combination with cutting of stems near the base with pruning shears, machete, or chainsaw. Stems and roots should be removed or burned on site to avoid re-rooting, or a chipper can be used to reduce material, although clogging by the fibrous material makes chipping difficult (Dale, pers. comm.). For larger infestations on accessible terrain, heavier tools (rotary brush-cutter, chainsaw, or tractor-mounted mower) may facilitate biomass reduction, followed by rhizome removal or chemical treatment. Such methods may be of limited use on complex or sensitive terrain or on slopes over 30 percent, and may interfere with reestablishment of native plants and animals.

Mechanical methods: Mechanical eradication is extremely difficult, even with a backhoe, as rhizomes buried under three to ten feet (1-3 m) of alluvium readily resprout (R. Dale, pers. comm., Else *et al.* 1996). Removal of all such material is infeasible, especially where extensive soil disturbance would be disruptive.

Prescribed burning: In most circumstances burning of live or chemically treated material should not be attempted, as it cannot kill the underground rhizomes and probably favors giant reed regeneration over native riparian species (Gaffney and Cushman 1998). Burning in place is problematic because of the risks of uncontained fire, the possibility of damage to beneficial species, and the difficulties of promoting fire through patchily distributed stands. There may be some cases where burning of attached material can be done, but only if other means of reducing biomass cannot be carried out. Cut material is often burned on site, subject to local fire regulations, because of the difficulty and expense involved in collecting and removing or chipping all material.

Biological Control

Insects and fungi: No biological control agents against *Arundo donax* have been approved by the USDA, although some invertebrates are known to feed on the grass in Eurasia/Africa (Tracy and DeLoach 1999). The green bug (*Schizaphiz graminum*) has been observed to feed on giant reed in winter (Zuniga *et al.* 1983). In France *Phothedes dulcis* caterpillars may feed on it. The insect *Zyginidia guyumi* uses giant reed as an important food source in Pakistan (Ahmed *et al.* 1977). A moth borer (*Diatraea saccharalis*) has been reported to attack it in Barbados. A USDA evaluation of the potential benefits of biological control against giant reed ranked it as a promising candidate and suggested several insects and pathogens as possible control agents (Tracy and DeLoach 1999).

Grazing: Vertebrate grazers such as cattle and sheep may be useful in controlling giant reed, and Angora goats have been partially successful in reducing this plant and other brush in southern California (Daar 1983). Grazers are unlikely to reduce population size sufficiently to eliminate the risks posed. Likewise, management of native plants to increase competition with giant reed probably provides insufficient control, and in fact seems to offer little resistance against the invading reeds.

Chemical Control

In many, if not all, situations it may be necessary to use chemical methods to achieve eradication, especially in combination with mechanical removal. The most common herbicidal treatment against giant reed is glyphosate, primarily in the form of Rodeo®, which is approved for use in wetlands (Round-Up® can be used away from water). Because glyphosate is a broad-spectrum herbicide, care should be taken to avoid application or drift onto desirable vegetation. The standard treatment is a foliar spray application of 1.5 percent by volume glyphosate with a 0.5 percent non-ionic surfactant (Monsanto 1992). Most effective application is post-flowering and pre-dormancy, usually late August to early November when plants are translocating nutrients into root and rhizomes (TNC 1996). Foliar uptake and kill may be achieved by spray application during active growth periods, primarily late spring through early fall (Monsanto 1992). Small patches can be treated from the ground using backpack or towed sprayers, and major infestations have been aerially sprayed using helicopters.

Direct treatment to cut culms can reduce herbicide costs and avoid drift onto desirable plants, with fair results year-round and best kill in fall, although it appears to be more successful in shaded sites (Else *et al.* 1996, Vartanian, pers. comm.). Concentrated glyphosate solution (50 percent to 75 percent Rodeo, or 27 percent to 40 percent glyphosate is applied to stems, cut within two to four inches (5-10 cm) of the substrate, by painting with a cloth-covered wand or a sponge or spraying with a hand mister. It may be helpful to add a dye or food coloring to the solution to identify treated material. Solution must be applied immediately following cutting because translocation ceases within minutes of cutting; a five-minute maximum interval is suggested (TNC 1996).

New growth is sensitive to herbicides, so a common alternative is to cut or mow a patch and allow regeneration, returning three weeks to three months later when plants are three to six feet (1-2 m) tall to treat new growth by foliar spraying of glyphosate. Promoting regrowth causes nutrients to be drawn from the roots, potentially reducing the movement of glyphosate to the roots (TNC 1996). With all methods, follow-up assessment and treatment should be conducted, and some professional applicators suggest six return spot treatments over six months (Van Diepen, pers. comm.). Other chemical control methods have been tested, including paraquat and triclopyr compounds (Arnold and Warren 1966, Horng and Leu 1979, Franklin 1996), but are not recommended near water.

Atriplex semibaccata R. Brown

Common name: Australian saltbush

Synonymous scientific name: none known

Closely related California natives: 31

Closely related California non-natives: 6

Listed: Cal EPPC A-2; CDFA nl

by Jonathan J. Randall and Marc C. Hoshovsky

HOW DO I RECOGNIZE IT?

Distinctive Features

Australian saltbush (*Atriplex semibaccata*) is a drought-tolerant, low-growing shrub with silvery gray evergreen foliage and small red fruit. It forms a dense groundcover that is fire retardant. The dense mat is less than one foot tall, spreading up to six feet or more across.

Description

Chenopodiaceae. Perennial or subshrub usually less than 13 in (33 cm) tall, from deep woody taproot; often mounded. Stems: usually several stems that range from 1-5 ft (30-150 cm), spreading on the ground or slightly ascending. Stems usually white-scaly or becoming glabrous. Leaves: alternate, subsessile. Blade 0.32-1.2 in (8-30 mm), oblong to narrowly elliptic, entire to wavy-toothed, and often scaly, especially below blade. Inflorescence: monoecious, pistillate flowers axillary, staminate flowers in terminal spikes, each subtended by 2 bracts. Inflorescence ovate to almost diamond-shaped, fleshy, reddish, net-veined, sub-entire. Fruit: with bracts approximately 0.2-0.3 in (4-8 mm), fused to the mid-section or slightly above, becoming thick and fleshy in fruit. Seed: 0.06-0.08 in (1.5-2 mm) (Hickman 1993).

WHERE WOULD I FIND IT?

In California, Australian saltbush is found mostly in waste places, shrubland, or woodland below 3,280 feet (1,000 m) elevation in the Mojave and Sonoran deserts and arid parts of the South Coast, Central Coast, San Francisco Bay Area, and Central Valley as far north as Glenn County. It also inhabits coastal areas and coastal salt marshes from San Diego County to Mendocino County. Australian saltbush is especially fond of heavy saline soils, particularly areas that have been heavily grazed or disturbed. It is quick to invade newly developed lands, roadsides, coastal marshes, and the margins of cultivated fields (Halvorson *et al.* 1988, Hickman 1993).

WHERE DID IT COME FROM AND HOW DOES IT SPREAD?

Australian saltbush is native to Australia and was originally introduced to the United States as livestock forage. It was introduced to California as forage in alkaline areas, starting in Tulare County in 1901. Seeds were distributed throughout the state from there. By 1916 Australian saltbush was abundant in San Diego. By 1940 it was common on the South Coast and found infrequently inland to Imperial and San Bernardino counties and the Salinas Valley (Robbins 1940).

Australian saltbush was promoted as a groundcover in arid landscapes, for erosion control, and to attract birds. Birds eat the red, berry-like fruits and may act as a means of dispersal (Sanders, pers. comm.). The plant escaped cultivation and has become a common weed.

WHAT PROBLEMS DOES IT CAUSE?

As a ground-spreading plant, Australian saltbush displaces native plants.

HOW DOES IT GROW AND REPRODUCE?

Australian saltbush reproduces by seed only. The plant flowers from April to December. Male and female flowers are borne on the same plant. Other similar *Atriplex* species are self-compatible and wind-pollinated, suggesting this also may be true of this plant. Seeds are produced in large numbers and are surrounded by fleshy bracts when mature (Sanders, pers. comm. 1997). These fleshy bracts are attractive to fruit eaters, which may help disperse the seeds. Seeds have been found in the stomach contents of foxes and lizards on Santa Cruz Island (Valido and Nogales 1994, Crooks 1994). Degree of persistence of seeds in soil and germination conditions are unknown.

Seed germination can take place on saline soils, which provides a competitive advantage over other native species (De Villiers *et al.* 1995). Halvorson *et al.* (1988) reported that Australian saltbush was one of the first species to colonize eroded lands on Santa Barbara Island.

This low-spreading subshrub can form mats up to four feet in diameter. It grows in full sun and requires little soil moisture (Plant Advisor 1997). Australian saltbush is believed to be short-lived (two to five years), but in favorable situations it may survive for up to ten years (Sanders 1997). It is deep-rooted (Sunset 1996). Plants will survive winter temperatures well below freezing (Plant Advisor 1997).

HOW CAN I GET RID OF IT?

Physical Control

Manual methods: Australian saltbush is easy to control by hand pulling because of its diminutive size. If it is pulled before it bears seeds, it can be effectively controlled, but any residual seed pool will remain to propagate the local population.

Prescribed burning: The effectiveness of burning as a control method is not known.

Biological Control

Although the subject has not been researched in detail, it has been observed that the larvae from pygmy blue butterflies (*Brephidium exile*) feed on the foliage (Sanders, pers. comm.).

Chemical Control

Chemical control of Australian salt-bush has not been reported, although chemicals that control similar species, such as kochia (*Kochia scoparia*) and Russian thistle (*Salsola tragus*), likely will work on this saltbush. Russian thistle can be controlled with dicamba, 2,4-D, and picloram plus 2,4-D at 1-1.5 fl oz/acre + 0.75 pt/acre. Picloram is currently not registered in California. Kochia can also be controlled with dicamba plus MCPA amine at label

strengths. 2,4-D at 1.0 pt/acre gives good kochia control, but good spray coverage is essential because 2,4-D does not translocate readily in kochia. Treatment should be to plants less than three inches tall or large spray volumes should be used to penetrate the kochia foliage. The esters of 2,4-D generally are more effective than the amines for both weeds. MCPA is not as effective as 2,4-D in controlling either weed. However, MCPA at 1.0 pt/acre will control small kochia plants. Picloram is not effective on kochia; but control is good when combined with 2,4-D ester at 0.75 pt/acre (North Dakota State University Extension Service 1998).

Bassia hyssopifolia (Pallas) Kuntze

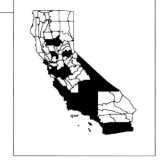

Common names: bassia, five-hook bassia, thorn orache, five-horn smotherweed

Synonymous scientific names: *Echinopsilon hyssopifolia, Salsola hyssopifolia*

Closely related California natives: 0

Closely related California non-natives: 0

Listed: CalEPPC B; CDFA nl

by Marc C. Hoshovsky and Guy B. Kyser

HOW DO I RECOGNIZE IT?

Distinctive Features

Bassia (*Bassia hyssopifolia*) is a grayish annual up to three and one third feet tall; with inconspicuous flowers and younger stems that are densely covered with long, soft, straight hairs. Branches angle out at thirty to sixty degrees from the stem. The small fruit has five distinctive

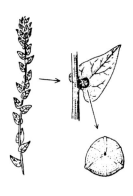

hooked structures on each seed, looking and adhering like a five-legged tick. Overall, this plant looks similar to lambs' quarters (*Chenopodium album*), but it has smaller, elongated, pointed leaves. Bassia is sometimes confused with other members of the Chenopodiaceae, such as Russian thistle (*Salsola tragus*) or kochia (generally *Kochia scoparia*). Russian thistle is more profusely branched and spiny than bassia (Fischer *et al.* 1979). Bassia more closely resembles plants of the genus *Kochia*, within which some taxonomists believe it should be included (Hickman 1993). Bassia is hairier than kochia, produces a pronounced, woolly-looking flowering spike unlike kochia's small clusters of flowers, and has a characteristic five-spined fruit.

Description

Chenopodiaceae. Leaves: linear to lanceolate; largest leaves, found toward base of plant, are 1.6-2.4 in (40-60 mm) long, 0.04-0.14 in (1-3.5 mm) wide, flat, untoothed, and alternate. Bassia flowers between July and October. Inflorescence: spike 0.2-2 in (5-50 mm) long, with oblong, leaf-like bracts 0.08-0.2 in (2-5 mm) long that often wither in fruit. Flowers: tiny, without petals, in axillary groups of one to a few; 5 stamens and 2 or 3 stigmas. Calyx is tan, densely woolly, becoming leathery; attached to the calyx are 5 incurved, hooked spines about 0.04 in (1 mm) long. Fruit: including the persistent calyx, is 0.04-0.06 in (1-1.5 mm) in diameter, containing a single dark brown seed about 0.04 in (1 mm) long (Munz and Keck 1973, Hickman 1993).

WHERE WOULD I FIND IT?

Hickman (1993) reports that bassia occurs widely in California, except in the Klamath and

northern Coast Ranges and in the Sierra Nevada above 1,200 meters (about 3,900 ft). It appears to do well on basic or saline soils. Robbins *et al.* (1970) report the occurrence of bassia in the "spiny salt bush association" of the San Joaquin Valley. It is also common in abandoned agricultural fields in the Owens Valley, the Mojave Desert (Lancaster), the Colorado (Imperial Valley) Desert, the South Coast (Santa Ana River), and northward through the Sacramento Valley (Robbins *et al.* 1970, Sanders, pers. comm.). It is also known from the Sierran foothills, such as near Mount Lassen in the north and Lake Isabella (Kern River) in the south (Ahart, pers. comm., Hewett, pers. comm.). It may also occur in extreme northern California, near Tulelake, on the Modoc Plateau.

WHERE DID IT COME FROM AND HOW DOES IT SPREAD?

Bassia is native to parts of Europe and Asia, particularly around the Caspian Sea. It was first recorded in North America near Fallon, Nevada, about 1915 (Collins and Blackwell 1979). It probably was introduced as a seed contaminant, possibly with Turkestan alfalfa (*Medicago sativa*) seed (Alex 1982). It was found as early as 1921 near Los Banos in the San Joaquin Valley. By 1940 it could be found in spiny saltbush (*Atriplex confertifolia*) and mixed lowland associations of the

San Joaquin, Owens River, Santa Ana River, Imperial, Coachella, and Palo Verde valleys (Robbins 1940). It had also spread to neighboring Arizona and as far as British Columbia and Wyoming, growing well in soils too alkaline for crops. By this time bassia was also established on the East Coast, where it has maintained a limited distribution from Maine to New York City.

Considering the external structure of the fruit, bassia seeds probably disperse by attaching to the fur or feathers of passing animals (Collins and Blackwell 1979). Human disturbances, such as road building or ditch clearing, help to establish bassia and likely contribute to dissemination as well. The seeds do not survive well in fresh water for extended periods (Bruns 1965).

WHAT PROBLEMS DOES IT CAUSE?

Bassia occasionally may displace native species, but there is no evidence that it alters other ecosystem processes (e.g., fire cycles, hydrological cycles, soil chemistry, etc.). On The Nature Conservancy's Kern River Preserve, in the southern Sierra, bassia covers five to ten acres (2 to 4 ha) in a multitude of small clusters, becoming a monospecific stand in the densest areas. Once established, it is somewhat persistent, although it does not appear to be on the increase at the Kern River Preserve. In some areas native species are replacing bassia (Hewett, pers. comm.), suggesting that it is possibly ruderal or stress-tolerant rather than competitive. Because it is toxic to sheep, bassia can be a threat to livestock (James *et al.* 1976).

HOW DOES IT GROW AND REPRODUCE?

Bassia is an annual, reproducing by seeds (Muenscher 1955). Its germination and growth patterns have not been extensively studied (Collins and Blackwell 1979). However, the following can be inferred from the plant's environmental preferences and from the habits of its close relatives: seed dormancy is relatively short; germination requires warm, high-light conditions; germination and seedling growth are not hindered by moderately saline/alkaline conditions; and initial growth is rapid, especially below ground.

HOW CAN I GET RID OF IT?

Physical Control

Mechanical removal: Muenscher (1955) recommends hand pulling of bassia, done most easily after a rain when soil is loose. Plants should be pulled as soon as they are large enough to grasp but before they produce seed. Pieces of root remaining in the soil will not sprout again. Plants can be destroyed readily while they are still small by hand hoeing, either by cutting off the tops or by stirring the surface soil to expose seedlings to drying by the sun.

Prescribed burning: This might be a useful control strategy for bassia, though it has not been tried. Because bassia produces flowers and seeds later in the season than do most range-land plants, there may be a period in mid-summer during which desirable plants have senesced and dropped seed but bassia has not. If such a window exists, dry senesced plants could provide enough fuel to kill bassia before it produces seed. However, most reported infestations of bassia are sparse and limited in extent, so that presence of this weed alone may not justify burning.

Biological Control

Insects and fungi: The only mention of insect herbivory in the literature is of *Lygus* sp. leafhoppers feeding on bassia in late summer (Parker 1972). The degree to which these insects affect the growth of bassia was not reported. No program currently exists for biological control of bassia.

Grazing: Livestock readily graze on bassia, although sheep have died after a single feeding (James *et al.* 1976). Goats have not yet been used to control bassia.

Plant competition: Experience at the Kern River Preserve suggests that minimizing disturbance in non-crop settings may allow more desirable plants to outcompete and replace this weed.

Chemical Control

Chemical control of bassia has not been reported, although it might be similar to control of other similar species, such as kochia and Russian thistle. Kochia and Russian thistle are well controlled by metsulfuron, triasulfuron, thifensulfuron + tribenuron, or tribenuron at label-recommended concentrations. Herbicide-resistant populations of kochia are known to arise, and this can be minimized by tank-mixing these sulfonylurea herbicides with other broadleaf herbicides with differing modes of action.

Russian thistle can also be controlled with dicamba, 2,4-D, and picloram plus 2,4-D at 1-1.5 fl oz/acre + 0.75 pt/acre. Kochia can also be controlled with dicamba plus MCPA amine at label strengths. 2,4-D at 1.0 pt/acre gives good kochia control, but good spray coverage is essential because 2,4-D does not translocate readily in kochia. Treatment should be to small plants (less than three inches tall), or large spray volumes should be used to penetrate the kochia foliage. The esters of 2,4-D generally are more effective than the amines for both weeds. MCPA is not as effective as 2,4-D in controlling either weed. However, MCPA at 1.0 pt/acre will control small kochia plants. Picloram is not effective on kochia; but control is good when it is combined with 2,4-D ester at 0.75 pt/acre (North Dakota State University Extension Service 1998). Because bassia infestations occur in non-agricultural areas and tend to be limited in scope, spot spraying (not broadcast) is the preferred method of herbicide application in most cases. Check herbicide labels for current registered uses in California.

Bellardia trixago L.

Common names: Mediterranean linseed, garden bellardia

Synonymous scientific name: *Bartsia trixago*

Closely related California natives: 0

Closely related California non-natives: 0

Listed: CalEPPC B; CDFA nl

by John M. Randall and Carla C. Bossard

HOW DO I RECOGNIZE IT?

Distinctive Features

Mediterranean linseed (*Bellardia trixago*) is a glandular-hairy annual in the snapdragon family, with erect stems six to twenty inches tall and small, opposite, narrow leaves with coarsely toothed or scalloped margins. The small flowers are arranged in spike-like racemes with leaf-like bracts between the individual flowers, which have a purple hood-like upper lip (galea) and white lower lip. Mediterranean linseed is a green, photosynthetic plant that must parasitize other plants for carbohydrates in order to develop normally and flower (Carafa *et al.* 1980).

Description

Scrophulariaceae. Small annual herb with erect, generally simple stems. Plants glandular-hairy 6-29 in (15-70 cm) tall at maturity. Leaves: opposite above, sessile, linear to lanceolate, 0.6- 3.6 in (1.5-9 cm) long, with coarsely dentate to crenate margins. Inflorescence: spike-like raceme with leaf-like bracts decreasing in size upward; upper cordate with entire margins. Flowers: fused calyx, 0.25-0.33 in (0.8-1 cm) long, with unequal, triangular lobes, 0.06 in (1–1.5 mm) long. Corolla fused, 2-lipped, 0.75-1 in (2-2.5 cm) long, upper lip hood-like, 3-lobed, pale purple, lower lip much longer than upper, two-lobed, usually white or yellow, throat with 2 ridges. On some plants corollas are all-white. Four stamens in 2 pairs, anthers hairy, awned at base. Stigma club-shaped. Fruit: ovoid, loculicidal capsule. Seeds many, ridged, more or less oblong, <0.1 in (0.5-1 mm) long (from Hickman 1993; Tutin *et al.* 1976).

WHERE WOULD I FIND IT?

Mediterranean linseed may be found in disturbed areas and coastal grasslands in places below 700 feet (300 m) elevation in Mendocino County, the North Coast Ranges, the Central Coast, and the San Francisco Bay Area (Hickman 1993; Smith and Wheelor 1990). It can grow on serpentine. It appears to be especially abundant in wet years.

WHERE DID IT COME FROM AND HOW DOES IT SPREAD?

Mediterranean linseed is native to the Mediterranean region from Portugal to Turkey (Tutin *et al.* 1976). It spreads through dispersal of seeds.

WHAT PROBLEMS DOES IT CAUSE?

This species displaces native vegetation. It parasitizes native plants to obtain water and nutrients.

HOW DOES IT GROW AND REPRODUCE?

Mediterranean linseed is a green, photosynthetic plant, but growth studies indicate that it must parasitize other plants for carbohydrates in order to develop and flower (Carafa *et al.* 1980).

It may also obtain water and mineral salts from its host. Plants flower in spring and reproduce by seed.

HOW CAN I GET RID OF IT?

Chemical Control

Only one article was found with information on how to control *Bellardia trixago*, and it focused on control in wheat crops on marginal land in Chile (del Pardo and Encina 1977). None of the herbicides this study found effective are currently registered for use in rangeland or wildland situations in California. Because of its pubescent leaves, this plant is not controlled effectively by foliar herbicides. Del Pardo and Encina found that best results were obtained with terbutryne + simazine + MCPA (as Agren 3614®) at 1.5-2 kg product/ha, cyanazine + MCPA at 4.5 litres product/ha and terbutryne (Igran 50®) at 2 kg product/ha. Agren 3614® was applied at the six-leaf stage of wheat, Igran 50® at pre-emergence and the other treatments at tillering.

No information on other control methods was found.

Brassica tournefortii Gouan

Common names: Sahara mustard, Asian mustard

Synonymous scientific names: none known

Closely related California natives: 0

Closely related California non-natives: 8

Listed: CalEPPC A-2; CDFA nl

by Richard A. Minnich and Andrew C. Sanders

HOW DO I RECOGNIZE IT?

Distinctive Features

Sahara mustard (*Brassica tournefortii*) is an annual herb with stems four to forty inches tall. Plants flower early, but the flowers are small and dull yellow, making them inconspicuous compared to most other true mustards. Petals are less than one-quarter inch. Individual flower stalks are longer than the sepals and spread away from the stem. Fruits have an obvious beak at the tip. The pedicels of the fruits are one-half to four-183
fifths of an inch long and diverge stiffly from the stem at a forty-five degree angle.

Description

Brassicaceae. Annual. Stems: 4-40 in (10-100 cm) tall. Leaves: usually in a moderately well developed basal rosette and quickly reduce in size upward on the stems, so that in the inflorescence only minute bracts are present. Basal leaves 3-12 in (7 to 30 cm) long and deeply lobed. Lobes further toothed. Inflorescence: on a typical well developed branch consists of racemes, which

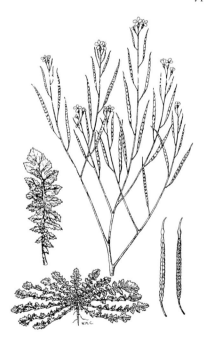

elongate greatly in fruit, of 6 to 20 flowers. Flowers: inconspicuous and dull yellow with 0.6 in (1.5 cm) wide and 0.2-0.3 in (5-7 mm) long petals, only slightly longer than the 0.12 in (3 mm) length of sepals; flowering pedicels 0.2-0.4 in (4-10 mm) long, much longer than the sepals of the open flowers. Fruits: pedicels of fruits 0.4-0.8 in (10-20 mm) long and diverge from the stem at about a 45 degree angle. Fruit a dehiscent silique about 1.4-2.6 in (3.5-6.5 cm) long and 0.088-0.12 in (2-3 mm) in diameter with an obvious terete beak capsule 0.4-0.8 in (1-2 cm) long, but with no stipe; 2 locules with a single row of 7 to 15 seeds in each locule. Seeds: red, globose, and 0.04 in (1 mm) in diameter.

WHERE WOULD I FIND IT?

Sahara mustard is an abundant annual weed at low elevations throughout southwestern deserts of North America, including southern California, southern Ne-

vada, Arizona, New Mexico, west Texas, and northwestern Mexico. In California it occurs in both the
Mojave and Sonoran deserts, but is more common in
the latter. It is becoming increasingly frequent in semi-
arid south coastal California. It is found as high as 3,300
feet (1,000 m) elevation, but is especially abundant below 1,000 feet (305 m).

Apparently uncommon at mid-century, this plant
attracted little attention in southern California until recent decades when it began to be collected widely,
mostly by botanists and students. In the late 1950s it
was known only from Riverside, Imperial, and southwestern San Bernardino counties (Munz 1959). It was
not reported in the Sonoran Desert flora until the mid-
1960s (Shreve and Wiggins 1964). It seemed to have
experienced a population explosion from 1977 to 1983,
during successive years of above-normal precipitation, becoming well established in all counties
of southern California. It was described as being a "rare adventive, usually in desert waste areas"
as late as the mid-1980s in San Diego County (Beauchamp 1986).

Now abundant in the Coachella, Borrego, and Imperial valleys of southeastern California, it is especially common in areas with wind-blown sediments. It is also invading exotic
annual grassland and coastal sage scrub on the coastal slope of southern California. Plants

were observed in Colton Dunes in southwestern San Bernardino County in the 1950s. First noticed in the Riverside area in 1988, it has spread throughout the city, including Box Spring Mountain and Mt. Rubidoux. It is abundant in Baja California, on both coasts. It often forms almost pure stands on abandoned sandy fields. Anderson (pers. comm.) observed a dense stand of the species near San Felipe on the northeast coast in about 1990. It had reached the extreme southern edge of Sonora by 1993 as an uncommon weed along roadsides. As far north as Coso, near the southern Owens Valley of the Mojave Desert, it is an uncommon weed of roadsides.

Sahara mustard is most common in wind-blown sand deposits and in disturbed sites such as roadsides and abandoned fields. It is scarce on alluvial fans and rocky hillslopes.

WHERE DID IT COME FROM AND HOW DOES IT SPREAD?

Sahara mustard is native to semi-arid and arid deserts of North Africa and the Middle East, as well as Mediterranean lands of southern Europe (Townsend and Guest 1980, Tutin *et al.* 1964, Zohary 1966) in habitats similar to those it now occupies in North America. The plant apparently was first collected in North America at Coachella in Riverside County by J.B. Feudge (#1660, RSA) on 25 February 1927. This collection was incorrectly identified as *Brassica arvensis* and was only recently corrected (by Andrew Sanders). Sahara mustard probably was introduced with date palms brought from the Middle East in the early part of this century with the development of the date industry in the Coachella Valley.

During rains, a sticky gel forms over the seed case that permits seeds to disperse long distances by adhering to animals. The rapid spread of *Brassica tournefortii* through the Sonoran Desert, with first occurrences along roadsides, may be related to its ability to adhere to automobiles during rare periods of wet weather.

WHAT PROBLEMS DOES IT CAUSE?

Dense stands in the Coachella and Imperial valleys appear to suppress native wildflowers. Because of its early phenology, it appears to monopolize available soil moisture as it builds canopy and matures seed long before many native species have begun to flower. In coastal southern California, it locally dominates exotic grasslands in dry, open sites, especially disturbed areas. It expands over larger areas when drought suppresses other exotic annuals such as *Bromus rubens*, *Avena fatua*, *Brassica geniculata*, and *Erodium cicutarium*.

Sahara mustard increases fuel loads and fire hazard in desert scrub and coastal sage scrub. It also establishes from a soil seedbank after fire.

HOW DOES IT GROW AND REPRODUCE?

Plants flower or fruit as early as December or January and set seed by February. Most plants are in fruit or dead by April. Time of flowering probably is controlled by the onset of the rainy season. Early flowering may be triggered by hot spells during winter. During warm or dry winters, plants mature at a small size, ripen seeds, and perish by February.

Sahara mustard appears to be self-compatible or autogamous, as there is virtually 100 percent fruit set on most plants. A well developed plant produces between 750 and 9,000 seeds. Seed longevity is unknown, but based on observations of other species of *Brassica*, it is probably several years. There is little evidence of herbivory or seed parasitism.

Sahara mustard may be the most rapidly developing annual in the winter and spring flora of southern California. Once soils have chilled in fall, rains as small as 1.5 in (4 cm) cause mass germination. The period of most rapid growth is from the first winter rains or February to April. Within two to three months, plants can grow to a biomass of 3.0 tons/ha-1, but usually less than 0.5 tons/ha-1. Total biomass does not correlate with annual precipitation because hot, dry spells frequently cause plants to reach premature flowering and fruiting in early winter. Abundance in the Sonoran Desert is sensitive to rainfall. Sahara mustard was found throughout the Sonoran Desert from the Coachella Valley to the Colorado River after heavy (300 percent of normal) rains during the winter of 1991-92. It was virtually absent in the same region after the dry (less than 25 percent of normal) winters of 1995-96 and 1996-97.

The density of Sahara mustard plants can vary with annual climate and fire history. For example, two years of drought during 1989-91 in Riverside County killed off existing red brome (*Bromus rubens*) cover on a dry southern exposure. Sahara mustard populations in this area subsequently increased by almost thirty-five times. During the wet winters of 1991-92 and 1992-93, while plant densities increased, overall biomass decreased, apparently reduced by intraspecific competition. A hot spell in February 1993 caused Sahara mustard to flower early, aborting a potentially productive growing season in moist soils over the following months. After a fire in November 1993, plants continued to decline in both density and biomass, apparently because of the proliferation of other exotics, especially red brome.

HOW CAN I GET RID OF IT?

There appear to have been no efforts to control this weed, so effective control methods are unknown. Experimentation with biological control and mechanical control (such as mowing) would be desirable.

Minimizing soil disturbance by off-road vehicles or construction activities will reduce disturbed habitat for this plant and thus the extent of seed rain on neighboring areas. Populations on roadsides should be suppressed before they can expand into neighboring areas.

Physical Control

Manual methods: Hand pulling might be effective in limited areas when seed pools have been suppressed.

Prescribed burning: The occurrence of this annual in harsh deserts of the Old World has no doubt selected it to survive long periods in soil seedbanks. Therefore, planned burns may not be a useful option. Although fires cause high seed loss, stem densities reach pre-burn levels within one or two growing seasons. Partial seed survival after fire may be related to its hard seed coat.

Biological Control

Insects and fungi: Sahara mustard is closely related to a number of important vegetable crops (broccoli, cauliflower, brussels sprouts, etc.), so it will be difficult to find an agent that will attack this plant but not damage food crops. Even the possibility of transfer of a control agent to a valuable food crop may create political pressures that could prevent importation of the agent.

Grazing: Since Sahara mustard establishes from a seedbank, it is doubtful that grazing could suppress the spread of this annual. Experiments could be undertaken to determine whether foraging interferes with recruitment and growing season biomass by placing livestock in fields of Sahara mustard during early winter (e.g., January).

Plant competition: Establishment of dense cover of exotic annual grasses apparently suppresses this species.

Chemical Control

The extremely early development of this species might make early chemical control a possibility, especially when desirable native species have not yet begun to develop. This should be investigated experimentally.

Bromus madritensis ssp. *rubens* (L.) Husnot

Common names: foxtail chess; red brome, compact brome, Spanish brome

Synonymous scientific name: *Bromus rubens*

Closely related California natives: 9

Closely related California non-natives: 13

Listed: CalEPPC Red Alert; CDFA nl

by Matthew L. Brooks

HOW DO I RECOGNIZE IT?

Distinctive Features

Foxtail chess (*Bromus madritensis* ssp. *rubens*) is characterized by a brush-like inflorescence that becomes a distinctive purplish color at maturity. Plants growing in arid regions are gener-

ally less robust with a more open and rigid panicle than those growing in mesic regions. Seedlings are bright green and hairy, much like those of cheatgrass (*B. tectorum*), the other common brome grass of arid and semi-arid regions in southwestern North America. At maturity foxtail chess is erect with a panicle of ascending florets on short pedicels arranged roughly equally around the peduncle. Cheatgrass is more spreading, with a panicle of florets that are attached to the peduncle by long, thin pedicels hanging loosely to one side, nodding toward the ground. These characteristics persist for many years and are useful in differentiating these species even after the florets have fallen from the plant. Young plants are green, but foliage and inflorescences become purplish at maturity, fading to light tan during the months following senescence. Areas of foxtail chess infestation can be recognized at a distance by their purplish color. Until recently, foxtail chess was considered a distinct species, *B. rubens*, but it is now generally considered a subspecies of its congener, *B. madritensis* (Wilken and Painter 1993, Sales 1994).

Description

Poaceae. Annual. Stems: erect or ascending culms to 16 in (40 cm), bearing small soft hairs. Leaves: blades and sheaths with small, soft hairs; blade to 4 in (12.5 cm) x 0.2 in (5 mm); ligule to 0.2 in (5 mm), whitish. Inflorescence: to 3 in (75 mm) x 1.5 in (37 mm), stiffly erect, dense, ovoid at the top, wedge-shaped at the base; rachis internodes generally <0.1 in (3 mm); reddish brown to purplish at maturity. Flower: spikelet: 4 to 11-flowered, uppermost sterile and reduced; to 2 in (50 mm); densely crowded; wedge-shaped; soft or stiff hairs; cylindric to slightly compressed; sessile; lower glume 1-veined, to 0.3 in (8 mm); upper glume 3-veined, to 0.4 in (10 mm); both glumes narrow, gradually tapering to a short point, smooth or with soft hairs, and translucent. Floret: lemma to 0.6 in (16 mm) x 0.1 in (3 mm); short stiff hairs rough to the touch; back rounded, lance-shaped, 5-veined; awn to 0.9 in (22 mm) and generally straight; palea shorter than lemma, very narrow. Seeds: caryopsis to 0.4 in (11 mm). Seed lanceolate with short stiff hairs, bounded by palea and lemma with long, rough awn (Munz 1959, Wilken and Painter 1993, Bor 1968, Davis *et al*. 1985).

WHERE WOULD I FIND IT?

Foxtail chess is considered a weedy grass in cultivated lands and waste places in its native range (Davis *et al.* 1985; Bor 1968a) and occupies similar habitats in California (Munz 1959, Wilken and Painter 1993). It is a common weedy species of grassland and scrub habitats in arid and semi-arid regions of California, especially those that have been disturbed by wildfire, livestock grazing, off-road vehicles, or agriculture. It is known to occur throughout most of the state except the Sierra Nevada, Modoc County, and northwestern parts. Foxtail chess is supplanted as the dominant weedy brome grass at higher elevations by cheatgrass, and in more mesic lowland regions by ripgut grass (*Bromus diandrus* Roth). Distribution of foxtail chess in California deserts is limited by mineral nutrients and water, but where moist, nutrient-rich microhabitats occur, populations can be quite large (Brooks 1998). Such microhabitats include areas beneath perennial shrubs, crevices on rocky outcrops, and margins of roads and washes.

WHERE DID IT COME FROM AND HOW DOES IT SPREAD?

Foxtail chess is native to southern Europe, northern Africa, and southwestern Asia, where it occurs from sea level to 4,260 feet (1,300 m) on stony or sandy soils of cultivated fields and rangelands in arid to mesic scrub and steppe regions (Bor 1968a, Jackson 1985). It was established in California by 1848 (Frenkel 1977) and appears to have naturalized there by the 1890s (Davidson 1907). By 1904 it could be found in Kern, Mendocino, Orange, Amador, and Contra Costa counties (Robbins 1940). Foxtail chess was common in the Mojave Desert by 1950 (Hunter 1991) and subsequently spread across the deserts of California into southern Baja California and eastward to Texas. Its spread into the Great Basin appears to be limited by sensitivity to low winter temperatures (Hulbert 1955) or competitive exclusion by cheatgrass.

Long-distance dispersal of foxtail chess is accomplished by seeds that lodge in animal fur and in loosely woven clothing. Short-distance dispersal is aided by wind, which blows seeds along the ground until they settle in eddies behind shrubs or rocks or in depressions in the ground.

WHAT PROBLEMS DOES IT CAUSE?

Altered patterns of wildfire, microhabitat characteristics, and nutrient cycling caused by foxtail chess and competition for soil nutrients and light negatively affect native annual plant populations and revegetated plants (D'Antonio and Vitousek 1992). Foxtail chess is highly flammable and promotes wildfires in desert plant communities where fires historically have been infrequent (Brooks in press, Brooks 1998). Wildfires convert woody perennial scrub into non-native annual grassland, which in turn promotes further wildfires. In southern California both coastal and desert plant communities are being "type-converted" into annual grassland dominated by foxtail chess and other exotic annual plants (Brooks unpubl. data). Native reptiles such as snakes and desert tortoises are sometimes killed in rapidly moving fires (Fisher and Esque unpubl. data) fueled by this species.

Although foxtail chess is sometimes grazed by livestock, it is not considered a good forage plant and is generally regarded as having no economic value (Bor 1968). Dried florets become entangled in wool, reducing its value, and lodge in digestive tracts of some livestock, sometimes causing death.

HOW DOES IT GROW AND REPRODUCE?

Foxtail chess reproduces by seed only. It is generally considered a winter annual, emerging in early winter following rainfall and remaining largely quiescent until early spring, when rainfall and higher temperatures stimulate growth and flowering. Plants continue to flower until water stress kills them, typically by the middle of May. Populations increase during years of average to high rainfall. During years of low rainfall a high percentage of seedlings die prior to reproducing, thereby depleting the seedbank. Localized populations of foxtail chess can be virtually wiped out following a few years of drought, suggesting that seed dormancy may not last more than two to three years.

Like many successful annual weeds, foxtail chess can survive up to a year near human habitations and agricultural fields, where it receives enough supplemental water to survive through the summer. Biomass of this species can accumulate over many years, producing annual grasslands with a gray thatch of litter under a mass of tan and red erect and bent-over stems. In desert regions stems can remain rooted and upright for one to three years following death of the plant.

HOW CAN I GET RID OF IT?

Physical Control

Manual methods: Seedlings can be pulled before they produce seeds, but this is practical only on a small scale. Hand pulling may be an option to help revegetated plants become established during the initial stages of restoration projects, but seed rain from plants in adjacent areas will recolonize any open habitat.

Prescribed burning: Burning aids the establishment of foxtail chess in most cases. One exception is fire occurring in spring before seeds are fully mature or have otherwise dispersed to the ground. Naturally occurring spring wildfires can reduce the above-ground biomass of foxtail chess while enhancing that of native forbs in both coastal and desert regions of southern California. Temperatures in fires in grassland and scrub habitats easily kill foxtail chess seeds suspended in the flame zone, but often are not high enough to kill seeds located at or below the soil surface (Brooks 1998). Some perennial plants are more vulnerable to fire in spring than in other seasons. However, the high water content of perennials during spring can provide some protection if the intensity of the fire is low.

Biological Control

Insects and fungi: Some species in the genus *Bromus* are susceptible to both viral and fungal infections. A black smut that destroys the inner part of the spikelet, thereby reducing or preventing seed production, is naturally present in wild populations of foxtail chess in California. Unfortunately, this fungus does not reach levels of infestation that significantly affect population size.

Grazing: Livestock grazing may be used in lieu of hand pulling. Unfortunately, desirable native species are eaten as well, and alterations to the soil caused by livestock may promote further establishment of foxtail chess.

Chemical Control

Various herbicides, including glyphosate, have controlled foxtail chess in agricultural applications, but they are either not practical to use over the large expanses typically infested by foxtail chess or not currently registered for wildland use.

Bromus tectorum L.

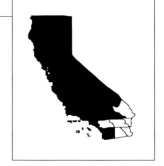

Common names: cheatgrass, downy brome, downy cheat, downy chess, early chess, drooping brome, cheatgrass brome, wild oats, military grass

Synonymous scientific names: none known

Closely related California natives: 9

Closely related California non-natives: 13

Listed: CalEPPC Red Alert; CDFA nl

by Jim Young

HOW DO I RECOGNIZE IT?

Distinctive Features

Cheatgrass (*Bromus tectorum*) typically is a short grass. Seedlings are bright green with conspicuously hairy leaves, which suggests the alternate common name, downy brome. At maturity the foliage and seedheads often become reddish. After maturity the fine herbage is characterized by a light tan reflectance. The nodding open panicles with moderately awned seeds (caryopses) are distinctive. Seeds readily penetrate clothing of passersby.

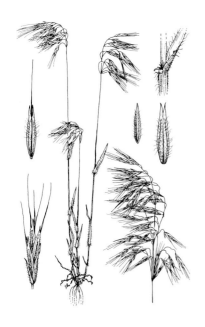

Description

Poaceae. Phenotypically extremely variable annual grass. Can mature at 1 in (2.5 cm) high with single floret or at 24 in (60 cm) with multiple tillers and fertile florets. Leaves: leaf sheaf is densely soft-hairy; blade 1/16-1/8 in (0.1-0.5 mm) wide. Leaf blade can be nearly glabrous to dense soft-haired, but is generally softly cillulate near the base. Inflorescence: open to more or less compact panicle with branches usually nodding. Flower: spikelet subcylindric to slightly compressed; glumes glabrous to short-hairy, lower 0.25-0.5 in (5-8 mm), 1-veined, upper 0.33-0.75 in (7-12 mm), 3-veined. Floret: 3 to 7 per spikelet; lemma body 0.33-0.5 in (9-13 mm) long, 5 to 7 veined, glabrous to short-hairy, tip with 2 teeth, 0.07-0.13 in (1-3 mm) long, awn 0.33-0.75 in (8-18 mm) long. Description adapted from Wilken and Painter (1993) and Hitchcock (1950).

WHERE WOULD I FIND IT?

Cheatgrass is widespread throughout California. It is the dominant annual grass on sagebrush (*Artemisia* sp.) rangelands on the Modoc Plateau in northeastern California and along the eastern Sierra Nevada to Owens Valley. It is relatively rare in the annual range communities of California west of the Sierra Nevada, but widespread throughout the Great Basin, Snake River Plain, and the Columbia Plateau. Cheatgrass is a weed of croplands, especially winter wheat and alfalfa. In wildlands it is most commonly found in sagebrush/bunchgrass communities, although its distribution extends to higher-elevation juniper, pinyon-juniper, and pine woodlands.

Cheatgrass grows in many climatic areas. It is found primarily in locations that receive 6-22 in (15-56 cm) of precipitation. Cheatgrass will grow in almost any type of soil. Research has shown that it is most often found on coarse-textured soils and does not grow well on heavy, dry, or saline soils. Cheatgrass has been found growing on B and C soil horizons of eroded areas and areas low in nitrogen. It grows in a narrow range of soil temperatures. Growth starts at just above freezing and stops when temperatures exceed 60 degrees F (15 degrees C). Litter promotes germination and establishment of seedlings.

WHERE DID IT COME FROM AND HOW DOES IT SPREAD?

Cheatgrass is native to southern Europe, northern Africa, and southwestern Asia, where it occurs from sea level to 5,000 feet (1,500 m). This grass has been in the shadow of livestock production since ruminants were first domesticated in southwestern Asia (Young *et al.* 1972). It has spread to Europe, southern Russia, west central Asia, North America, Japan, South Africa, Australia, New Zealand, Iceland, and Greenland. It was first identified in the United States in 1861 in New York and Pennsylvania and was accidentally introduced to northeastern California late in the nineteenth century. In 1900 it was found in Yosemite National Park, and by 1920 it could be found along the Klamath River, near Yreka (Siskiyou County), Santa Barbara, and Upland in the South Coast (Robbins 1940). It now occurs throughout the United States (including Hawaii and Alaska), except for portions of Alabama, Georgia, South Carolina, and Florida. The hairy seed heads are spread by wind, attachment to animal fur or human clothing, or by small rodents. Contaminated grain seed probably was the early method of dispersal. Seeds can also be dispersed as a contaminant in hay and straw or by mud clinging to machinery.

WHAT PROBLEMS DOES IT CAUSE?

Cheatgrass displaces native vegetation. It outcompetes the seedlings of native and desirable species for soil moisture. In a classic paper Robertson and Pearce (1944) determined that cheatgrass closed communities to the establishment of seedlings of perennial herbaceous species. Subsequently, it has been determined that cheatgrass also interferes with seedling establishment of shrubs such as antelope bitterbrush (*Purshia tridentata*) and with pine (*Pinus* sp.) transplants.

Cheatgrass changes the frequency, extent, and timing of wildfires. The early-maturing fine-textured herbage of cheatgrass increases the chance of ignition and the rate of spread of wildfires. Repeated wildfires lead to the loss of native shrubs and continued cheatgrass dominance (Young and Evans 1978, Young *et al.* 1987). Slow-moving fauna such as desert tortoises also are sometimes killed in the rapidly moving fires (Lovich, pers. comm.).

HOW DOES IT GROW AND REPRODUCE?

Cheatgrass establishes by seeds only. The plant typically flowers from mid-April through June. It has a tremendous seed production capacity, with a potential in excess of 300 seeds per plant, depending on plant density. Plants as small as one inch (2.5 cm) in height may produce seed. Seed production is so abundant that many seeds do not find safe sites for germination. These seeds can remain dormant in the soil for two to three years. Seeds can withstand extremely high soil temperatures. As a general rule, there are twice as many viable cheatgrass seeds in seedbanks as there are plants established in a given year. The dormancy can be broken by gibberellin treatment or nitrate enhancement of the seedbed. Through the mechanism of acquired seed dormancy cheatgrass enjoys the ecological benefits of continuous germination.

In keeping with its flexible flowering traits, cheatgrass can germinate in fall and act as a winter annual. The primary limit to germination is adequate fall, winter, and/or spring moisture. If fall precipitation is limiting and spring moisture is adequate, germination may be delayed until the following spring. Seeds germinate best in the dark or in diffuse light, and they readily germinate at a wide range of temperatures. They do not need to be in contact with bare soil to germinate, and a litter cover generally will improve germination. However, seeds will germinate more quickly when covered with soil, and seedlings rapidly emerge from the top one inch (2.5 cm) of soil. No emergence occurs from seeds buried four inches (10 cm) below the surface.

Shoot growth occurs in early spring and continues until soil moisture is exhausted. Cheatgrass grows rapidly and may produce dry matter at a rate of 2.9 g/mm^2/day. However, growth varies widely from year to year, with practically nothing one year and tons per acre in subsequent years.

Extensive root growth can occur during fall and winter. Cheatgrass will produce roots to depths of seven to eight inches (18-20 cm) before sending out far-reaching lateral roots. These lateral roots are one key to survival of this plant. One study showed that cheatgrass has the capability to reduce soil moisture to the permanent wilting point to a depth of twenty-eight inches (70 cm), reducing competition from other species.

HOW CAN I GET RID OF IT?

Physical Control

Mechanical methods: Mechanical fallows are effective in controlling cheatgrass and establishing herbaceous perennial seedlings. The fallow process accumulates moisture and nitrate to aid in seedling establishment. Tillage in spring after cheatgrass is established is effective if sufficient moisture remains for perennial seedling establishment. Mowing has been shown to reduce seed production when the stand is mowed within one week after flowering. This reduces seed production, but does not eliminate it because plants that develop later and escape mowing will produce seed.

Prescribed burning: Burning of pure cheatgrass stands enhances cheatgrass dominance. This is because wildfires often occur in late summer or fall, a poor time for perennial plants to reestablish. Open ground created by fires is readily colonized by annuals such as cheatgrass. However, burning of mixed shrub-cheatgrass stands generates enough heat to kill most cheatgrass seeds and offers a one-season window for the establishment of perennial seedlings. This is why prompt revegetation after wildfires in sagebrush communities is so important. Because cheatgrass is a cool-season annual, prescribed fire in late spring might help to control this species, especially in areas where native warm-season grasses are desired. A prescribed fire should kill seedlings and further reduce the surface seedbank. Spring burning of the closely related Japanese brome (*Bromus japonicus*) showed that consecutive annual burns reduced brome density and standing crop (Whisenant and Uresk 1990).

Biological Control

Insects and fungi: No insects or fungi have been approved by the USDA for use on cheatgrass. Research into the biological control of cheatgrass is limited. Cheatgrass is often infected with a head smut fungus (*Ustilago bulleta* Berk.) that, when severe, may reduce seed yield. Some research has been conducted on pink snow mold (*Fusarium nivale*) as a biological control agent, but information has yet to be released. In addition to these molds and smuts, over twenty diseases of cheatgrass have been reported.

Grazing: Grazing management systems that favor perennial herbaceous species are excellent tools in the suppression of this pest. This is a good means to avoid the risk of extensive wildfires that cause severe ecological degradation. Late fall and early spring grazing has been shown to significantly reduce plant numbers. However, heavy grazing will promote cheatgrass invasion. Encouraging the reestablishment of native plants improves the effectiveness of grazing as a control method.

Plant competition: Biological suppression is the most cost-effective and least ecologically intrusive method of controlling cheatgrass. Cheatgrass is not competitive with established perennials, particularly grasses. Establishing native perennials is easiest after cheatgrass is removed by other control methods.

Chemical Control

Several effective herbicide techniques used in the past are no longer available. The registrations for these herbicides have either been lost or not renewed because of cost to the manufacturing companies. Glyphosate (as Roundup®, Rodeo®) applications control cheatgrass, but its effectiveness is limited by the environmental conditions during the cold early spring when glyphosate should be applied. Several newer herbicides are being tested for selective control of cheatgrass in perennial broadleaf seedling stands.

Most of the work on the chemical control of cheatgrass has focused on infestations in agricultural crops. Chemical control research in prairies has been primarily limited to atrazine. Herbicides active on cheatgrass in various crops include diclofop, atrazine, simazine, amitrole, imazapyr, sulfometuron, paraquat, and glyphosate. Many herbicides are not specific to cheatgrass or may not be specifically licensed for this use.

Cardaria chalepensis (L.) Hand-Mazz. and
C. draba (L.) Desv.

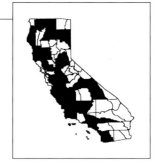

C. chalepensis

Common name: lens-podded hoary cress

Synonymous scientific names: none known

Listed: CalEPPC A-2; CDFA B

by Carla Bossard and David Chipping

C. draba

Common names: white top, perennial pepper-grass, heart-podded hoary cress, pepperwort, white-top, pepperweed whitetop, hoary cress, whitetop, white weed

Synonymous scientific names: *Cochlearia draba, Lepidium draba, Nasturtium draba*

Closely related California natives: 0

Closely related California non-natives: *Cardaria pubescens*

Listed: CalEPPC A-2; CDFA B

by David Chipping and Carla Bossard

HOW DO I RECOGNIZE IT?

Distinctive Features

Three species of hoary cress are found in California: heart-podded (*Cardaria draba*), lens-podded (*C. chalepensis*), and globe-podded (*C. pubescens*). Globe-podded hoary cress is not listed as noxious in California. These *Cardaria* species are single-stemmed, upright perennial herbs, less than knee high, with upper lobed leaves clasping the stem. The inflorescence is typically flat-topped and generally dense with white flowers. Lower leaves form a basal rosette and are somewhat hairy and lance-shaped; higher on the plant the leaves clasp the stem with two ear-like lobes and have fewer hairs. Flowers have four sepals with white margins and four petals narrowing to a claw at the base. Plants tend to form dense monospecific mats with shoots connected by white underground rhizomes (Hickman 1993, Robbins *et al.* 1974)

Cardaria chalepensis

These three *Cardaria* species are differentiated mainly by the shape of their fruit pods. The pods of heart-podded hoary cress are heart-shaped at the base. The pods of the lens-podded hoary cress are circular. The pods of both heart-podded and lens-podded are flattened in cross-section and have no hairs. The pods of globe-podded hoary cress are globular or spherical and are covered with fine hairs (Hickman 1993, Robbins *et al.* 1974).

C. draba

Description

Cardaria chalepensis: Brassicaceae. Perennial herb with rhizomes. Stems: 1 to several, generally erect; 7-14 in (20-40 cm) hairs below, sparse to none above. Leaves: widely oblanceolate to obovate; basal, short petioled, +or- toothed to entire; middle and upper cauline sessile, obovate, elliptic-oblong, or lanceolate, irregularly toothed to entire, clasping stem with cordate-sagittate bases. Inflorescence: variable, elongate with spreading racemes to shortened with racemes forming a corymb, generally dense. Flowers: sepals glabrous; petals 1/8 in (3 mm), white. Fruit: 0.2-0.3 in (2.5-7.5 mm), +or- round to widely ovate, or widely obovate, inflated, not narrowed at septum, glabrous; mature silicles with 1-4 seeds, indehiscent or tardily dehiscent (Hickman 1993).

Cardaria draba: Brassicaceae. Perennial herb typically 6-20 in (15-50 cm) high. Stem: generally solitary and erect, often procumbent in older plants, hairy at the base, sparsely hairy above. Leaves: 1.5-3 in (4-8 cm) long, broadly ovate to lens-shaped, with edges entire and sometimes toothed. Basal leaves more slender but larger than stem leaves, narrow into short petiole, and form basal rosette. Stem leaves sessile, clasping stem with ear-like lobes. Leaves

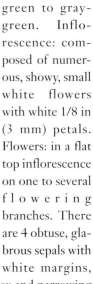

and stem blue-green to gray-green. Inflorescence: composed of numerous, showy, small white flowers with white 1/8 in (3 mm) petals. Flowers: in a flat top inflorescence on one to several f l o w e r i n g branches. There are 4 obtuse, glabrous sepals with white margins, and 4 white petals, broad at the apex and narrowing to a claw at the base. Fruit: somewhat heart-shaped, consisting of two valves and a small persistent style at the apex; becomes inflated and veined at maturity. Seeds: 0.08 in (2 mm) long, 0.06 in (1.5 mm) wide, dark reddish brown with 1-2 seeds in each half, which are finely pitted, slightly flattened, rounded at one end and narrowed to a blunt point at the other. The whitish rhizome-like rootstocks can penetrate downward several feet and laterally as far as 6 ft (1.8 m)

Cardaria chalepensis

(Hickman 1993, Cook 1987, Mulligan and Frankton 1962).

WHERE WOULD I FIND IT?

Cardaria draba and *C. chalepensis* grow in many habitats and areas of the state, except in the Mojave and Colorado deserts (Barbe 1990). *C. draba* occurs in wet and dry grasslands, scrubs, and arid areas with alkali soils. It is most often found in open, unshaded areas on disturbed, generally saline soils, but can grow on almost any soil. It generally is found at elevations of less than 4,000 feet, but it is known from elevations of over 8,000 feet in Utah and over 6,000 feet in Montana. It survives heavy frosts and snowfall, but may favor wetter sites in harsher climates (Cook 1987, Robbins *et al.* 1974).

Cardaria chalepensis concentrations exist in grainfields and hayfields and along roadsides in Siskiyou County, the Sacramento-San Joaquin Delta, and in Kings and Tulare counties (Barbe 1990). It is particularly common in the north on red-brown soils and in disturbed, generally saline soils and fields (Mulligan and Findley 1974).

WHERE DID IT COME FROM AND HOW DOES IT SPREAD?

Hoary cress is native to central Europe and western Asia, probably centering on Turkey,

Georgia, Syria, Iraq, Iran, and Armenia. It has now spread to all continents and is particularly common in many parts of North America.

Cardaria species occur as crop weeds throughout the Middle East, Europe, Australia, and New Zealand. Cook (1987) reports that *C. draba* has been considered England's most serious weed pest since 1949, and the most serious weed pest for the Wimera region of Australia.

Cardaria draba

Hoary cress appeared in New York in 1898, apparently introduced from ship ballast. In 1910 it was introduced into alfalfa fields in the southwestern United States from imported seed from Turkestan (South Australia Dept. of Agriculture 1973; Chipping 1992). It was first reported in California near Yreka (Siskiyou County) in 1876 (Robbins 1940).

Seed is commonly spread in hay and forage such as cut alfalfa, in soil attached to livestock and farm equipment, and by flowing water. Seed may be spread by wind along highways (South Australia Dept. of Agriculture 1973; Chipping 1992). Despite prolific seed production, spread by seed is likely not the most important means of spread. Many infestations remain virtually the same size year after year in spite of annual seed production (Parsons 1992).

Plants also spread by means of extremely persistent root systems, which consist of extensive rhizomes from which shoots emerge (Mulligan and Findley 1974). Another method of dispersal is through movement of root fragments in mud carried by livestock and vehicles, spread by highway maintenance, carried in streams, and spread by tillage (Cook 1987, Robbins *et al.* 1974). Even very small pieces of root are capable of growth. Infestations in areas with frequent disturbance, such as cultivation, regularly increase in size and density (Parsons 1992).

WHAT PROBLEMS DOES IT CAUSE?

Cardaria draba establishes monospecific mats that exclude most or all other herbaceous vegetation. *C. chalepensis* forms dense infestations that crowd out forage plants in meadows and fields. By displacing native vegetation used by wildlife, both species negatively affect native fauna as well. These *Cardaria* species are strong competitors for nutrients and moisture. In Australia

C. chalepensis slowed water drainage, increasing flooding. Gophers increased and, as a result, the amount of soil disturbance also increased, increasing colonization rates of *Cardaria* species. Australians consider *C. chalepensis* toxic to grazing stock (Chipping 1972, South Australia Dept. of Agriculture 1973).

HOW DOES IT GROW AND REPRODUCE?

Growth and reproduction are better understood for *Cardaria draba* than for *C. chalepensis*, although the two may be quite similar. *C. draba* reproduces by seed and expands by creeping roots. Flowering is generally from March to June, but may occur as early as December in mild coastal climates. In large stands that are close to the water table a few flowering plants may be encountered year round. In early spring, infestations may resemble carpets of snow (Cook 1987, Robbins *et al.* 1974, Mulligan and Findlay 1974). Under stressful conditions flowers may develop on stems just four to six inches (10-15 cm) high with just one branch, but in well watered conditions flowering may start when the plant is as small, but continue until many flowering branches have developed and plant height approaches twenty inches (50 cm). The plant is self-incompatible, is pollinated by insects, and can produce 1,000 to 5,000 seeds per stem, with seed viability of about 80 percent. Seeds are small, with 550,000 seeds per kilogram. Plants appear to produce few seeds in dry years and are prolific seeders in wet years. Seedbanks are generally depleted in three years under both irrigated and non-irrigated conditions. Seed survives in uncomposted cattle dung.

Germination in these *Cardaria* species typically occurs in autumn, with the plant overwintering as a rosette and flowering the following spring. Plants can also germinate in spring or early summer and over-summer as a rosette, but flowering still is delayed until the following spring. Seeds can germinate thirty-five to forty-two days after they are released from the fruit, and in California generally do so after the first rains. *C. draba* sometimes appears after a grass fire; the heat presumably breaks the dormancy of seeds lying in or on the soil (Parsons 1992).

Basal rosettes of *Cardaria draba* are formed three to four weeks after sprouting. The rosettes may be formed from seed or, more commonly within infested areas, from sprouts that arise from rootstocks. Plants usually do not flower the first year. After twenty-five days a plant may be rooted to a depth of 10 inches (25 cm) and may have up to six lateral roots with buds. Shallowly buried buds may form rosettes, while deeper buds form new rhizomes.

Hoary cress has a deep, penetrating root system, numerous underground buds, and large food reserves. The extensive root system spreads horizontally and vertically with frequent shoots arising from the rootstock. Within three months roots can extend a foot from the stem, with nearly fifty new shoots and over eighty buds. In the absence of competition, one plant can produce 455 shoots the first year (Cook 1987). Energy is stored in the rootstock during the growing season, and new plants are produced from joints in the roots. The roots may survive complete removal of shoots for a period of one season without noticeable loss in vigor; plants suffer visibly the second consecutive year. Plants must be cut off at depths greater than 20 inches (50 cm) below the soil surface to prevent regeneration from underground parts. Even root fragments will readily regenerate, allowing *C. draba* to be spread by any vector that can carry a root fragment.

HOW CAN I GET RID OF IT?

Physical Control

Manual/mechanical methods: Tillage may control infestations if started at flower bud time and continued every ten days throughout the growing season. Slightly longer intervals may be possible at different times of year, but it is essential that no green leaves be allowed to form. This deprives rootstock fragments of energy, but the process may have to be continued for at least three growing seasons to deplete the seedbank. Care should be taken not to spread fragments of the plant out of the infested area on tillage equipment.

Prescribed burning: *Cardaria* species apparently are favored by fire through removal of competition.

Flooding: For control of *Cardaria draba*, flooding to a depth of six to ten inches (15-25 cm) for about three months can produce 90 percent control of the plant (Cook 1987, Fryor and Makepeace 1978, Pryor 1959, Robbins *et al.* 1974). However, short-term submergence lasting a week has no effect on the plant (Chipping, pers. observation).

Biological Control

Insects and fungi: No USDA recommended biological control agents exist, and potential introductions from the native range are complicated by the large numbers of cruciferous crops. Although the mite *Acerea draba* is effective in sterilizing plants of *Cardaria draba*, it is also found on commercial crops, as is the aphid, *Aphis armoracea* (Cook 1987).

Grazing: Grazing is not effective on *Cardaria draba*, as it survives and resprouts using energy stored in its rhizome-like rootstock. In *C. chalepensis* young plants may be grazed to the ground by cattle and sheep, which also ingest seed heads. Although *C. chalepensis* contains glucosinolates and can be mildly toxic, nutritional levels are adequate to meet the requirements of most livestock, especially in early growth stages. Problems arise as the foliage becomes coarse and bitter as it matures, when plants have low nutritive value compared to other forages (Cook 1987, Robbins *et al.* 1974).

These plants actually are spread by grazing, as cattle ingest seed heads and may become vectors for plant fragments.

Chemical Control

Most research on chemical control of *Cardaria* species has focused on cropland, usually alfalfa, clover, or wheat fields. Experiments commonly include combinations of herbicides and other non-chemical methods in association with the herbicide treatment. Check with your county agricultural agent to determine which of the possible chemical means of control are currently registered for use in wildlands in California.

Different forms of 2,4-D have been tried with limited success in northern California, although Canadian trials have had success with applications at 1 to 2 lbs/acre, repeated for three years to remove the seedbank.

Mixes of 2,4-D ester and dicamba have been applied by aircraft, and mixes of 0.50 2,4-D and 0.25 each dicamba and R-11 surfactant have worked in roadside applications of one gallon of the mix in 100 gallons of water. Airplane application inevitably affects non-target plants and carries with it the danger of drift of the herbicide to non-target areas and surface water.

Chlorsulfuron, which is selective for broadleaf plants, has been used on California rangeland at 0.33-1 oz/acre with limited success, but has a half-life of four to six weeks and affects non-target species. Chlorsulfuron at 0.50-2 oz/acre has been successful in roadside applications in central coastal California. Glyphosate at 1 pt/acre produces 80 percent control at the budding or flowering stage, but is also non-selective (Cook 1987, Fryor and Makepeace 1978, Pryor 1959, Robbins *et al.* 1974).

Carduus pycnocephalus L.

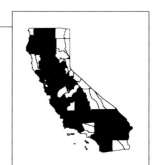

Common names: Italian thistle, slender thistle, shore thistle, Italian plumeless thistle

Synonymous scientific name: none known

Closely related California natives: 0

Closely related California non-natives: *Carduus acanthoides, C. nutans, C. tenuiflorus*

Listed: CalEPPC B; CDFA C

by Carla Bossard and Rich Lichti

HOW DO I RECOGNIZE IT?

Distinctive Features

The annual Italian thistle (*Carduus pycnocephalus*) varies in height from ankle to head high. Its leaves are white-woolly below, hairless-green above, and deeply cut into two to five pairs of spiny lobes. The terminal lobe spine grows longer and more rigid than the other spines. Stems are slightly winged. Flower heads are covered with densely matted, cobwebby hairs. The thimble-

sized, rose to pink to purple flowers are clustered in groups of two to five. The flowerheads are smaller and fewer than those of bull thistle or Canada thistle, and Italian thistle has narrow bracts under its heads with many tiny, firm, forward-pointing hairs on them (Roche 1992). The closely related slender thistle, *C. tenuiflorus*, differs in having stems with continuous wings, leaves with twelve to twenty lobes, and five to twenty heads per cluster.

Description

Asteraceae. Winter annual. Stems: 8 in-6.6 ft (2-20 dm), glabrous or slightly woolly, narrowly spine-winged. Leaves: basal 4-6 in (10-15 cm), 4-10 lobed, cauline +/- tomentose. Inflorescence: heads 2-5 per cluster, sessile or short peduncled, involucre 0.5-0.8 in (1-2 cm) diameter, cylindric to elliptic, phyllary bases loosely tomentose, margins not scarious, tips ascending, linear-lanceolate, spiny, scabrous. Flowers: corollas 0.4-0.6 in (1-1.4 cm), pink to purple, tube 0.2-0.3 in (5-8

mm), throat 0.1 in (2-3 mm) lobes 0.2 in (4-5 mm). Fruit: 0.3 in (4-6 mm) golden to brown, veins 20, pappus 0.4-0.6 in (1-1.5 cm) (Hickman 1993).

WHERE WOULD I FIND IT?

In California, Italian thistle infests areas below 3,000 feet (1,000 m) throughout most of the state except for the Great Basin and northern Mojave Desert. It is common in chaparral and oak savanna in the inner Coast Ranges from Solano County

north. It also occurs in meadows, pastures, and ranges, on roadsides, and in disturbed wildland areas. It is partial to warm, dry mediterranean climate areas, basalt soils, soils of naturally high fertility or soils with a relatively high pH > 6.5 (Bendall 1975). It commonly colonizes disturbed habitats with less intense interspecific competition (Parsons 1977).

WHERE DID IT COME FROM AND HOW DOES IT SPREAD?

Native to the Mediterranean, southern Europe, and North Africa to Pakistan, Italian thistle is now widespread in temperate zones and a major pest in Australia, New Zealand, South America, and South Africa. It was accidentally introduced into United States (Batra *et al.* 1981) and California (Goeden 1974) in the 1930s. Robbins (1940) reports it as early as 1912 near Fort Bragg in

Mendocino County. It is spread by seeds on wind, vehicles, and animals. The seeds are mucilaginous, which aids in dispersal (Goeden 1974, Evans *et al*. 1979). Seeds can disperse by wind an average of seventy-five feet (23 m) from the parent plant and can travel more than 325 feet (108 m) in strong winds. Ants may also play a role in dispersing seed (Uphof 1942). This species can also spread through seed-contaminated hay and soil from infested quarries.

WHAT PROBLEMS DOES IT CAUSE?

Italian thistle dominates sites and excludes native species, crowding out forage plants in meadows and pastures. The blanketing effect of overwintering rosettes can severely reduce the establishment of other plants, as the leaves of the rosette can become erect in dense stands (Parsons 1973). Most animals avoid grazing on it because of its spines. The spines also discourage grazing on neighboring forage species (Parsons 1992). Italian thistle is troublesome in meadows and grasslands, along roadways, in firebreaks, and in utility and railway rights-of-way where control is uneconomical or impractical (Batra *et al*. 1981). It grows well in oak savanna and can carry grass fires to tree canopies.

HOW DOES IT GROW AND REPRODUCE?

Italian thistle reproduces by seed only; it has no means of vegetative reproduction. An annual, it flowers from mid-September to December, and plants die early the following summer. Italian thistle is known to hybridize with *Carduus tenuiflorus* (Pitcher and Russo 1988), although Olivieri (1985) reported low levels of successful seed set in these hybrids. Italian thistle is bisexual, self-compatible, and pollinated by many different insects (Bendall 1975, Evans *et al*. 1979, Olivieri *et al*. 1983). A single plant can produce 20,000 seeds in one season (Wheatley and Collett 1981). Seeds are produced in two forms: brown seeds and silver seeds. Brown seeds generally remain in the flowerheads, falling with them to the ground at the end of the season. These seeds can germinate at lower temperatures than the silver seeds. Silver seeds are dispersed by wind and can remain dormant in the soil longer than the brown seeds, up to eight to ten years (Evans *et al*. 1979, Parsons 1992).

Germination generally occurs in autumn with the first substantial rains. In New Zealand, in a few areas with cold winter temperatures, this species was found to germinate as late as June (Kelly 1988). Seeds can germinate from depths of three inches (8 cm) but usually germinate at less than one inch (0.5–2 cm) (Evans *et al*. 1979). Partly because of its germination requirements and timing, Italian thistle has been rapidly spreading on rangelands previously dominated by alien annual grasses (Evans *et al*. 1979). It germinates under temperature and moisture regimes and in seedbed environments that would inhibit germination of the alien annual grass species that currently dominate California grasslands.

Drought favors an increase in Italian thistle (Wheatley and Collett 1981). Any disturbance of vegetative cover encourages establishment of this thistle. Seedlings establish best on bare disturbed soil; in areas with dense groundcover they cannot establish (Harradine 1985). Plants overwinter as rosettes and produce flowering stalks in late spring before summer drought. The rosettes can be so dense that they blanket the soil, inhibiting germination of all other plants. Italian thistle has a branched, slender taproot.

HOW CAN I GET RID OF IT?

Physical Control

Manual/mechanical methods: Hand pulling is used at Golden Gate National Recreation Area for small patches, but the root must be severed at least four inches (10 cm) below ground level so the plant does not regrow (Alverez, pers. comm. 1997). Plants should be pulled well before seed is set.

Mowing or slashing is not reliable because plants can regrow and still produce seed. A significant amount of seed can be produced even if thistles are consistently mowed at three inches (8 cm) (Tasmanian Department of Agriculture 1977). Cultivation before seed production may eventually eliminate this thistle, but only if repeated until the seedbank is depleted (up to ten years) (Wheatley and Collett 1981).

Biological Control

Insects and fungi: Biological control methods offer limited options for containment of Italian thistle. The subject has been extensively researched, but there are no USDA approved biocontrol agents recommended for use in California. Many insects feed on Italian thistle, but the few that effectively control infestations also feed on economically valuable species (Sheppard *et al.* 1991). Only three insect species, *Psylloidas chalcomera*, *Rhinocellus conious*, and *Ceutorhynchuys trimaculatus*, tested host-specific and caused injury sufficient to decrease reproductive potential of Italian thistle (Goeden 1974). Concern that these insects may prey on several of California's endangered native thistles in the genus *Cirsium* has limited the use of these insects for control of Italian thistle.

Several species of rust fungi infest Italian thistle. *Puccinnia cardui-pycnocephali* is apparently restricted to Italian thistle, although *P. cauduorum*, *P. centaureae*, and *P. galatica* also are found on the plant. Rust fungus reduces growth, especially during the rosette and vegetative phase, but it has insignificant effects on flower or fruit production. Optimal conditions for rust infection and decline of host plants are 18 to 20 degrees C and 90 to 100 percent humidity (Batra *et al.* 1981, Olivieri 1984, Bruckart 1991).

Grazing: Grazing management showed some promising results in control of Italian thistle populations in Australia. Sheep or goats must be used. Infested areas are closed when thistles start to germinate in autumn and not grazed until plants reach a height of four to six inches (10-15 cm). Areas are then heavily grazed at twice normal stocking rate for three weeks (Bendall 1973).

Chemical Control

Clopyralid at label-recommended concentrations was effective in controlling Italian thistle in trials in Australia. Clopyralid can be used on advanced or early-flowering plants and causes seed to abort or be sterile (Sindel 1991). Glyphosate (as Roundup®) applied with a rope wick has resulted in good control in New Zealand. Diquat at label-recommended rates kills thistle seedlings but not seedlings of legumes and pasture grasses (Parsons 1992). Picloram, applied in February or March at concentrations of 1/8 to 1/16 lb acid equivalent per acre was recommended for control of Italian thistle in California (Pitcher and Russo 1988). The herbicides 2,4-D ester and MCPA have been used extensively for control of this thistle, but in the rosette stage plants are not as susceptible to these herbicides as during other life stages. Application of 2,4-D ester should be applied when plants are no more than ten inches tall (25 cm) (Wheatley and Collett 1981). Herbicides are likely also to damage desirable native plants and animal species, and consequently are not appropriate for some infested sites, especially those near water. An integrated, long-term plan with persistent follow-up and twice-yearly monitoring is needed to eliminate this thistle. See herbicide table in appendix for California-registered herbicides.

Carpobrotus edulis (L.) N. E. Br.

Common names: highway iceplant, Hottentot fig, iceplant

Synonymous scientific name: *Mesembryanthemum edule* L.

Closely related California natives: 0

Closely related California non-natives: 8

Listed: CalEPPC A-1; CDFA nl

by Marc Albert

HOW DO I RECOGNIZE IT?

Distinctive Features

Highway iceplant (*Carpobrotus edulis*) is a ground-hugging succulent perennial that roots at the nodes, has a creeping habit, and often forms deep mats covering large areas. Shallow, fibrous roots are produced at every node that is in contact with the soil. Highway iceplant has been widely planted for soil stabilization and landscaping, and is well known by most Californians for its succulent three-sided leaves and its propensity to form deep mats and monospecific stands. In

California flowering occurs throughout the year, peaking in late spring and early summer; flowers do not appear to require specific pollinators.

Highway iceplant is easily confused with its close relative, the more diminutive and less aggressive *Carpobrotus chilensis* (sea fig), and the two species hybridize readily throughout their ranges in California. The large, two and a half- to six-inch-diameter, solitary flowers of highway iceplant are yellow or light pink, whereas the smaller, one and a half- to two and a half-inch-diameter sea fig flowers are deep magenta. Hybrid flowers are pink and intermediate in size. *C. edulis* hybrids also appear to be invasive pest plants in wildlands (Albert *et al.* in press).

Description

Aizoaceae. Succulent perennial. Stems: root at the nodes and have a prostrate, creeping habit. Leaf and stem tissue is medium green with red, orange, or purple along margins and sometimes

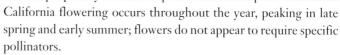

on leaf surfaces, possibly depending upon exposure to sun. Lateral branches can emerge from all nodes along primary axes, and shoots can grow rapidly and overtop each other to form dense mats. Stem tissues become woody after several years' growth. Leaves: opposite and much longer than wide, 2.4-4 in (6-10 cm) in length and 0.2-0.5 in (5-12 mm) in width, with 3 acutely angled edges, one of which is partially to completely serrated. Flowers: terminal, solitary 2.5-5 in (6-12 cm) in diameter with pink, yellow, or yellowish white petals, 1.2-1.6 in (3-4 cm) in length, aging to pink. Flowers have dozens of petals and stamens. Unequal, fleshy sepals are 0.5-1.5 in (1-4 cm) in length. Inferior ovary contains 10-20 chambers. Fruits: fleshy and indehiscent, remain on plants for several months, and are eventually eaten by native mammals or dry out on the plant (D'Antonio 1993, Hickman 1993).

WHERE WOULD I FIND IT?

Highway iceplant is found in coastal habitats from north of Eureka, California, south at least as far as Rosarita in Baja California. It has been planted (along with *Carpobrotus* hybrids) and is still abundant along highways, on military bases, and in other public and private landscapes. It spreads beyond landscape plantings and has invaded foredune, dune scrub, coastal bluff scrub, coastal prairie, and most recently maritime chaparral communities. Establishing readily after disturbance, its seedlings are often seen along roads and on trails and gopher mounds, as well as in areas of open sand and recently burned areas. It is intolerant of frost, and is not found far inland or at elevations greater than approximately 500 feet (150 m).

WHERE DID IT COME FROM AND HOW DOES IT SPREAD?

Native to coastal areas of South Africa, a region with a Mediterranean climate similar to that of coastal California, highway iceplant was brought to California in the early 1900s for stabilizing soil along railroad tracks. It was later used by Caltrans for similar purposes, and until the 1970s thousands of acres were planted with iceplant. For several decades it was also widely promoted as an ornamental plant for home gardens, and it is still available at some nurseries. It will spread easily to natural areas via a number of mammals (D'Antonio 1990a).

Highway iceplant spreads both vegetatively and by seeds. Individual clones can grow to at least 165 feet (50 m) in diameter, and shoot segments can continue to grow if they are isolated from the parent plant. This form of reproduction is important for survival in beach and dune areas in which burial by sand occurs regularly. The abundant seeds are dispersed by generalist mammalian frugivores. Seeds have been found in deer scat more than a kilometer from the nearest clone (D'Antonio 1990a). Its ability to establish and grow in native plant communities differs from one community to another (D'Antonio 1993). In coastal prairie it requires rodent disturbance to provide suitable open soil and is usually outcompeted by grasses at the seedling stage. Once established, however, highway iceplant can spread rapidly by vegetative means. In foredune and dune scrub areas, establishment is limited by herbivory (probably mostly by rabbits) but not by competition, although growth is slow in the dry, low-nutrient conditions. In the less harsh conditions of backdune scrub areas, seedling mortality is high as a result of herbivory, but a moderate rate of growth allows for fairly rapid vegetative spread.

WHAT PROBLEMS DOES IT CAUSE?

Highway iceplant tolerates a range of soil moisture and nutrient conditions and can establish and grow in the presence of competitors and herbivores. These qualities and others have meant that in many natural areas it has formed nearly impenetrable mats that dominate resources, including space. It has invaded foredune, dune scrub, coastal bluff scrub, coastal prairie, and maritime chaparral communities, and competes directly with several threatened or endangered plant species for nutrients, water, light, and space (State Resources Agency 1990). It can suppress the growth of both native seedlings (D'Antonio 1993) and mature native shrubs (D'Antonio and Mahall 1991). In addition, it can lower soil pH in loamy sand (D'Antonio 1990a) and change the root system morphology of at least two native shrub species (D'Antonio and Mahall 1991).

An indirect effect of highway iceplant on the communities it invades can be the build-up of organic matter in normally sandy beach and dune soils, especially in areas where dieback and regrowth have occurred or in areas where iceplant has been treated with herbicide. This can result in invasion by non-native plants that normally would not be able to establish in sandy soils. Another indirect effect is the stabilization of dune sands, resulting in a change in the natural processes that sustain dune community formation over time.

HOW DOES IT GROW AND REPRODUCE?

Highway iceplant can reproduce both vegetatively and by seed. Flowering occurs almost year round, beginning in February in southern California and continuing through fall in northern California, with flowers present for at least a few months in any given population. Seed production is high, with hundreds of seeds produced in each fruit. Fruits mature on the plant and are eaten by mammals such as deer, rabbits, and rodents. Germination is enhanced by passing through animal digestive systems. Seeds in scat were found to have a higher germination rate than seeds from fruits that were not eaten (D'Antonio 1990a, Vila and D'Antonio 1998). Because of the ability to produce roots and shoots at every node, any shoot segment can become a propagule. This allows for survival of individual branch segments when they are isolated from the rest of the plant by being severed or buried by sand. For this reason it is important to remove all material from the site when attempting to eradicate this species.

Active growth appears to occur year round, with individual shoot segments growing more than three feet (1 m) per year (D'Antonio 1990b). All segments can produce roots at the nodes when in contact with soil, allowing for the formation of broad, thick mats. The impact on native competitors changes with the availability of water throughout the year, with the greatest impact occurring in times of drought (D'Antonio and Mahall 1991).

HOW CAN I GET RID OF IT?

Physical Control

Manual methods: Highway iceplant is easily removed by hand pulling, making it a good target for community or school group restoration projects. Because the plant can grow roots and shoots from any node, all live shoot segments must be removed from contact with the soil to prevent resprouting. If removal is not possible, mulching with the removed plant material is adequate to prevent most resprouting, but requires at least one follow-up visit to remove resprouts.

Mechanical methods: Mechanical removal by bobcat or tractor is efficient for areas in which there are no sensitive resources, although in order to prevent significant soil removal, the use of a brush rake attached to the scoop is recommended (Pickart, pers. comm.). Mechanical removal is effective at any time of year.

Prescribed burning: Because of the high water content of shoot tissues, burning of live or dead plants is not a useful method of control or disposal. Attempts to control *C. edulis* by solarization or freezing also have been found to be ineffective (Theiss and Associates 1994).

Biological Control

Insects and fungi: There are currently no biological controls for *Carpobrotus edulis*. The iceplant scale insects, *Pulvinariella mesembryanthemi* and *P. delottoi*, have a small impact on some individuals (Washburn and Frankie 1985), but would likely not be useful as a control tool. In addition, occasional parasitism by dodder (*Cuscuta* sp.) can be seen, but its impact appears to be minimal.

Grazing: Because of the salty and astringent quality of the leaves and the fibrous to woody quality of stems, grazing is unlikely to be an effective control for highway iceplant.

Chemical Control

The herbicide glyphosate has been effectively used to kill *Carpobrotus edulis* clones at concentrations of 2 percent or higher. The addition of 1 percent surfactant to break apart the cuticle on the leaves increases mortality (Moss, pers. comm.). Mortality reportedly is greater when the water utilized is more acidic. Adding an acidifier to hard water before mixing with glyphosate can increase the effectiveness of the treatment (Gray, pers. comm.). It takes several weeks for the clones to die off, and resprouting can occur from apparently dead individuals for several months afterward. Spraying should be avoided in areas in which native species are interspersed with highway iceplant clones. Impacts to native species can be reduced by treating iceplant in early or mid-winter when most native plants are dormant (Moss, pers. comm. 1998). Subsequent growth from seedlings needs to be controlled.

Centaurea calcitrapa L.

Common names: purple starthistle, red star thistle, red starthistle, St. Barnaby's thistle, golden starthistle

Synonymous scientific names: none known

Closely related California natives: 0

Closely related California non-natives: 11

Listed: CalEPPC B; CDFA B

by John M. Randall

HOW DO I RECOGNIZE IT?

Distinctive Features

Purple starthistle (*Centaurea calcitrapa*) is an annual to perennial thistle with a mounding growth habit and heads of purple flowers surrounded by long, stout, sharp-pointed spines. Plants form rosettes in their first growing season, the leaves deeply pinnately lobed and gray-hairy with light-colored midribs; older rosettes have a circle of spines in the center. Mature plants are

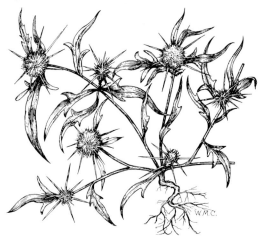

one to four feet high, densely and rigidly branched, and have numerous flowerheads (Roche and Roche 1990). Purple starthistle is similar to Iberian starthistle (*Centaurea iberica*), which is also found in California, but which differs in having seeds that are topped by a crown of bristles. The young heads of purple starthistle are reportedly edible like an artichoke (Allred and Lee 1996).

Description

Asteraceae. Usually behaves as a biennial, but may be an annual or a short-lived

perennial under some conditions. Plants form rosettes in their first growing season. Leaves of young rosettes more or less gray-tomentose with light-colored midribs; leaves of older plants glabrous and resin-dotted. Older rosettes often develop a ring of stout spines at the center before bolting. Plants usually bolt during second growing season. Mature plants 0.5 to 4 ft (20-130 cm) tall, often more or less mounded and densely and rigidly branched. Lower leaves on bolted plants are 4-8 in (10-20 cm) long

and more or less deeply lobed. Flower heads many, each surrounded by leaves, involucres 0.25-0.3 in (6-8 mm) in diameter, 0.6-0.8 in (1.5-2 cm) long, ovoid in outline, main phyllaries greenish or straw-colored and tipped with a stout spine 0.4-1 in (1-2.5 cm) long, spine with a fringe of small spines at base. There are 25-40 flowers in each flowerhead. Corollas 0.6-1 in (1.5-2.5 cm) long, purple. Flowers at margins of flowerheads not enlarged as in some species of *Centaurea*. Achenes about 1/8 in (2.5-3.5 mm) long, white or brown streaked, smooth, with no pappus. The species epithet *calcitrapa* is derived from the word caltrop, a weapon with protruding spikes used in ancient times to obstruct the movement of cavalry. (Description from Hickman 1993, Roche and Roche 1990, Tutin *et al.* 1976).

WHERE WOULD I FIND IT?

Purple starthistle is most troublesome in recently or repeatedly disturbed areas such as pastures and overgrazed rangelands and along roads, ditches, and fences, usually below 3,000 feet (1000 m) elevation. It is most prolific on fertile soils and seems to prefer heavier bottomland and clay soils (Roche and Roche 1990). Found along the coast from San Diego to Humboldt County, in the northern and southern Coast Ranges, and across the Central Valley to the Cascade and Sierra foothills, it is particularly abundant in the San Francisco Bay Area and north into Marin, Solano, Napa, and Sonoma counties (Amme 1985, Havlik 1985).

WHERE DID IT COME FROM AND HOW DOES IT SPREAD?

Purple starthistle is native to the Mediterranean region of southern Europe and northern Africa (Roche and Roche 1990). It was first detected in California near Vacaville in 1886 (Robbins 1940) and has recently become established as a rangeland and pasture pest as far north as Washington. Purple starthistle reproduces only by seed. Since the seeds have no pappus, it is likely that they have been dispersed long distances in hay and straw and on farm and ranch machinery. Some seeds may also be dispersed moderate distances as are those of tumbleweeds, remaining in the flowerheads until after the plants have died, broken off at the soil, and rolled before strong winds.

WHAT PROBLEMS DOES IT CAUSE?

Purple starthistle is a pest of pastures, and in the San Francisco Bay Area it is regarded as a major problem (Roche and Roche 1990). It has also invaded some grassland preserves, notably within the East Bay Regional Park District and at the Jepson Prairie Preserve northwest of Rio Vista. It is not clear whether purple starthistle can form dense infestations in grasslands not subject to heavy grazing or other disturbances, but it is suspected that it can and that it will replace desirable native species.

Purple starthistle's stiff, sharp spines and bitter taste discourage feeding by cattle, deer, and rodents (Amme 1985). It replaces palatable species in some grazed areas, and dense stands of mature plants can make areas inaccessible to livestock and humans (Roche and Roche 1990). Its spines are thicker and stronger than those of yellow starthistle and do not fall from the plants in autumn as do those of yellow starthistle. Because of this, forage that may grow in infested areas during fall and winter after purple starthistle has senesced may be inaccessible to grazers.

HOW DOES IT GROW AND REPRODUCE?

Purple starthistle reproduces only by seed. It is a rosette-forming herb. Most plants remain in the rosette stage for one year, bolt, flower, and set seed in the second growing season, and then die. Some individuals may complete their life cycles in one year in extremely favorable circumstances (annual) or only after several years in unfavorable circumstances (monocarpic perennial) (Roche and Roche 1990). The seeds have no pappus, and most are deposited below or near the parent plant. Viable seeds can be found in the heads of senesced plants, which may break and be blown long distances, scattering seeds as they go (Amme 1985). Longevity of seed in the soil is unknown.

HOW CAN I GET RID OF IT?

Physical control

Manual methods: Grubbing or digging can control small infestations. Amme (1985) reported that purple starthistle populations were sharply reduced after three years of hand grubbing efforts at the Las Trampas site in the East Bay Regional Park system. Plants should be cut at least two inches below the soil surface early in the growing season. They are easiest to see after they have begun to bolt, but they should be cut before they begin to flower in order to prevent the release of viable seed. If plants are cut after they have begun to flower, they should be removed from the site and destroyed. Follow-up treatments will be necessary as field tests indicated that 10-15 percent of plants cut below the root crown resprouted (Roche and Roche 1990).

Mechanical methods: Mowing is not an effective method of control. The rosettes are too low to be cut and plants that have already bolted often respond to mowing by producing multiple rosettes. Mowing plants that have begun to flower will spread the cut flowerheads, which may still be capable of dropping mature seed.

Biological control

Insects and fungi: There is no biological control program for purple starthistle. Two species of *Bangasternus* seed head weevils that have been introduced to control yellow starthistle (*Centurea. solstitialis*) are reported to have 'biotypes' that feed on purple starthistle in Europe but there are no plans to introduce these to North America.

Grazing: Conventional grazing by sheep or cattle will not control purple starthistle and in fact can promote it, because grazing animals usually avoid this plant and selectively feed on species that would otherwise compete with it. It has been suggested, however, that rotational grazing practices with short graze periods followed by recovery periods may reduce purple starthistle and promote grasses and other species that compete with it (DiTomaso pers.comm.).

Chemical control

Clopyralid, 2,4-D and dicamba provided effective control of purple starthistle but had little or no effect on grasses (Whitson *et al.* 1987). Late winter or spring application is recommended because the seedlings and rosettes are most sensitive at this time (Roche and Roche 1981). Amme (1985) reported that a 1 percent solution of glyphosate killed all purple starthistle along a rocky road shoulder.

Centaurea melitensis L.

Common names: tocolote, Maltese star thistle, Napa star thistle, Malta starthistle

Synonymous scientific names: none known

Closely related California natives: 0

Closely related California non-natives: 11

Listed: CalEPPC B; CDFA nl

by Joseph M. DiTomaso and John D. Gerlach Jr.

HOW DO I RECOGNIZE IT?

Distinctive Features

In California, tocolote (*Centaurea melitensis*) grows as a winter annual, producing one to several solitary or clustered, spiny, yellow-flowered heads during spring and early summer. The pre-bolting vegetative characteristics of this species are similar to those of yellow starthistle. Bolting occurs during early spring. The stem leaves extend downward, giving the stems a winged appearance. Flowerheads are generally produced from April through June (approximately four to six weeks before yellow starthistle begins flowering). Flowering plants range from two to thirty-six inches in height and may change from green to bluish green as they senesce. The main phyllaries are pinnately spined with an apical, needle-like spine and a few, much smaller, lateral spinelets. The apical spine bears a characteristic pair of spinelets approximately one-eighth inch from its base. The heads produce only one type of fruit or achene, which is usually light brown with faint tan stripes and always bears a white pappus.

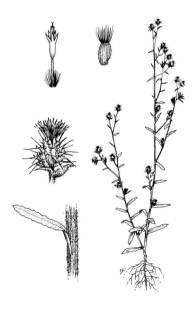

Description

Asteraceae. Stems: at flowering stems range from 2-36 in (5-100 cm) in height. Leaves: basal are entire to lobed, green, resin dotted, often with scabrous hairs, 0.8-6 in (2-15 cm) long; cauline are long, entire, narrow, extending downward on stem; leaf blades 0.4-1.2 in (1-3 cm) long; initially green but sometimes becoming bluish green and lightly covered with cobwebby hairs later in the season. Inflorescences: produced April-June; heads 1 to several, solitary or in groups of 2-3; involucre 0.4-0.6 in (10-15 mm) tall, ovoid, often cobwebby; outer phyllaries with apical appendage pinnately spiny with a few pairs of spinelets; spine 0.2-0.4 in (5-10 mm) long, slender, with a characteristic pair of tiny spinelets about 1/8 in (2-4 mm) from the base. Flowers: many; corollas 0.4-0.5 in (10-12 mm) long, usually equal, yellow; outer ring of flowers sterile, ascending to erect; remaining interior flowers fertile, erect. Fruits: achenes 1/8 in (2-3 mm) long; typically light brown with tan stripes; all with white pappus, bristles usually 1/8 in (2-3 mm) long, but can be as short as 0.04 in (1 mm) (Gerlach unpubl. data), pappus bristles covered with rows of minute barbs; achene attachment scar acute, achene base narrow and hook-like (Hickman 1993, Roché and Roché 1988, Gerlach unpubl. data).

WHERE WOULD I FIND IT?

Tocolote is widely distributed in California, but the largest populations are found in central-western and southwestern regions of the state (Hickman 1993). Scattered small to medium-sized populations occur in the San Francisco Bay Area, North Coast Ranges, and Sierra Nevada foothills.

WHERE DID IT COME FROM AND HOW DOES IT SPREAD?

Tocolote was brought to California during the Spanish mission period. The earliest record of its occurrence was seed found in adobe bricks of a building constructed in 1797 in San Fernando (Hendry 1931). It appears to have been a contaminant in wheat, barley, and oat seed and was widely distributed in dry-farmed grain fields. In one instance its seed was found embedded in an oat floret (Stanton and Boerner 1936). Seed is transported by humans, animals, or wind, similar to starthistle (Gerlach unpubl. data)

WHAT PROBLEMS DOES IT CAUSE?

Dense infestations of tocolote displace native plants and animals, threatening natural ecosystems and nature reserves. It significantly reduces seed production of the endangered plant, *Acanthomintha ilicifolia* (Bauder unpubl. data). Long-term ingestion by horses causes a chewing disease, a lethal lesion of the nigropallidal region of the brain (Kingsbury 1964).

HOW DOES IT GROW AND REPRODUCE?

Tocolote generally flowers from April through June. Inflorescences can produce 1 to 100 heads with 1 to 60 seeds per head (Gerlach unpubl. data). Nothing is known about its pollination biology. Wild oat (*Avena fatua*) litter has been implicated as producing allelopathic compounds that significantly reduce seed germination (Tinnin and Muller 1972). A study of European plants found that young seedlings are resistant to the effects of fall drought (Espigares and Peco 1995). Little is known about the biology of tocolote. Seeds germinate after fall rains, bolting occurs in early spring, and plants flower from spring to early summer.

HOW CAN I GET RID OF IT?

Little work has focused on the control of tocolote. However, it is likely that the strategies used to successfully control yellow starthistle (*Centaurea solstitalis*) will also provide effective control of this species. A possible difference in control would be methods for biological control.

Biological Control

Insects and fungi: No specific biological control agents have been released to control tocolote, and none of the agents introduced to control yellow starthistle have reproduced on tocolote. In a field test of alternate plant hosts, the weevil *Bangasternus orientalis*, which was

Centaurea solstitialis on left; *C. melitensis* on right

C. melitensis

introduced to control starthistle, did not reproduce on tocolote (Woods *et al.* 1995). While no biological control agents have been released to control this species, a small beetle inadvertently introduced into California has been found to destroy mature seeds in the seedhead (Pitcairn and Gerlach unpubl. data). This beetle, *Lasioderma haemorrhoidale*, was first collected in Santa Clara and Fresno counties during 1981, 1982, and 1983 (White 1990). In 1997 the beetle was found on *Centaurea melitensis* seed heads collected in San Diego and Colusa counties. The home range of the beetle is the entire Mediterranean region, and it appears to specialize on species in the thistle tribe (White 1990).

Centaurea solstitialis L.

Common name: yellow starthistle

Synonymous scientific names: none known

Closely related California natives: 0

Closely related California non-natives: 11

Listed: CalEPPC A-1; CDFA C

by Joseph M. DiTomaso and John D. Gerlach, Jr.

HOW DO I RECOGNIZE IT?

Distinctive Features

In California, yellow starthistle (*Centaurea solstitialis*) grows as a deep-taprooted winter annual, or rarely as a short-lived perennial. It produces one to many solitary, spiny, yellow flowerheads during late spring, summer, and fall. Seeds begin to germinate soon after fall rains, and young plants grow as prostrate to ascending taprooted rosettes until bolting occurs in late spring or early summer. Stem leaves of bolted plants extend downward, giving the stems a winged appearance. Flowering plants range from ankle to shoulder height and change color from green to bluish green in summer. Flowerheads are generally produced from June through September. The heads are initially produced on branch tips, but robust plants may produce heads in the branch axils later in the season. The main phyllaries (flowerhead bracts) are palmately spined with a single stout, apical spine and a few much smaller, lateral spines. Some individuals produce shorter apical spines. The heads contain two types of fruits or achenes. Most are cream to tan with a white pappus or plume; achenes in the outer ring are darker and lack a pappus.

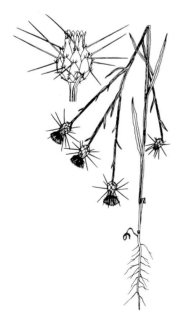

Description

Asteraceae. Annual. Stems: 6-72 in (15-200 cm) in height. Leaves: basal, earliest, entire to slightly toothed; subsequent, lobed to deeply lobed; bright green and scabrous-bristly in seedling and rosette stages; 2-6 in (5-15 cm) long; cauline leaves long, entire, narrow, decurrent; initially green, becoming bluish green and densely covered with cobwebby hairs later in the season; leaf blades 0.4-1.2 in (1-3 cm) long. Inflorescence: produced late May-December; heads 1 to many, always solitary; involucre 0.5-0.7 in (13-17 mm) tall, ovoid; outer phyllaries with apical appendages palmately spiny, central spine 0.4-1 in (10-25 mm) long, generally stout; tips of inner phyllaries with membranous winged tips about 1 mm wide. Flowers: many; corollas 0.5-0.8 in (13-20 mm) tall, unusually equal, yellow; marginal florets sterile, corollas 2-4 lobed,

spreading to ascending; inner florets fertile, 5 lobed. Fruits: achenes 0.08-0.12 in (2-3 mm) long; those produced by outer ring of flowers dull, dark brown to blackish, without pappus; those produced by interior flowers glossy, grayish to mottled light brown; pappus white with bristles 0.08-0.166 in (2-4 mm) long, pappus bristles covered with rows of minute barbs; achene attachment scar obtuse, achene base broad (Hickman 1993, Gerlach unpubl. data).

WHERE WOULD I FIND IT?

Yellow starthistle is most widely distributed in the Sacramento and northern San Joaquin valleys, Inner North Coast Ranges, northern Sierra Nevada foothills, Cascade and Klamath ranges, and the central-western regions of the state (Hickman 1993). There are many

small to large relict populations in the southwestern region of California. It is currently spreading in mountain regions of the state below 7,500 feet (2,250 m) and in the central-western region. It is uncommon in deserts and at moist coastal sites. Primarily it is a problem in moderately warm, exposed areas on fertile, drier soils, including disturbed sites, grasslands, rangeland, hay fields, pastures, roadsides, and recreational areas (DiTomaso *et al.* 1999).

WHERE DID IT COME FROM AND HOW DOES IT SPREAD?

Yellow starthistle is native to southern Europe and western Eurasia and was first collected in Oakland, California, in 1869. It was most likely introduced after 1848 as a contaminant of alfalfa seed. Introductions prior to 1899 were most likely from Chile, while introductions from 1899 to 1927 appear to be from Turkestan, Argentina, Italy, France, and Spain (Gerlach in prep., Hillman and Henry 1928). By 1917 it had become a serious weed in the Sacramento Valley and was spreading rapidly along roads, trails, streams, ditches, overflow lands, and railroad rights-of-way (Newman 1917). In 1919 Willis Jepson observed its distribution near Vacaville and stated: "It is 1,000 times as common as ten years ago, perhaps even six years ago" (Jepson 1919).

Yellow starthistle had spread to over a million acres of California by the late 1950s and nearly two million acres by 1965. In 1985 it was estimated to cover eight million acres in California (Maddox and Mayfield 1985) and perhaps ten to twelve million acres a decade later. It is equally problematic around Medford in southwestern Oregon and in Hell's Canyon in Oregon and Idaho (Maddox *et al.* 1985). It also infests, to a lesser degree, areas in eastern Oregon, eastern Washington, and Idaho (Roché and Roché 1988).

Human activities are the primary mechanisms for the long-distance movement of yellow starthistle seed. Seed is transported in large amounts by road maintenance equipment and on the undercarriage of vehicles. The movement of contaminated hay and uncertified seed is also an important long-distance transportation mechanism. Once at a new location, seed is transported in lesser amounts and over short to medium distances by animals and humans. The short, stiff, pappus bristles are covered with microscopic, stiff, appressed, hair-like barbs that readily adhere to clothing and to hair and fur (Gerlach unpubl. data). The pappus is not an effective long-distance wind-dispersal mechanism as wind moves seeds only short distances, with maximum wind dispersal being sixteen feet (<5 m) over bare ground with wind gusts of twenty-five miles per hour (40 km/hr) (Roché 1992).

WHAT PROBLEMS DOES IT CAUSE?

Dense infestations of yellow starthistle displace native plants and animals, threatening natural ecosystems and nature reserves. Yellow starthistle also significantly depletes soil moisture reserves in annual grasslands in California (Gerlach unpubl. data) and in perennial grasslands in Oregon (Borman *et al.* 1992). Long-term ingestion by horses causes a neurological disorder known as chewing disease, a lethal lesion of the nigropallidal region of the brain. This disease is expressed as a twitching of the lips, tongue flicking, and involuntary chewing. Permanent brain damage is possible, and affected horses may starve to death (Kingsbury 1964). Yellow starthistle interferes with grazing and lowers yield and forage quality of rangelands, thus increasing the cost of managing livestock (Roché and Roché 1988). It can also reduce land value and limit access to recreational areas.

HOW DOES IT GROW AND REPRODUCE?

Plants reproduce only by seed and generally flower from May to September. When adequate moisture is available, yellow starthistle can survive as a short-lived perennial and flower throughout fall, winter, and spring. However, the flowers produced during winter are often killed by frost (Gerlach unpubl. data). Almost all plants are self-incompatible and require pollen from a genetically compatible plant to produce seed (Maddox *et al.* 1996).

European honeybees are an important pollinator, and in some populations are responsible for 57 percent of seed set (Barthell unpubl. data). Seeds produced per head (30-80) and flowerhead production per plant (1-1,000) are variable, depending on soil moisture levels and intensity of competition (DiTomaso, unpubl. data). Large plants can produce nearly 75,000 seeds. Seed production in heavily infested areas varies between fifty to 200 million seeds per acre. Studies of seed survival in soil have found significant survival to ten years (Callihan *et al.* 1993). Seeds typically germinate in late fall or early winter, when soil moisture is present (Maddox 1981) and overwinter as basal rosettes.

Germination responses in yellow starthistle are greatly reduced in dark environments and by exposure to light enriched in the far-red portion of the spectrum (Joley 1995). The two types of achenes also differ in response to light (Joley 1995). During early seedling establishment, root growth is vigorous and can extend deeper than one meter (3.3 ft) (Roché *et al.* 1994, DiTomaso unpubl. data), providing plants with access to deep soil moisture reserves during dry summer months. Reduced light levels cause the rosettes to produce fewer but larger leaves and to assume a more upright growth form (Roché *et al.* 1994). Reduced light levels also significantly reduce root growth and flower production (Roché *et al.* 1994). Consequently, survival and reproduction are significantly reduced in shaded areas, and the plant is probably less competitive in dense stands of established perennials. Bolting occurs from late spring to early summer, and spiny flowerheads generally are produced from early summer to late summer or fall. The spines on the flowerheads may protect them from herbivory by large animals, but they do not prevent significant herbivory by grasshoppers or seed predation by birds (Gerlach unpubl. data).

HOW CAN I GET RID OF IT?

It is important to prevent large-scale infestations by controlling new invasions. Spot eradication is the least expensive and most effective method of preventing establishment of yellow starthistle. In established stands, any successful control strategy will require dramatic reduction or, preferably, elimination of new seed production, multiple years of management, and follow-up treatment or restoration to prevent rapid reestablishment.

Effective control using any of the available techniques depends on proper timing. Combinations of techniques may prove more effective than any single technique. For example, prescribed burning followed by spot application of post-emergence herbicides to surviving plants can prevent the rapid reinfestation of the treated area. Similarly, combining mowing and grazing, revegetation and mowing (Thomsen *et al.* 1996a, Thomsen *et al.* 1996b), or herbicides and biological control may provide better control than any of these strategies used alone. Effective combinations may depend on location or on the objectives and restrictions imposed on land managers.

Physical Control

Mechanical methods: Tillage can control this thistle; however, this will expose the soil for rapid reinfestation if subsequent rainfall occurs. Under these conditions, repeated cultivation is necessary (DiTomaso *et al.* 1998). During dry summer months, tillage practices designed to detach roots from shoots prior to seed production are effective. For this reason, the weed is rarely a problem in agricultural crops. Weedeaters or mowing can also be used effectively. However, mowing too early, during the bolting or spiny stage, will allow increased light penetration and more vigorous plant growth and high seed production. Mowing is best when conducted at a stage where 2 to 5 percent of the seed heads are flowering (Benefield *et al.* 1999). Mowing after this period will not prevent seed production, as many flowerheads will already have produced viable seed. In addition, mowing is successful only when the lowest branches of plants are above the height of the mower blades. Under this condition, recovery is minimized. Results should be repeatedly monitored, as a second or perhaps a third mowing may be necessary to ensure reduced recovery and seed production (Thomsen *et al.* 1996a, 1996b).

Prescribed burning: Under certain conditions, burning can provide effective control and enhance the survival of native forbs and perennial grasses (Robards, unpubl. data, DiTomaso *et al.* 1999a). This can be achieved most effectively by burning after native species have dispersed their seeds but before yellow starthistle produces viable seed (June-July). Dried vegetation of senesced plants will serve as fuel for the burn. At Sugarloaf Ridge State Park in Sonoma County, three consecutive burns reduced the seedbank by 99.5 percent and provided 98 percent control of this weed, while increasing native plant diversity and perennial grasses (DiTomaso *et al.* 1999a). No additional control method was used in the fourth year. In that year, unfortunately, the seedbank of yellow starthistle increased by thirty-fold compared to the previous year (DiTomaso unpubl. data).

Biological Control

Insects and fungi: Six USDA approved insect species that feed on yellow starthistle have become established in California (Pitcairn 1997a and 1997b). These include three weevils, *Bangasternus orientalis*, *Eustenopus villosus*, and *Larinus curtus*, and three flies, *Urophora sirunaseva*, *Chaetorellia australis*, and *C. succinea* (Woods *et al.* 1995). All of these insects attack yellow starthistle flowerheads, and the larvae utilize the developing seeds as a food source. The most effective of these species are *E. villosus* and *C. succinea* (Balciunas and Villegas 1999). With the possible exception of a few sites, the insects do not appear to be significantly reducing starthistle populations, but success may require considerably more time for insect numbers to increase to sufficient levels.

Current evidence indicated a 50 to 75 percent reduction in seed production in areas with significant bioagent populations (Pitcairn and Ditomaso unpubl. data). A root-attacking flea beetle (*Ceratapion brasicorne*) is also being studied (Pitcairn, pers. comm.). Researchers are seeking other starthistle-specific foliar- and stem-feeding insects in Asia Minor. Research is also currently being conducted on three native or naturalized fungal pathogens, *Ascochyta* sp., *Colletotrichum* sp., and *Sclerotinia sclerotiorum* for the control of yellow starthistle seedlings (Woods and Popescu 1997).

Grazing: Intensive grazing by sheep, goats, or cattle before the spiny stage but after bolting can reduce biomass and seed production in yellow starthistle (Thomsen *et al.* 1996a, 1996b). To be effective, large numbers of animals must be used for short durations. Grazing is best between May and June, but depends on location. This can be a good forage species.

Plant competition: Revegetation with annual legumes capable of producing viable seed provides some level of control in pastures (Thomsen *et al.* 1996a, 1996b). In some areas subterranean clover (*Trifolium subterraneum*) proved to be the best of sixty-six legumes tested. In other sites rose clover (*T. hirtum*) and/or perennial grasses may be the preferred species. Control was enhanced when revegetation was combined with repeated mowing (Whitson *et al.* 1987).

Chemical Control

Although several non-selective pre-emergence herbicides will control yellow starthistle, few of these can be used in rangeland or natural ecosystems. The exception is chlorsulfuron, which provides good control in winter when combined with a broadleaf selective post-emergence compound. However, chlorsulfuron is not registered for use in rangelands or pastures.

The primary options for control in non-crop areas are post-emergence herbicides; 2,4-D, triclopyr, dicamba, and glyphosate (DiTomaso *et al.* 1998). All but glyphosate are selective and preferably applied in late winter or early spring to control seedlings without harming grasses. Once plants have reached the bolting stage, the most effective control can be achieved with glyphosate (1 percent solution). The best time to treat with glyphosate is after annual grasses or forbs have senesced, but prior to yellow starthistle seed production (May-June). The most effective compound for yellow starthistle control is clopyralid (as Transline®), a broadleaf selective herbicide (DiTomaso *et al.* 1998). Clopyralid provides excellent control, both pre-emergence and post-emergence, at rates between 1.5-4 acid equivalent or 4-10 oz formulated product per acre. Although excellent control was achieved with applications from December through April, earlier applications led to significant increases in quantity of other forage species, particularly grasses.

Cirsium arvense (L.) Scop.

Common names: Canada thistle, California thistle, creeping thistle, corn thistle, perennial thistle, field thistle

Synonymous scientific names: *Cirsium lanatum, Serratula arvense*

Closely related California natives: 20

Closely related California non-natives: 3

Listed: CalEPPC B; CDFA B.

by David E. Bayer

HOW DO I RECOGNIZE IT?

Distinctive Features

Canada thistle (*Cirsium arvense*) is a persistent perennial thistle that grows vigorously, forming dense colonies and spreading by roots growing horizontally that give rise to aerial shoots. Plants generally grow one to four feet tall, but on occasion may grow more than six feet tall and branch freely. Stems are smooth, mostly without spiny wings, green and glabrous. Flower heads are numerous, small, and almost spineless. Flowers are purplish lavender or, less commonly, white.

Description

Asteraceae. Perennial, vigorous plant that spreads by horizontally growing root stalks. Stems: smooth, mostly without spiny wings, green and glabrous. Plants 1-4 ft (30-120 cm) tall with green foliage. Leaves: alternate, margins variable from entire to deeply lobed, 2-8 in (5-20 cm) in length, sessile, slightly clasping or shortly decurrent. Spines along margins of leaf 1/8-3/8 in (3-5 mm) in length. Inflorescence: flowerheads small, 0.4-1 in (1-2.5 cm) diameter and approximately 1 in (2.5 cm) tall, rounded or flat-topped; involucre hemispheric to ovoid, outer phyllaries ovate, tipped with stout spines 1/16 in (1 mm), inner phyllaries progressively longer with tips flattened. Flowers: plants dioecious with all flowerheads either pistillate (female) or staminate (male). Florets all tubular, generally purplish lavender to, less commonly, white. Staminate flowers (male) with corolla 1/2 in (12-13 mm) long, tube 1/3 in (8 mm) and lobes 1/8 in (3-4 mm). Pistillate flowers (female) with corolla 3/5-3/4 in (14-20 mm) long, tube 3/8-5/8 in (10-16 mm), and lobes 1/16-1/8 in (2-3 mm). Pappus abundant, white, feathery (branched), to 3/4-1 3/4 in (20-30 mm) long and fragile, easily broken off at the achene. Fruits: achenes light brown, approximately 1/8 in (3 mm) long, slightly grooved along long axis, flattened, and occasionally curved. There is an inconspicuous yellow rim at the apex where the pappus joins the seed with a small conical point in the center. Base of seed slightly rounded (Hickman 1993).

WHERE WOULD I FIND IT?

Canada thistle is common throughout northern California, including the Sierra Nevada, Modoc Plateau, Central Valley, Coast Range, and San Francisco Bay Area. Although less vigorous in southern California, it has been found along the coast as far south as Orange and Riverside counties. Inland it is also known to occur in Kern County and in montane areas of Fresno, Inyo, and Mono counties (Barbe 1990). The southern distribution is probably limited by high temperatures.

Canada thistle grows on a variety of soil types. It does well on deep, well aerated, moist loam soils, but is known to grow in dry habitats and on sandy soils. It may also grow on stream

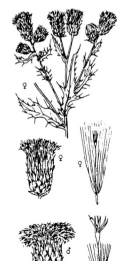

banks, in meadows, and even in wet ditches, but it will not survive in saturated soil. It is intolerant of shade, requiring good light conditions for aggressive growth (Moore 1975, White *et al.* 1993).

Canada thistle infests many habitats such as cultivated fields, roadsides, pastures and rangeland, railway embankments, and lawns. It is a major pest in streamside grasslands from the Pacific Northwest eastward to the plains. It also invades moist prairies.

WHERE DID IT COME FROM AND HOW DOES IT SPREAD?

Canada thistle is native to southeastern Europe and the eastern Mediterranean area. It has spread to most temperate parts of the world and is considered an important weed in thirty-seven countries (Parsons 1992). It is particularly troublesome in cooler areas of North America, extending from coast to coast in both Canada and the United States. It was introduced into North America in the seventeenth century by French settlers as a contaminant in crop seed (Moore 1975).

It was probably introduced into California the same way.

Canada thistle spreads by seed, either by wind or as a contaminant in crop seed. The pappus is fragile and easily separated from the seed, so most seeds stay in the vicinity of parent plants. The seeds float and are easily distributed by water. Seeds can be spread in mud attached to farm equipment. Infested packing material has also contributed to its spread.

Once established, Canada thistle spreads rapidly by horizontal roots, up to several meters per year. The extensive horizontal root system assures long-term persistence and spread by vegetative means. A segment of root as small as 1/8 to 3/8 inch (3-6 mm) in length and 1/16 inch (1 mm) in diameter is able to propagate a new plant and is easily spread with plant material or by equipment (Rogers 1928, White 1979).

WHAT PROBLEMS DOES IT CAUSE?

Canada thistle is considered a noxious weed in California, and in many states it is considered one of the most serious pests to agriculture. In North America heavy infestations reduce yields of spring cereals by 40 to 70 percent. Canada thistle occurs on 40 percent of cultivated lands of the Canadian prairie provinces, and loss of wheat in Saskatchewan alone was estimated to be worth $4 million in 1980. Approximately one million hectares of grassland are infested in England and Wales (Parsons 1992). Once established, Canada thistle is a fierce competitor for nutrients and water needed by crops or native plants. It produces allelopathic chemicals that assist in displacing competing plant species (Stachion and Zimdahl 1980). It has been reported to accumulate nitrates that cause poisoning in animals (Fuller and McClintock 1986). The spiny leaves scratch animal skin, sometimes causing infection, or, at a minimum, restrict animal grazing in heavily infested areas (Moore 1975).

HOW DOES IT GROW AND REPRODUCE?

In established plants of Canada thistle, carbohydrates move from the root system up to the newly forming shoots as growth starts in spring. As leaves on the shoots develop, photosynthates start moving to newly developing roots and flowerheads. The developing flowerheads take more and more of the energy (photosynthates) produced by the leafy stems and stored in the roots. Carbohydrates in the root system are at their lowest when the plant begins flowering. As seeds develop, photosynthates start moving down into the root system again and, after the seed reaches maturity, carbohydrate movement to the root system continues until frost kills the foliage.

Canada thistle typically flowers from late June through August. Male and female flowers are borne on separate plants. Seed production thus depends on having both male and female plants present within 330 feet (100 m) of each other, which often requires at least two introductions. It is rare to find a clone of Canada thistle that has both female and male parts in the same flower or that produces more than the occasional apomictic seed (Hodgson 1964, 1968). Pollination is almost exclusively by insects. Male and female plants growing in close proximity result in high rates of seed production, with some plants producing over 5,000 seeds per plant.

Viable seeds are formed eight to ten days following pollination. Seeds mature in late summer or early fall. They may germinate immediately after falling from the plant if conditions are favorable, or they may remain dormant in the soil for up to twenty-one years (Detmers 1927, Moore 1975, Lalonde and Roitberg 1994). Seeds that germinate immediately form a rosette that overwinters and flowers the following summer. However, germination of most seeds is delayed until spring (Moore 1975).

Approximately 90 percent of seeds germinate within one year. Some seeds remain dormant in the soil for several years. Viability is a function of age of seed and depth in the soil. The deeper the seed is buried, the longer the viability. Ideal conditions for germination are abundant soil moisture and temperatures averaging 20 to 30 degrees C (White 1979).

Seedling survival is poor. Seedlings do not establish in areas with existing groundcover and survive only on disturbed or bare areas in unshaded situations (Parsons 1992). Seedlings develop rapidly, first developing a taproot. Approximately six to eight weeks later, plants develop lateral roots that grow more-or-less horizontally. These lateral roots last approximately two years before they die and disintegrate. In the meantime, they produce new shoots and plants, resulting in a clonal infestation that spreads outward rapidly, approaching twenty feet (6 m) in a single season (Donald 1990).

Despite the taproot, over half of the root system grows in the top twelve inches (30 cm) of soil. Roots go much deeper than that, however, with some vertical roots penetrating fifteen to twenty feet (5-7 m) deep (Haderlie *et al.* 1987). New plants readily emerge from small root fragments. Shoots can emerge from root fragments at a depth of at least twenty inches (50 cm).

Within two years, plants can produce over sixty-six feet (20 m) of new roots (Parsons 1992). Growth is prolific, and patches exist with 130 shoots per square meter.

HOW CAN I GET RID OF IT?

The most effective method of control depends on the site, the extent of the infestation, the presence or absence of both male and female flowering plants, growth characteristics of the plant, and the impact of the control method on non-target species. The many ecotypes respond differently to control measures. There is no easy method of control, and all methods require follow-up. Combinations of mechanical, cultural, and chemical methods are more effective than any single method used alone (Trumble and Kok, 1982).

Regardless of the method or methods selected, competition from other plants should always be considered in the control program. Because of the longevity of viable seeds in the seedbank and the short time from germination to perennation, monitoring of control sites to locate and eliminate new seedlings and resprouts is essential. The site should be monitored at least twice a year for the first four years, and should follow rain or irrigation by a month to six weeks. Monitoring should be more frequent at the beginning of a control program and can become less frequent with time.

Physical Control

Mechanical methods: Cultivation is not generally recommended unless it is carried out with care and persistence, because it often increases the problem by spreading root fragments to new locations. Repeated cultivation at regular intervals of twenty days is effective in exhausting the remaining root fragments but does not kill ungerminated seeds. Care needs to be taken not to remove or disturb other desired native plants in the area. Cultivation must start as soon as the plants first emerge in late winter. Shallow cultivation in hot, dry weather is best.

Repeated mowing at three-week intervals will weaken the plant, prevent flowering, and seed production, and generally can be timed to avoid major impacts on desirable plants growing in the infested area.

Prescribed burning: Repeat burning has shown some reduction in old, established stands of Canada thistle, but overall control generally is less than satisfactory. Removal of old plant residue resulting from fire may promote earlier seed germination of native species (Olson, 1975).

Biological Control

Insects and fungi: There are no effective biological control organisms available at this time. There are several insects, such as *Cassida rubiginosa*, *Cleonus piger*, *Orellia ruficauda*, and *Vanessa cardui*, that feed on Canada thistle and cause some damage, but none effectively control Canada thistle populations (Moore 1975, Maw 1976). *O. ruficauda* has been reported to be the most effective. The rust species *Puccinia obtegens* has shown some promise for controlling Canada thistle, but it must be used in conjunction with other control measures to be effective (Turner *et al.* 1980).

Grazing: Sheep and cattle graze on Canada thistle when the plants are young and tender, helping to deplete the root reserves. Recent New Zealand studies show that goat grazing can be effective. Grazing and trampling can increase stress on plants, enhancing the effectiveness of other control measures such as herbicides. Livestock tend to avoid grazing in and around dense patches of older plants. If young sheep eat older plants, their tender mouth parts may be damaged by the spines and become infected (Parsons 1992).

Plant competition: Vigorous competition from native plants is essential in achieving lasting control.

Chemical Control

Chemical control should be used cautiously and in strict compliance with the label, realizing that some non-target plant species may be killed or seriously affected by the herbicide. Herbicides applied to the foliage will enter the plant and be translocated or move with water in the transpiration stream. Those herbicides that land on the soil may be taken up by the roots and transported in the plant by water in the transpiration stream. Herbicide use must be timed to the growth stage and physiology of Canada thistle (Tworkoski, 1992) and is most effective when used in combination with competition from other plants. Thorough coverage of all foliage of an infestation (clones) is essential because all shoots from infestations over two years old may not be interconnected. Properly timed repeat applications of selected herbicides will always be necessary to achieve complete control of Canada thistle. Applications may be made either in spring or in fall, but fall applications have provided the most control from a single application.

Glyphosate is a non-selective foliar applied herbicide that will kill or seriously injure all growing vegetation with which it comes in contact. Once it penetrates a plant, it is transported efficiently from cell to cell. It has little or no soil residual except on light, sandy soil low in organic matter. Applications should be made soon after flowering when photosynthates are moving from the foliage to the roots. When the first application is properly timed, extensive injury will occur to the developing root system. A second application or another herbicide will be necessary to kill the crown and new shoots arising from old roots that were not affected by the first application.

Triclopyr, dicamba, and 2,4-D are all foliar-applied herbicides that may seriously affect grasses and other monocots as well as many dicotyledon plants. Like glyphosate, they are efficiently transported from cell to cell once they enter the plant's tissue. They all have some soil residual, but dicamba has the longest (up to six months). Application should be made when Canada thistle is actively growing and when photosynthates are being translocated from leaves to roots.

Clopyralid or clopyralid + 2,4-D has shown good control of Canada thistle where the root system has been disturbed. These herbicides are foliar applied and generally have a soil residual of one to two months. Repeat applications for two to four years generally have provided complete elimination of established root systems. Caution should be exercised because many broadleaf plants may be injured. Clopyralid leaches readily, especially in open sandy soil.

Cirsium vulgare (Savi) Tenore

Common names: bull thistle, spear thistle

Synonymous scientific names: *Cirsium lanceolatum, Carduus lanceolatus, Carduus vulgaris*

Closely related California natives: 20

Closely related California non-natives: 3

Listed: CalEPPC B; CDFA nl

by John M. Randall

HOW DO I RECOGNIZE IT?

Distinctive Features

A coarse biennial, bull thistle (*Cirsium vulgare*) is distinguished from other thistles by the following combination of characteristics. Leaf blades, especially those that are larger and deeply lobed, are rough to the touch like medium sandpaper and dark green. Stems of bolted plants appear winged because leaf blades continue along the petioles and several inches down the stems. Flowerheads are one to two inches wide and one and a half to two and a half inches high with deep purple flowers. The bristles on the pappus (thistledown) are feathery. This characteristic distinguishes thistles in the genus *Cirsium* from those in the genus *Carduus*, which have un-branched, thread-like bristles.

Description

Asteraceae. Biennial thistle, 2-6 ft (60-200 cm) tall when mature. In the first year plants form a rosette that may grow to over 3 ft (1 m) in diameter. Rosette leaves oblanceolate to elliptic in

outline and coarsely toothed on older plants. Plants usually bolt in the second year and may have a single stem, a branched stem, or many stems from a single root crown. Stems lightly covered with fine, white, cobwebby hairs and become woody with age. Leaves on stems of bolted plants alternate, up to 1 ft (30 cm) long, lanceolate and deeply lobed; green and rough with coarse hairs above and finer hairs on the underside, the midribs and veins extending beyond the leaf blades to form long, fierce spines. Leaf blades extend down petiole and along stem, forming long, prickly wings. Plants may have one to many flowerheads (inflorescences or clusters of flowers), each with an involucre composed of several overlapping series of spine-tipped bracts (phyllaries). Bracts overlap, are lanceolate to linear in outline, 1-1.5 in (3-4 cm) high, and tightly clasp the flowerhead when it is immature, but begin to

curl away after flowering begins and spread widely when seeds are released. Flowers bisexual, purple (a white-flowered form exists but is rare), 1-1.5 in (3-4 cm) long, producing abundant nectar. Flower has a tubular corolla with 5 long, narrow lobes. Tube is about 1 in (23 mm) long with a <1/4 in (4-6 mm) long bell at its base where nectar collects. Stamens extend well beyond petals. Seeds roughly 1/8 in (3.5-4.0 mm) long and roughly 1/16 in (1.2-1.7 mm) wide, yellowish brown, streaked with black or purple, and with a narrow yellow band at the top. Each seed topped with a white pappus up to 1 in (2-2.6 cm) long with many feathery bristles. The pappus is easily broken off and usually is absent from seeds found on or in soil (description from Delorit 1970, Hickman 1993, Moore and Frankton 1974).

WHERE WOULD I FIND IT?

Bull thistle is widespread in California and is most common in coastal grasslands, along edges of fresh and brackish marshes, and in meadows and mesic forest openings in the mountains below 7,000 feet (2,120 m). It is most troublesome in recently or repeatedly disturbed areas such as pastures, overgrazed rangelands, recently burned forests and forest clearcuts, and along roads, ditches, and fences. Even small-scale disturbances

such as gopher mounds promote bull thistle establishment and survival (Klinkhamer and de Jong 1988, Randall 1991). It can also colonize areas in relatively undisturbed grasslands, meadows, and forest openings.

WHERE DID IT COME FROM AND HOW DOES IT SPREAD?

Native to Europe, western Asia, and North Africa, bull thistle is now naturalized and widespread throughout the United States (including Hawaii and southeast Alaska) and southern Canada and on every other continent except Antarctica (Grime *et al.* 1988). It probably was introduced to eastern North America during colonial times as a contaminant in seed and/or ballast (Moore and Frankton 1974) and to scattered locations in western North America in the late 1800s or early 1900s (Jepson 1911). By 1925 it had been reported in California from the San Francisco Bay Area, Central Valley, Klamath region, North Coast, and the northern Sierra Nevada (Robbins 1940).

Bull thistle reproduces only by seed, and individual plants set seed only once before dying. Large individuals may produce tens of thousands of wind-dispersed seeds.

WHAT PROBLEMS DOES IT CAUSE?

Bull thistle invades a variety of wildland habitats, where it competes with and displaces native species, including forage species favored by native ungulates such as deer and elk. In addition to outcompeting native plant species for water, nutrients, and space, the presence of bull thistle in hay decreases feeding value and lowers market price (McDonald 1994). In pastures and irrigated rangeland it may interfere sufficiently with livestock grazing so that liveweight gain is significantly reduced (Hartley 1983).

McDonald and Tappeiner (1986) noted that bull thistle often dominates recently clearcut forest areas in the Sierra Nevada of California, and experimental work in a replanted clearcut indicated that infestations limit growth of ponderosa pine saplings (Randall and Rejmánek 1993). Bull thistle also colonizes and maintains high population densities for up to six years in clearcuts in redwood and mixed evergreen forests in northwestern California (Glusenkamp, pers. comm.).

HOW DOES IT GROW AND REPRODUCE?

Bull thistle is a rosette-forming biennial. Seedlings grow slowly until soil temperature rises in the spring. Absolute growth rates for thistle rosettes are slow for two months after sowing, even under ideal conditions. Only after two months do seedlings expand their leaves and accumulate dry weight rapidly. Seedlings quickly grow into rosettes, nearly stemless plants with leaves clustered radially at the soil surface or pressing against adjacent vegetation. They typically remain in the rosette phase through the first growing season and the following winter. Some studies indicate that rosettes must be exposed to cold temperatures before they will bolt and flower and may also require a minimum daily period of light following exposure to cold (Klinkhamer *et al.* 1987a, 1987b).

Most plants remain in the rosette stage for one year, then bolt, flower, and set seed in the second growing season, although some individuals may bolt and flower in the first year and others may not bolt until their third, fourth, or fifth year. Seeds are released in late summer or fall and may germinate after rains in fall or the following spring. Seed viability is high, often above 90 percent, and germination is stimulated by soil moisture. Flowering may occur from early June until the first snowfall or hard frost; in California there is a pronounced peak in July and early August. Seeds ripen and are released from early July through October, occasionally later along the coast. Studies indicate that most seeds fall within three feet of the parent plant, but up to 10 percent may travel distances of more than ninety feet, even on days with little wind (Klinkhamer *et al.* 1988, Randall unpubl. data). A single flowerhead can produce from forty to over 250 seeds, and individual plants may have anywhere from one to 475 flowerheads or more. Bull thistle reproduces only by seed, and individual plants die after setting seed.

HOW CAN I GET RID OF IT?

Physical Control

Mechanical methods: Bull thistle can be controlled by mowing or hand cutting shortly before plants flower (Harris and Wilkinson 1984, Randall 1991). If cut too early in the season, plants will resprout and flower before the first frost. The uneven flowering times may make more than one treatment necessary. Gill (1938) reported that bull thistles that were cut while flowering were not capable of producing and releasing viable seeds. Plants from Yosemite Valley that were cut at the root crown a few days after their first flowers appeared and then laid on the ground produced abundant viable seed (Randall pers. observation). Thus it may be important to remove cut stems from the area.

Less than 5 percent of adult thistles cut at the soil surface resprouted, while over 80 percent of adult thistles in control plots survived and flowered (Randall 1991). Mean height and number of inflorescences were lower for plants that resprouted (twenty-five inches or 63 cm and 3.7 flowerheads) than for adults in control plots (thirty-three inches or 85 cm and 15.8 flowerheads). Subsequent work (Randall unpubl. data) indicated that about 4 percent of the thistles cut two to four inches (5-10 cm) above the soil surface a month before flowering will resprout.

Variations in seed production have more influence than variations in adult mortality on population fluctuations of biennials (de Jong and Klinkhamer 1988b). De Jong and Klinkhamer (1988a) found seed input limited establishment of bull thistle seedlings on coastal dunes in the Netherlands. Even if some plants resprout, manual control may reduce bull thistle populations by limiting seed production. The distance seeds are dispersed is positively correlated with the height at which they are released (Sheldon and Burrows 1973), so cutting may reduce the spread of seed even when plants resprout and produce seed.

Biological Control

Two USDA approved insects, *Urophora stylata* and *Rhinocyllus conicus*, have been released for bull thistle control in California. Neither has been successful in controlling bull thistle populations in California to date, but *U. stylata* shows some promise in coastal sites where it was released in 1997 and 1998 (Villegas and Coombs 1999). *U. stylata* is a gall-forming fruit fly (Tephritidae) with a narrow host range that has been released at sites in Canada and the United States (Harris and Wilkinson 1984, Julien 1987). Harris and Wilkinson (1984) found that over 90 percent of the flowerheads examined in some British Columbia populations had been attacked by the flies, and the number of larvae per flowerhead increased with the percentage of flowerheads attacked. They calculated that an infestation of this intensity reduced the population's seed output by over 65 percent.

Rhinocyllus conicus is a weevil (Coleoptera, Curculionidae) that attacks species of *Cirsium*, *Carduus*, and *Silybum*. It has been used as a biocontrol agent for several species in these genera. Studies indicate that there are local populations or strains with strong preferences for particular host species (Zwolfer and Harris 1984). One strain collected from bull thistle in France was released in 1984 in British Columbia, where it bred well through 1986 (Julien 1987). A strain of

R. conicus introduced to southern California to control *Carduus pycnocephalus* L. and *Silybum marianum* (L.) Gaertn. was rarely found on bull thistle in the release area, although it destroyed an estimated 55 percent of its intended hosts' seeds (Goeden and Ricker 1986a). Unfortunately, the weevils were occasionally found feeding on two native *Cirsium* species (*C. californicum* Gray and *C. proteanum* T. Howell) (Goeden and Ricker 1986b). *R. conicus* also attacks several other native *Cirsium* species in the Rockies and the Great Plains and has been found to limit populations of at least one of these native thistles (Louda *et al.* 1997, Louda 1998).

Chemical Control

Bull thistle is relatively easily controlled with herbicides. Several Agricultural Extension bulletins recommend 2,4-D at 0.5 kg/ha; dicamba at 0.15 kg/ha; picloram (not registered in California) at 1 kg/ha, and various tank mixes of these chemicals for control of bull thistle (Fawcett and Nelson 1981, Harris and Wilkinson 1984). Timing of herbicide application is important, with the exact date dependent upon life cycle stage. Autumn or spring application is recommended to control rosettes (Fawcett and Nelson 1981).

Conicosia pugioniformis (L.) N.E. Br

Common names: narrow-leafed iceplant, false iceplant, conicosia

Synonymous scientific name: *Mesembryanthemum elongatum*

Closely related California natives: 0

Closely related California non-natives: 8

Listed: CalEPPC A-2, CDFA nl

by Marc Albert and Carla D'Antonio

HOW DO I RECOGNIZE IT?

Distinctive Features

Conicosia (*Conicosia pugioniformis*) is a short-lived succulent with prostrate to ascending shoots and a central, thickened taproot. The taproot intergrades with a short, thick root crown, allowing for shoot growth after inundation with sand. Leaves are slender, bright gray-green to green, somewhat irregularly scattered along stems, and clustered near stem tips. Flowers are large and solitary with numerous shiny, light yellow petals. The fruit is a cone-shaped capsule that splits open when drying and is easily dispersed by wind, spilling seeds as it tumbles. Conicosia is easily distinguished from highway iceplant (*Carpobrotus edulis*) and sea fig (*C. chilensis*) by its narrow, long, bright green leaves, lack of rooting along trailing shoots, and absence of large clonal mats.

Description

Aizoaceae. Perennial, usually short-lived. Roots: taproot, 1.6-3.3 ft (0.5-1 m) in length,

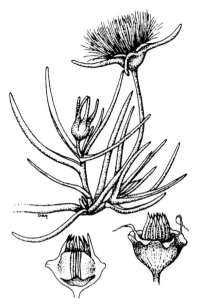

swells extensively, especially in first year. Caudex: 0.8-8 in (2-20 cm) in length and up to 2 in (5 cm) in diameter, vertical growth apparently related to burial; intergrades with taproot. Stems: above-ground shoots decumbent, 1.66-5 ft (0.5-1.5 m) in length, moderately branched. Leaf arrangement: basal leaves form a loose rosette; cauline leaves alternate but somewhat irregularly arranged along stem. Leaves: narrowly linear to cylindrical, rounded triangular in cross section, 6-8 in (15-20 cm) in length and 0.4-0.6 in (1-1.5 cm) in diameter. Inflorescence: 1-flowered axillary peduncle, <4.8 in (<12 cm) in length. Flowers: bright in appearance; 2-3.2 in (5-8 cm) in diameter; the 5 sepals irregular, succulent to papery; the several dozen petals shiny yellow and linear; multiple stamens; ovary +/- inferior, many chambered, multi-ovulate. Fruit: cylindrical capsule with conical top, dehiscent, with ovary chambers opening individually and capsule sometimes separating into individual valves; dried capsule dispersed easily by wind. Seeds: smooth and round, approximately 0.04 in (1 mm) in length (description from Hickman 1993).

WHERE WOULD I FIND IT?

Conicosia is found in coastal dunes and adjacent disturbed areas from the San Francisco peninsula to Point Conception, California. It is locally abundant, with particularly large populations occurring at Vandenberg Air Force Base and in the Guadalupe-Nipomo dunes. It is most abundant in open patches on dunes and in recently disturbed areas (e.g., along trails, roads, and railroad rights-of-way) with sandy soil. It seems to require well drained, sandy soils. It colonizes foredune, dune scrub, and, to a lesser extent, coastal scrub, coastal prairie, and maritime chaparral communities.

WHERE DID IT COME FROM AND HOW DOES IT SPREAD?

Conicosia is native to South Africa, probably introduced to the United States as an ornamental in the early 1900s. This species has only recently been noted as an important component of coastal California habitats (Shmalzer and Hinkle 1987). Spread occurs by seeds. Wind-dispersed capsules spill seeds from individual valves as they roll across the ground. Capsules have been observed several meters from adult plants. This low-growing, short-lived perennial has shoots that spread outward from a central stem that remains at soil level. It can cover or intersperse with other vegetation, and can be locally abundant. Spread does not occur by vegetative propagules. Any soil disturbance (e.g., by hikers, bicycles, or off-road vehicles) in established dune or coastal scrub areas could contribute to its spread, as would intentional planting in areas near the coast.

WHAT PROBLEMS DOES IT CAUSE?

Conicosia invades disturbance corridors within coastal scrub or maritime chaparral and has the potential to inhibit regeneration of native species in restoration sites. It colonizes recently

disturbed sandy areas, forming high-density populations that can preclude establishment of other vegetation. In addition, it readily invades openings in foredune
and dune scrub communities, often becoming established under shrubs and alongside subshrubs,
graminoids, and herbs. It is likely that this results in
interference with the growth and establishment of native plants. Individual conicosia plants produce many
seeds in fruit capsules that are easily dispersed; thus,
once established in an area, it is difficult to stop further colonization. Conicosia also may invade maritime
chaparral following fire (Odion *et al.* 1992). Other invasive members of this family have been shown to alter soil chemistry (Vivrette and Muller 1972, D'Antonio
1990a). It is not currently known if this plant can influence soil chemistry.

HOW DOES IT GROW AND REPRODUCE?

Flowering occurs in late summer through autumn. Plants may flower in the first year, but
flowering often begins in the second year. Plants can live ten years or longer (Bleck, pers. comm.).
Reproduction occurs by seed. Plants flower in summer or fall of their first or second year and
every subsequent year. As capsules dry, tissue from the upper surface of individual valves pulls

away, allowing seeds to drop. Capsules separate from the plant when dry. Whole dried capsules are readily moved by wind. Individual capsules produce tens to hundreds of seeds. Plants can resprout from the buried root crown after above-ground tissue is removed.

Conicosia plants can grow to several feet in diameter in a single growing season. Seeds from dense populations can spread rapidly into adjacent areas. There are currently no known studies of the physiological ecology of this species. Seedlings appear to devote much of their photosynthate to the thickened taproot; young individuals may have a taproot of one to three centimeters in diameter but only a few small leaves. Evergreen shoots may grow year round, with most vigorous growth occurring December to June. Like other succulents and other members of this family, conicosia may be able to alternate between CAM and C3 photosynthesis, which would allow for growth during periods of high temperature and low water availability, or in places with high salinity.

HOW CAN I GET RID OF IT?

Removal of live plants is easily accomplished by either mechanical or chemical means. Eradication of the species from any large area, however, is complicated by the wind dispersal of seeds and the propensity of conicosia to colonize disturbed areas. A critical component of any control effort will be follow-up removal of seedlings.

Physical Control

Manual methods: Plants are killed either by hand pulling or by slicing the taproot with a shovel, pulaski, pick and hoe tool, saw, or knife. Hand pulling is easy when plants are small (<0.5 m diameter) and soil is loose and/or moist, but can be difficult on large plants or in dry and/or compacted soils. Resprouting from basal shoots is possible, so care must be taken to ensure that the thick underground root crown is severed. Removal is recommended before fruits have set (Theiss 1994).

Prescribed burning: Because of the high water content of shoot tissues, burning live plants is not an effective method of control.

Biological Control

Insects and fungi: There are no known studies of biological control, and iceplant scale (*Pulvinariella mesembryanthemi*, *P. delottoi*) has not been observed on this plant. Plants at Vandenberg Air Force Base have appeared to be killed by dodder (*Cuscuta* sp.).

Grazing: Because of the astringent and often salty quality of the succulent tissue, grazing is not likely to be a control option.

Chemical Control

Glyphosate applied with a surfactant by foliar spray in concentrations as low as 0.5 percent has been effective in killing seedlings and mature plants (Mulroy, pers. comm.).

Conium maculatum L.

Common names: poison hemlock, carrot fern, poison parsley,
   spotted hemlock

Synonymous scientific names: none known

Closely related California natives: 0

Closely related California non-natives: 0

Listed: CalEPPC B; CDFA nl

by Jennifer Drewitz

HOW DO I RECOGNIZE IT?

Distinctive Features

   Poison hemlock (*Conium maculatum*) is a member of the carrot family. It is usually a biennial, with first-year plants producing ground-level rosettes. During the second year plants grow
from two to ten feet tall with a stem that is ribbed, hollow, and has purplish streaks or splotches
(Pitcher 1986). Small, white flowers grow in many umbrella-shaped clusters, each supported by
a stalk. Leaves have a somewhat fern-like appearance and are finely pinnately divided. When
crushed, they have a rank odor (Pitcher 1986).

Description

   Apiaceae. Biennial. Stems: hollow, ribbed with purple spots. Leaves: opposite, ovate, tapered, serrate, and finely pinnately divided. Inflorescence: umbels terminal and lateral. There
are 4-6 bracts, brown in color, and 5-6 bractlets present. Pedicels abundant. Flower: petals white
or yellowish, usually wide with narrow tips. Ray flowers number 10-20, 0.6-2 in (1.5-5cm).

Fruit: 0.08-0.12 in (2-3mm) wide, ovate with distinctively wavy ribs (Hickman 1993).

WHERE WOULD I FIND IT?

   Poison hemlock has spread throughout California in areas below 5,000 feet (1,500 m) elevation, excluding the Great Basin and Desert provinces (Pitcher
1986, Hickman 1993). It is commonly found in dense
patches along roadsides and fields. It also thrives in
meadows and pastures and is occasionally found in
riparian forests and flood plains (Goeden and Ricker
1982). It does best in disturbed areas where soil is
moist with some shade. Poison hemlock is also able
to form stands in dry, open areas (Parsons 1992).

WHERE DID IT COME FROM AND HOW DOES IT SPREAD?

Poison hemlock is native to Europe, North Africa, and Asia (Pitcher 1986). It was brought to the United States as a garden plant sometime in the 1800s and sold as a "winter fern" (Goeden and Ricker 1982, Parish 1920). The earliest poison hemlock collections were made in 1893 and 1897 in Berkeley and Truckee, respectively (Parish 1920). Poison hemlock has spread throughout the United States, Canada, Australia, New Zealand, and South America (Parsons 1992, Holm *et al.* 1979).

Poison hemlock reproduces only by seed, which is dispersed by water, mud, wind, animal fur, human clothing, boots, and machinery (Pitcher 1986, GGNRA 1989). It has no means of vegetative reproduction.

WHAT PROBLEMS DOES IT CAUSE?

Poison hemlock can spread quickly after the rainy season in areas that have been cleared or disturbed. Once established, it is highly competitive and prevents establishment of native plants by overshading (Serpa 1989). In agricultural areas it interferes with crops and production of feed for livestock (Jeffery and Robinson 1990).

Poison hemlock is best known for its toxicity to vertebrates, causing death primarily by respiratory paralysis after ingestion. The alkaloids in poison hemlock depress the central nervous system. Symptoms include nausea, vomiting, convulsions, loss of muscle power, dilation of pupils, slowing of heartbeat, and eventual death from respiratory failure (Parsons 1992). In livestock, symptoms appear immediately after ingestion, and death occurs within two to three hours. The recommended treatment is the same as that for nicotine poisoning: tannic acid followed by a purgative (Parsons 1992).

Poison hemlock is toxic to livestock, wildlife, and humans. Cattle, sheep, horses, pigs, goats, and fowl are all susceptible to its toxicity. Poisoning occurs in horses when they eat a quantity approximately equal to 0.25 percent body weight and in cows, 0.5 percent body weight (Kingsbury, 1964). Poisonings are usually not caused by direct foraging of poison hemlock, but by consumption when it is mixed in stock feed (Parsons and Cuthbertson 1992).

Wildlife is also susceptible to the toxic effects of poison hemlock. Ten percent of an elk population on Grizzly Island, California, died from ingesting poison hemlock in 1985 (Parsons and Cuthbertson 1992).

Field experiments have not established any allelopathic effects of poison hemlock (Serpa 1989).

Ancient Greeks used poison hemlock to carry out judicial executions, including the execution of Socrates (Parsons and Cuthbertson 1992). Most human poisonings occur when the leaves are eaten by people confusing them with edible seeds used as spices, or when children use the hollow stems as flutes (Kingsbury 1964). Seeds and young leaves are the most toxic parts of the plant. It is recommended that gloves be worn when handling the plant, as some people develop dermatitis, and that inhalation of particles be minimized (Parsons and Cuthbertson 1992). Children should be monitored when in areas containing poison hemlock.

HOW DOES IT GROW AND REPRODUCE?

Reproductive plant parts develop in mid-April, usually one year after germination (Amme

1988). In summer, once plants have set seed, they dry up and die leaving tall stalks to shade out other plants. The seed of poison hemlock is fully developed by mid-June. Plants disperse about 90 percent of their seed in September through December, with the remainder dispersed by late February (Baskin and Baskin 1990). This lengthy dispersal period allows poison hemlock to produce new seedlings continuously for several months (Baskin and Baskin 1990).

Poison hemlock has a large range of conditions in which it can germinate. It can germinate at temperatures greater than 9.4 C and lower than 33.8 C. It can germinate in darkness as well as in light. About 85 percent of seed produced is able to germi-

nate as soon as it leaves the parent plant (Baskin and Baskin 1990). The remainder is dormant and requires certain environmental conditions (thought to be summer drying) in order to germinate (Baskin and Baskin 1990). This ensures that some seed will remain in the seedbank until the following growing season. Seed can remain viable in the soil for up to three years (Baskin and Baskin 1990). It germinates most readily in soil, but can also germinate in sand. The combination of long seed dispersal period, seed dormancy, and non-specific germination requirements enable poison hemlock seedlings to emerge in almost every month of the year (Roberts 1979).

Germination takes place in all months of the year except April, May, and July, with late winter and early spring being the periods of greatest germination (Roberts 1979). Most vegetative growth occurs in winter months, with plants developing a deep taproot that is sometimes branched (Pitcher 1986).

HOW CAN I GET RID OF IT?

Physical Control

Manual methods: Hand pulling of poison hemlock is effective, especially prior to seed set, and easiest when the soil is wet (Parsons 1992). Because of the biennial nature of the plant, the entire root system does not need to be removed (Pitcher 1986).

Mechanical methods: Spring mowing has proven effective in killing mature plants, yet regrowth may occur and new seedlings may continue to establish (Amme 1988). A second mow in late summer is recommended to eliminate remaining or subsequent growth (GGNRA 1989). Because poison hemlock seed has been shown to germinate up to three years after dispersal, a third year of mowing may be necessary (Baskin and Baskin 1993).

Prescribed burning: Joseph DiTomaso, with the University of California, Davis, Weed Science Group, states that burning is probably not a good control option. In areas where poison hemlock is the dominant vegetation (usually moist environments), sufficient dried material would not be available to provide adequate fuel to control poison hemlock before fruit maturation. This method has yet to be tried on poison hemlock.

Biological Control

Insects and fungi: Although biocontrol is being examined, at this time there is no USDA approved biocontrol agent for use on poison hemlock in California. Goeden and Ricker (1982) found few insects attacking poison hemlock in California. Those that were found feeding on poison hemlock were unspecialized and polyphagous species (Goeden and Ricker 1982). In 1994 Berenbaum and Harrison published a paper noting a native European oecophorid caterpillar, *Agonopterix alstroemeriana*, was infesting poison hemlock in northeastern and western parts of the United States (Berenbaum and Harrison 1994). According to Berenbaum and Passoa (1983), "*C. maculatum* appears to be the sole host of *A. alstroemeriana*." It is not known how *A. alstroemeriana* was introduced into the United States, but this defoliating moth is now described as a biological control agent providing good to excellent control when present in a medium to heavy infestation in the Pacific Northwest (William *et al.* 1996).

Chemical Control

Effective post-emergent herbicides include 2,4 D ester, 2,4 D amine, and glyphosate plus surfactant, all to be applied in late spring. 2,4 D has been effective when sprayed at 1.0 lb ai/acre (1.1 kg/ha) and mixed with a wetting agent (Jeffrey and Robinson 1990). Glyphosate at a rate of 1.0 lb/acre (1.1 kg/ha) plus surfactant (as Roundup®) has also proved effective in killing poison hemlock, especially in the rosette stage (Amme 1988, Jeffery and Robinson 1990). Glyphosate plus surfactant (as Roundup®) at a rate of 1.0 lb/acre (1.1 kg/ha) has also proved effective in killing poison hemlock, especially in the rosette stage (Amme 1988, Jeffery and Robinson 1990).

Cortaderia jubata (Lemoine) Stapf

Common names: jubata grass, pampas grass, Andes grass, selloa pampas grass, cortaderia, pink pampas grass, purple pampas grass

Synonymous scientific name: *Cortaderia atacamensis*

Closely related California natives: 0

Closely related California non-natives: *Cortaderia selloana*

Listed: CalEPPC Listed A-1; CDFA nl.

by Joseph DiTomaso

HOW DO I RECOGNIZE IT?

Distinctive Features

Jubata grass (*Cortadaria jubata*) is a perennial grass six to twenty-three feet tall with long leaves arising from a tufted base or tussock. The inflorescence or flower cluster is a plumed panicle at the end of a long stem. Stems generally are at least twice as long as the tussock. Plumes consist of hairy female flowers, deep violet when immature, turning pinkish or tawny cream-white at maturity. Jubata grass is easily confused with, and often called, pampas grass (*Cortaderia selloana*). The two species are distinguished by stem height, leaf, plume, and spikelet color, florets, leaf tip, and presence of viable seed. The tussocks of jubata grass are less erect and more spreading and not fountain-like, when compared to tussocks of *Cortaderia selloana*.

CHARACTERISTIC	*Cortaderia jubata*	*Cortaderia selloana*
Stem (culm) height	2-2.5 times longer than tussock	equal to or slightly longer than tussock in female plants; two times longer in male plants
Leaf color	bright to deep green	glaucous-green
Plume color	pinkish to deep violet	light violet to silvery white; female plants with lighter plumes than males
Spikelet color	glumes purple	glumes white; males sometimes purplish near base
Florets	hairy at base; awn slightly extending beyond hairs	males sparsely or not at all hairy; females densely hairy at base, awns twice the length of hairs
Leaf tip	not bristly or curled	bristly and curled
Viable seed	yes	only when male and female plants are present

Description

Poaceae. Perennial grass. Leaves: blades 3-5 ft (1-1.5 m) long, 0.8-4 in (2-10 cm) wide, flat or slightly V-shaped in cross-section, deep green, upper and lower surfaces glabrous, occasionally with hairs near collar on upper surface, tips not setaceous (bristly) or curled, margins scabrous and sharp. Sheath: densely hairy. Inflorescence: dense panicle, 1-3 ft (3-10 dm) long, flexuous, deep violet when immature, pinkish turning cream-white or tawny at maturity. Spikelets: numerous, all female, 0.6 in (14-16 mm) long, 3-5 florets in each. Florets: 0.12-0.2 in (3-5 mm) long, glumes purple, lemma long-hairy, awns short <0.04 in (<1 mm), stigmas not exerted (Hickman 1993, Robinson 1984). Caryopsis: numerous seeds produced apomictically (without pollen transfer), easily separated from rachilla.

WHERE WOULD I FIND IT?

In California jubata grass occurs only in coastal areas (DiTomaso *et al.* 1999). It has become common in disturbed ditch banks, road cuts, cliffs, and cut-over areas, and eroded or exposed soil below 2,600 feet (800 m) elevation in the coastal fog belt from Santa Barbara County to Humboldt County, and less frequently in open habitats of southern California (Costas-Lippman 1977). Large infestations are common along US Highway 1 near Big Sur. Jubata grass nearly always occurs on open sites, such as roadside cuts, forest clearcuts, mudslides, or burned areas. Although typically found on sandy soils, jubata grass can survive on other soil types, including serpentine.

WHERE DID IT COME FROM AND HOW DOES IT SPREAD?

Jubata grass is native to northern Argentina and the Andes of Bolivia, Peru, Chile, and Ecuador (Costas-Lippmann 1977). In its native range it can be found from sea level to elevations

greater than 11,000 feet (3,400 m). It was first cultivated in France and Ireland from seed collected in Ecuador (Costas-Lippmann 1977). It is not clear how or when it was introduced into California, but it may have come through France via the horticultural trade (Madison 1992).

Jubata grass was first reported as a weed in California in logged redwood forests of Humboldt County in 1966 (Fuller 1976). Since infestations exist only in coastal areas of California, it is likely that the origin of this weed is a low-elevation biotype from South America. Because all seed production occurs without pollen transfer (apomictic), little genetic diversity exists within these plants (Connor 1973). This would explain its limited range in California. Spread occurs by wind-blown seed or by humans using mature inflorescences in decorative arrangements or using plants in landscaping. Seeds have been reported to disperse over twenty miles under windy conditions (Gadgil *et al.* 1984). Movement throughout the state also occurs when nurseries mistakenly sell this weedy species instead of *Cortaderia selloana* (Madison 1992).

WHAT PROBLEMS DOES IT CAUSE?

Large infestations threaten California's native coastal ecosystems by crowding out native plants, particularly in sensitive coastal dune areas (Cowan 1976). In addition to its effect on native plant diversity, jubata grass can reduce the aesthetic and recreational value of natural areas. In cut-over coastal redwood forests in northern California, jubata grass suppresses reestablishment of seedling conifers (Madison 1992). It is a significant weed problem in forestry operations and conservation areas in other countries, particularly New Zealand and Australia (Gadgil *et al.* 1984, Harradine 1991). In forests, jubata grass can outcompete seedling

trees and retard their establishment and growth. It creates a fire hazard with excessive build-up of dry leaves, leaf bases, and flowering stalks. Large clumps can complicate fire management activities by blocking vehicle and human access and by becoming fire hazards themselves. The sawtoothed leaves can cause injury to humans.

HOW DOES IT GROW AND REPRODUCE?

Reproduction of jubata grass is by asexual means only. Flowers typically are produced from late July to September (Madison 1992), even in the first year of growth. Plants can flower twice during the same season. Although all plants produce only female flowers, viable seeds develop from unfertilized ovules (apomixis). No pollination is necessary. All seedlings are genetically identical to the parent plant. This unusual form of reproduction is probably the most important characteristic responsible for the weediness of jubata grass, as well as its limited range. An individual inflorescence can produce 100,000 minute seeds (Harradine 1991), and large clumps can produce a million or more seeds (Cowan 1976). Like *Cortaderia selloana*, jubata grass can also reproduce vegetatively from fragmented tillers that produce adventitious roots in moist soils.

Germination generally occurs in spring and requires sandy soils, ample moisture, and light. The temperature range for germination is 55 to 70 degrees F (13-21 degrees C), with an optimum temperature of 63 degrees F (17 degrees C) (Drewitz and DiTomaso unpubl. data). Seeds do not appear to survive long in the soil, although no detailed studies have yet been conducted. Seedling growth and establishment are most rapid on bare, sandy soil and exposed road cuts, but typically require cool, foggy climate and moist soil (Cowan 1976). Seedling survival is low in shaded areas or in competition with grasses (Gadgil *et al.* 1990) or sedges. Growth initially is slow, but once established, plants grow rapidly. Roots are clustered in a shallow crown and can be fine to fairly thick. Technically, the roots are considered fibrous. Jubata grass does not tolerate winter frost (Costas-Lippman 1977), hot summer temperatures, intense sunlight, or drought. This may account for its inability to become established in the Central Valley of California. Plants are capable of surviving about fifteen years (Moore 1994).

HOW CAN I GET RID OF IT?

Control of jubata grass is similar to methods used to remove pampas grass. Because of the sensitivity of coastal sites occupied by jubata grass, few control strategies are available. Infestations sometimes can be averted by overseeding disturbed sites with desirable vegetation to prevent jubata grass seedling establishment.

Physical Control

Manual methods: Pulling or hand grubbing jubata grass seedlings is highly effective. Seedling leaves are shiny, stiff, and erect. Other more desirable grasses are not as stiff. For larger plants, however, a pulaski, mattock, or shovel are the safest and most effective tools for removing established clumps. To prevent resprouting, it is important to remove the entire crown and top section of the roots. Detached plants left lying on the soil surface may take root and reestablish under moist soil conditions (Harradine 1991). A large chainsaw or weedeater can expose the base of the plant, allow better access for removal of the crown, and make disposal of the detached plant more manageable (Moore 1994). Cutting and removing or burning the inflorescence is important to prevent seed dispersal during the operation (Cowan 1976, Harradine 1991). This is best accomplished prior to seed maturation. To reduce labor, the top of the foliage can be removed and the remaining crown treated with diesel oil (Cowan 1976).

Prescribed burning: Burning does not provide long-term control. The growing points of the grass are protected by surrounding leaves. This leads to rapid resprouting following a burn.

Biological Control

Insects and fungi: No insect or fungi control agents have yet been investigated.

Grazing: Successful control by grazing has not been reported in the United States, but cattle have been shown to be effective in controlling jubata grass in commercial forests of New Zealand (Harradine 1991, Gadgil *et al.* 1984).

Chemical Control

Control of jubata grass can be achieved by spot treatment with a post-emergence application of glyphosate at about 2 percent solution or eight quarts per 100 gallons. The addition of a non-ionic or silicone-based surfactant may be necessary to enhance foliar penetration of the herbicide. For most effective control, plants should be sprayed to wet, but not to the point of runoff. In one study, over 90 percent control was obtained during the first season, but continued spot applications were necessary to prevent rapid reestablishment (Madison 1992). Fall applications result in better control than do summer applications (Costello 1986) because photosynthetic assimilates are translocating downward at a faster rate late in the season. However, it may be necessary to apply the herbicide prior to maturation of viable seed in late summer. Low-volume (20 gal/ac) treatment with glyphosate at 4 percent can provide excellent control and reduce the amount of herbicide used as well as the cost of the treatment (Drewitz *et al.* unpubl. data).

Other registered post-emergence herbicides may also be effective in the control of jubata grass. These include the post-emergence graminicide fluazifop and the broad-spectrum herbicide imazapyr (Harradine 1991). Imazapyr at 1 percent low volume provides excellent control applied in spring or fall (Drewitz *et al.* unpubl. data). In forestry operations, hexazinone is a soil-residual root-absorbed compound also effective in the control of jubata grass (Harradine 1991), but only as a pre-emergence treatment. Once plants have been killed, clumps can be removed mechanically and left to decompose naturally.

Rope wick applications of glyphosate have also proven effective, but good coverage is essential (Drewitz *et al.* unpubl. data).

Cortaderia selloana (Schultes) Asch. & Graebner

Common names: pampas grass, Uruguayan pampas grass

Synonymous scientific names: *Arundo selloana, Cortaderia argentea, Gynerium argenteum*

Closely related California natives: 0

Closely related California non-natives: *Cortaderia jubata*

Listed: CalEPPC A-1; CDFA nl

by Joseph DiTomaso

HOW DO I RECOGNIZE IT?

Distinctive Features

Pampas grass (*Cortaderia selloana*) is a perennial grass six to thirteen feet tall with long leaves folded at the midrib and arising from a tufted base or tussock. The inflorescence or flower cluster is a plumed panicle at the end of a stiff stem. Stems are equal to or slightly longer than the tussock. Plumes nearly always consist of light violet to silver-white hairy female flowers that rarely produce seed. Pampas grass is easily confused with jubata grass (*C. jubata*). The two species are distinguished by several features, including stem height, leaf, plume, and spikelet color, florets, leaf tip shape, and presence of viable seed. The tussocks of pampas grass are more erect and fountain-like, not spreading, when compared to tussocks of jubata grass.

CHARACTERISTIC	*Cortaderia selloana*	*Cortaderia jubata*
Stem (culm) height	equal to or slight longer than tussock in female plants; two times longer in male plants	2-2.5 times longer than tussock
Leaf color	glaucous-green	bright to deep green
Plume color	light violet to silver white; female plants with lighter plumes than males	pinkish to deep violet
Spikelet color	glumes white; males sometimes purplish near base	glumes purple
Sheath	males sparsely or not at all hairy; females densely hairy at base, awns twice the length of hairs	hairy at base; awn slightly extending beyond hairs
Leaf tip	bristly and curled	not bristly or curled
Viable seed	only when male and female plants are present	yes

Description

Poaceae. Perennial grass. Leaves: blades to 6 ft (1.8 m) long, 1-3 in (3-8 cm) wide, V-shaped in cross-section, bluish green (glaucous), upper surface glabrous at base, lower surface

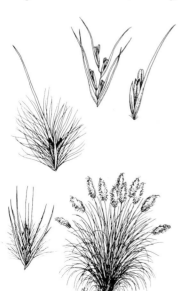

glabrous or hairy towards collar, tips bristly and curled, margins scabrous and sharp. Sheath: variable in hairiness, often glabrous, auricle-like outgrowth present at collar. Inflorescence: dense panicle, 1-4 ft (3-13 dm) long, stiff, light violet to silvery white when immature, white at maturity. Spikelets: numerous, 0.6 in (15-17 mm) long, typically with 6 florets in female plants and 3 in hermaphroditic plants. Florets: 0.16-0.3 in (4-8 mm) long, glumes white or membranous, lemma long-hairy, awns 0.1-0.2 in (2.5-5 mm) long, stigmas exerted (Hickman 1993, Robinson 1984). Caryopsis: viable seed rarely found and, when present, not easily separated from rachilla.

WHERE WOULD I FIND IT?

Pampas grass is common as an ornamental throughout California, including interior regions. It has escaped cultiva-

tion and spread along sandy, moist ditch banks throughout coastal regions of southern California (Costas-
Lippman 1977) below 1,000 feet (330 m). Its distribution is not as extensive as *Cortaderia jubata*, but it appears to be expanding (DiTomaso *et al.* 1998).

WHERE DID IT COME FROM AND HOW DOES IT SPREAD?

Pampas grass is native to Argentina, Brazil, and
Uruguay, where it grows in damp soils along river margins (Connor and Charlesworth 1989). It was first introduced to Europe by a Scottish horticulturist between 1775 and 1862. Samples were introduced to
California about 1848 by Joseph Sexton, a nurseryman from Santa Barbara. Commercial production began in 1874 in both California and Europe. Until 1895
nurserymen in the Goleta Valley near Santa Barbara were the primary producers of *Cortaderia
selloana* for ornamental use (Madison 1992). In 1946 it was planted by the Soil Conservation
Service throughout Ventura and Los Angeles counties to provide supplementary dryland forage and prevent erosion (Costas-Lippman, 1977). Pampas grass has escaped cultivation in many
coastal areas in California, presumably by fragmentation of the parent plant or, to a limited
extent, by seed.

WHAT PROBLEMS DOES IT CAUSE?

Although the more aggressive *Cortaderia jubata* is often called pampas grass, true pampas grass (*C. selloana*) can also be weedy in California. In other areas of the world, particularly New Zealand and Australia, *C. selloana* is an important weed problem in forestry operations and conservation areas (Gadgil *et al.* 1984, Harradine 1991). In forests it competes with seedling trees and can slow their establishment and growth. Pampas grass creates a fire hazard with excessive build-up of dry leaves, leaf bases, and flowering stalks. In addition, heavy infestations can block access to plantations and pose a significant fire hazard. In conservation areas pampas grass competes with native vegetation, reduces the aesthetic and recreational value of these areas, and also increases the fire potential.

HOW DOES IT GROW AND REPRODUCE?

Pampas grass is typically propagated for ornamental purposes through division of mature plants (Robacker and Corley, 1992). In nature it produces flowers two to three years after germination. Flowering usually occurs from late August though September (Madison 1992), but occasionally in winter. The species is considered gynodioecious, that is, flowers of some plants consist of both male and female parts on the same flower, but only the male parts are functional (Connor 1973). Other plants bear only female flowers. Thus, this species is functionally dioecious. Over the years, selection for ornamental plants in California has been for the showier plumes of the female plants. Consequently, few opportunities exist for seed production. This may account for the lack of spread of this species in California in past years.

In recent years some nurseries have propagated pampas grass from seed. Since it is impossible to distinguish male from female plants before they flower, the result is an increase in the proportion of male plants in the population. Consequently, there has been an increase in the amount of viable seed produced, and this species has escaped to become an invasive weed along the California coast. Populations that escaped from cultivation probably occurred in areas where seeds were produced. This can occur when both male and female plants are present in a population or when an occasional perfect flower (with both male and female parts) is produced on a typically male plant. In New Zealand, selection for female plants has not been as rigorous. As a result, more seed is produced, and pampas grass has become a significant weed problem (McKinnon 1984). Little is known of the germination of *Cortaderia selloana* from seed. Vegetative reproduction can occur when fragmented tillers receive adequate moisture and develop adventitious roots at the base of the shoot.

Establishment of seedlings generally occurs in spring and requires sandy soils, ample moisture, and light. Seedling survival is low in shaded areas or in competition with grasses (Gadgil *et al.* 1990) or sedges. Since few seeds are produced in California, little is known of the growth requirements. Unlike *Cortaderia jubata*, *C. selloana* can tolerate winter frost (Costas-Lippman 1977); it also tolerates warmer summer temperatures, more intense sunlight, and moderate drought. This accounts for its success as an ornamental in the Central Valley of California and its establishment as a weed along the American River near Sacramento. Once established, roots of a single plant can occupy a soil volume of about 1,100 square feet (103 m^2). Lateral roots can spread to thirteen feet (4 m) in diameter and eleven and one-half feet (3.5 m) in depth (Harradine 1991). Plants are capable of surviving about fifteen years (Moore 1994).

HOW CAN I GET RID OF IT?

Control of pampas grass is similar to that of jubata grass. Few strategies are available for the control of *Cortaderia selloana*. Burning does not provide long-term control, as plants resprout shortly thereafter. Infestations sometimes can be averted by overseeding open disturbed sites with desirable vegetation to prevent establishment of seedlings.

Physical Control

Manual methods: Pulling or hand grubbing *Cortaderia selloana* seedlings is highly effective. For larger plants however, a pulaski, mattock, or shovel are the safest and most effective tools for removing established clumps. To prevent resprouting, it is important to remove the entire crown and top section of the roots. Detached plants left lying on the soil surface may take root and reestablish under moist soil conditions (Harradine 1991). A large chainsaw or weedeater can expose the base of the plant, allow better access for removal of the crown, and make disposal of the detached plant more manageable (Moore 1994). Cutting and removing or burning the inflorescence prior to seed maturation in late summer may be important if seed production occurs in escaped populations of pampas grass. To reduce labor, the top of the foliage can be removed and the remaining crown treated with diesel oil (Cowan 1976).

Biological Control

Insects and fungi: No insect or fungal control efforts have been investigated for any species of *Cortaderia*.

Grazing: The success of grazing has not been reported in the United States, but cattle have been shown to provide effective control for pampas grass in commercial forests of New Zealand (Harradine 1991, Gadgil *et al.* 1984).

Chemical Control

Control of pampas grass can be achieved by spot treatment with a post-emergence application of glyphosate at about 2 percent solution or eight qts/100 gal. The addition of a non-ionic or silicone-based surfactant may enhance foliar penetration of the herbicide. For most effective control, plants should be sprayed to wet, but not to the point of herbicide runoff. In one study, over 90 percent control was obtained during the first season, but continued spot applications were necessary to prevent rapid reestablishment (Madison 1992).

Fall applications result in better control compared to summer applications (Costello 1986) because photosynthetic assimilates are translocating downward at a faster rate late in the season. However, if viable seeds are produced, it may be necessary to apply the herbicide prior to seed maturation. Although studies were conducted on jubata grass, it is likely that low-volume (20 gal/ac) treatment with glyphosate at 4 percent can also provide excellent control of pampas grass. The reduced volume can lower the amount of herbicide used as well as the cost of the treatment (Drewitz *et al.* unpubl. data). Rope wick applications of glyphosate have also proven effective, but good coverage is essential or tillers will recover (Drewitz *et al.* unpubl. data).

Other registered post-emergence herbicides useful for control of *Cortaderia jubata*, may also be effective in the control of pampas grass.

For large clumps, the top foliage can be removed by cutting or burning and the regrowth treated with a systemic post-emergence herbicide. This method reduces the amount of herbicide applied compared to herbicide treatment alone (Harradine 1991).

Cotoneaster spp.

Common names: cotoneaster, silverleaf cotoneaster, rockspray cotoneaster

Synonymous scientific name: *Cotoneaster buxifolius*

Closely related California natives: *Heteromeles arbutifolia*

Closely related California non-natives: 0

Listed: CalEPPC A-1; CDFA nl

by Jake Sigg

HOW DO I RECOGNIZE IT?

Distinctive Features

   Cotoneaster pannosa and *C. franchetii* are similar and frequently confused with each other, and some plants invading wildlands are not readily assignable to a particular species. Both species are evergreen shrubs, prostrate to erect, to ten feet tall, depending on species; many-branched from ground level, the branches laden with clusters of quarter-inch, white, rose-like flowers in summer followed by red berries in autumn and winter. The branches usually zig-zag, producing a complex, interwoven pattern.

Description

   Rosaceae. Fountain-shaped to 3m (10 ft). Leaves: 3/4 in (2 cm) leaves are dull gray-green above, felty beneath, and occur mostly in the upper part of plant. Flowers: those of *Cotoneaster pannosa* are white, with fully opened petals; those of *C. franchetii* have erect petals that are strongly tinged pink.

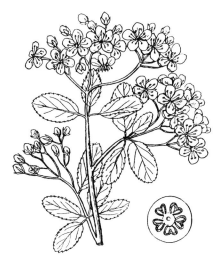

   Cotoneaster lacteus is profusely branched with leaves from the ground up. Leaves are 3 in (8 cm), ovate, shiny dark green above, densely hairy beneath, deeply pinnately veined. Fruits: orange-red berries.

   Cotoneaster microphyllus, rockspray cotoneaster. Prostrate with dense, arching, mounding branchlets arising from flat primary branch, which roots on contact with ground. Leaves: roundish and shiny deep green, 0.25 in (0.5 cm) wide. Fruit: bright red berries.

WHERE WOULD I FIND IT?

   Cotoneasters have escaped garden cultivation in widely scattered counties along the California coast

(Smith and Wheeler 1990). Some species have invaded coastal areas and forests of the north-western United States. They can be found in forests, shrublands, and grasslands and can tolerate a wide range of environmental conditions ranging from moist woodland and forest to open areas with thin, rocky soils that are dry for long periods. Their absence in wildlands of the interior indicates a need for coastal conditions, where frequent cool fogs reduce transpiration. *Cotoneaster pannosa* and *C. franchetii* are both naturalized in California. In Oregon and Washington only *C. franchetii* has been reported, and then only in waste places and roadsides.

WHERE DID IT COME FROM AND HOW DOES IT SPREAD?

Cotoneasters occur naturally in Eurasia, but the plants used in gardens emanate almost exclusively from China, with a few from the Himalaya (Hickman 1993). They were introduced into English gardens by the great plant explorers of the nineteenth century. Rockspray cotoneaster arrived in England in 1824 and in California in 1854. Collectors sent additional kinds during the next century, and by 1900 many species were available.

Considering that cotoneasters have been in California gardens for so long, it is surprising that they have been reported as a problem in wildlands only recently; the oldest shrubs in natural areas appear to be only fifteen to twenty years old. This may be a result of subtle changes in the environment that make ecosystems more vulnerable to invasion, to the build-up in numbers of plants (cotoneasters are more commonly seen in gardens today, in part because of natural seeding and in part because of their longevity), and to an increase in the numbers of seed-dispersing birds. Changes in genotype of plants may be another factor favoring invasiveness.

Although birds (cedar waxwings, robins, and their relatives) consume large numbers of berries, most are not eaten and fall to the ground, where many of them germinate. The numerous seedlings vigorously compete with each other. Birds facilitate dispersal of seeds away from the parent plant.

WHAT PROBLEMS DOES IT CAUSE?

Cotoneasters displace native plants by their rapid growth, competition for light, an aggressive, competitive root system, abundant seed production, and an effective seed-dispersal strategy. They may compete for the same ecological niche as the related native toyon (*Heteromeles arbutifolia*) in part of the toyon's range.

HOW DOES IT GROW AND REPRODUCE?

Natural propagation is almost exclusively by seed. Plants grow through the spring months, flower in summer, and set fruit in autumn; berries persist through winter. Although cotoneasters are apomictic (able to produce seed without benefit of fertilization), their flowers are attractive to wasps (especially yellow jackets), and this apparently can result in sexual reproduction. The plants self-sow abundantly. Many seedlings are of intermediate character from parent plants and presumably are hybrids. These indeterminate plants occasionally are found in the wild.

The showy fruits are produced in abundance and are consumed by birds. Long-range seed dispersal by birds and the ability of cotoneasters to establish in seemingly healthy native ecosystems make them weeds to take seriously along the coast. Seed longevity is not known, but multi-year follow-up is always advisable for any weed. Cotoneaster seeds can germinate readily at any time of year in cultivation, but in California wildlands the only window of opportunity is with

Cotoneaster pannosa

autumn and winter rains. Passage through a bird's digestive tract may facilitate germination but is not necessary, as shown by prodigious germination of seed of uneaten berries beneath parent shrubs.

A secondary means of self-propagation is by layering, that is, the rooting of branches that are in constant contact with the ground. The root system is fairly deep and strong, as is common with shrubs of the rose family. Layering is an important means of spreading in the case of prostrate plants such as rockspray cotoneaster. After being cut down, or in response to pruning, cotoneasters produce coppice shoots (Sunset 1996).

HOW CAN I GET RID OF IT?

Physical Control

Mechanical methods: Removal by a weed-whacker may be feasible at the seedling stage, but it is imperative to cut plants close to the ground, which risks hitting rocks. If herbicide is not applied, the stump will produce profuse coppice shoots. Effort required to kill the stump can be minimized by timing the initial cut to just after fruit set. This maximizes depletion of stored energy in the root system, thus weakening the plant. If plants are cut after fruit set but before fruit ripening, there is less chance of mature berries falling to ground and creating new plants. Frequent removal of coppice shoots will eventually starve the root, but if the

C. microphyllus

initial cut is not correctly timed, it could take two or three years to effect kill. Stump removal is difficult and labor-intensive because of the tenacious root system.

Biological Control

Plant competition: The numerous seedlings from a single parent plant vigorously compete with each other. Rather than spend the time pulling out hundreds of small plants, it is better to wait a year or two, when most are killed by sibling competition. The surviving few can then be hand pulled in the moist season.

Chemical Control

A cotoneaster can be killed by cutting down its branches, which, because of their dense zig-zag pattern, is not easy. Cut surfaces of the cambium-phloem layer should be painted with a 25 percent solution of triclopyr (as Garlon 4®) herbicide with 75 percent cottonseed or other light cooking oil as surfactant and inert ingredient. Glyphosate (as Roundup Pro®, 100% solution) may be substituted for triclopyr, but with less certain results. Frilling the bark to expose more phloem adds to the absorptive surface. Herbicide should be applied to cut surfaces immediately after cutting; delay of even a few minutes may reduce or prevent effectiveness.

Crataegus monogyna Jacq.

Common names: single-seed hawthorn, whitethorn

Synonymous scientific name: *Mespilus monogyna*

Closely related California natives: *Crataegus douglasii, C. suksdorfii*

Closely related California non-natives: 0

Listed: CalEPPC Red Alert; CDFA nl

by Jake Sigg and Ed Alverson

HOW DO I RECOGNIZE IT?

Distinctive Features

Single-seed hawthorn (*Crataegus monogyna*) is a deciduous shrub or small tree to twenty feet, with clusters of white flowers in mid-spring, small red berries in autumn, lasting to late winter. The small leaves are divided into three to seven lobes. The branches have stout spines, and the trunk has pale gray, smooth bark. It is usually found in forest understory. The plant has many garden forms and is capable of hybridizing with other species of *Crataegus*, including the native *C. douglasii* (Love and Feigen 1978); thus variation in characters is possible with hybridizing populations.

Description

Rosaceae. Shrub or small tree up to 20 ft (6 m) tall. Stems: with straight, stout spines on

branches. Leaves: 0.6-1.5 in (1.5-3.5 cm) on slender petioles, alternate on stem, ovate to obovate, 3-7-lobed, margins entire or sparingly serrate, smooth except for patches of hairs in axils of veins on the underside. Flowers: flat-topped clusters of many white (fading to pink) blossoms, petals 5, styles 1. Fruit: 0.5 in (1.25 cm) wide red berries, each with a single seed. Pacific Coast native hawthorn species bear purple-black fruit, but hybrids between natives and red-fruited *Crataegus monogyna* are also purple-black. Flowers of native plants have 5 styles and leaves are mostly unlobed; *C. monogyna* has a single style and prominently lobed leaves. Hybrids tend to have intermediate characters, 2-3 styles being usual (Bailey 1928; Willis 1973).

WHERE WOULD I FIND IT?

Incipient populations of single-seed hawthorn have been found on San Bruno Mountain, in the San Francisco Water Department's Crystal Springs Watershed (both in San Mateo County), and in eastern parts of the San Francisco Bay Area. It is well established in parts of the maritime Pacific Northwest and is a common invasive species in prairies and deciduous woodlands in the Willamette Valley, Oregon, where it has been present for over 100 years. Lack of reporting likely accounts for the apparent distribution gap between San Francisco Bay and the Pacific Northwest.

Single-seed hawthorn grows best in humid or subhumid temperate regions and does well on most soil types, including shallow, stony soils. Riparian areas, abandoned fields and pastures, oak woodlands, and other forested habitats from central California to Alaska must be considered potential habitat. Vigilance in spotting new infestations would be especially rewarding for controlling this plant, with its capability for long-range seed dispersal. Although forests are prime habitats, outlying plants of single-seed hawthorn can be found in shrubland or grassland, especially near the coast.

WHERE DID IT COME FROM AND HOW DOES IT SPREAD?

Single-seed hawthorn is indigenous to Europe and from North Africa to the Himalaya. It was introduced to gardens of North America (including the West Coast) from England and has been sold here since the 1800s. Fruit-eating birds are the primary agents of seed dispersal, and since they typically spend much of their time perched in trees and on fence posts, the forest understory and along fencerows are where seedlings, young plants, and older infestations are usually found. The berries are more attractive to some native birds such as American robins than are native hawthorns that grow in the same place (Sallabanks 1992, 1993). Seeds may also spread by clinging to farm machinery, vehicles, and animals or by contaminating agricultural produce. Fruit-eating mammals may also spread seed.

WHAT PROBLEMS DOES IT CAUSE?

Single-seed hawthorn displaces native plants, and dense thickets alter the structure of the forest understory and make movement of large animals difficult. Some species of *Crataegus* contain hydrocyanic acid in the leaves, which is poisonous to cattle. However, the presence of this toxin in single-seed hawthorn is currently unknown (Parsons 1992). Birds may prefer its berries to those of native berried plants, including native hawthorn, and this may lead to local extirpation of the latter.

Single-seed hawthorn has only recently been reported as a wildland weed in California, and it is not included in *The Jepson Manual* of 1993. Its rate of spread and degree of threat has not been quantitatively established, but every indication is that it is a plant to be taken seri-

ously. One nature preserve in Oregon's Willamette Valley has had to be abandoned because the hawthorn is beyond control of the available resources.

HOW DOES IT GROW AND REPRODUCE?

Single-seed hawthorn reproduces by seed, which is primarily dispersed by birds. Flowering occurs in spring. When pollinators are excluded, flowers do not develop fruits (Love and Feigen 1978). Single-seed hawthorn is a prolific fruit producer. At one site in western Oregon, the fruit crop was estimated at 2,721 fruits per plant. Seed longevity is unknown. Seed germination is aided by passing through a bird's digestive tract, but this is not necessary for germination.

Germination occurs in spring. Most vegetative growth occurs in spring and early summer, and normal growth rate is one to two feet a year. In the forest plants will shed lower branches and become tree-like, but in the open they will retain branches to ground level. The presence of single-seed hawthorn in forests indicates tolerance of shade and cool temperatures, but it will fruit more heavily if exposed to stronger light. Like most shrubs of the rose family, it has a tenacious root system, although probably not a deep one. Once established, single-seed hawthorn can withstand moderate drought. Established plants are capable of stump sprouting when cut or injured.

HOW CAN I GET RID OF IT?

Physical Control

Manual methods: Small infestations of young plants can be hand pulled or weed-wrenched out. The best time to cut plants is in early summer. At this time most of the plant's energy is in the growing tips and little is stored in the roots. Regrowth occurs unless the entire crown and the top few centimeters of the main roots are removed.

Biological Control

Insects and fungi: This species has no known effective biocontrol agents. In the British Isles, single-seed hawthorn is an important host of the fireblight bacterium, which also affects pears and apples. In some seasons in Australia, plants are severely attacked by the pear and cherry slug (*Caliroa cerasi*), which damages the foliage without any permanent effect on the plant (Parsons 1992).

Grazing: The spines deter grazing animals, and the plant is regarded as an impenetrable barrier to grazing.

Chemical Control

Larger plants can be cut to the ground and painted with a solution of 25 percent triclopyr (as Garlon 4) and 75 percent cottonseed or other light cooking oil as surfactant and inert ingredient. Glyphosate (Roundup at label-recommended strength) may be substituted for triclopyr, but results are less certain. A 2 to 3 percent solution of triclopyr or glyphosate has been sprayed on the foliage for control, but overall spraying has not generally been reliable and is more likely to affect non-target species than is painting with herbicide.

Cynara cardunculus L.

Common names: artichoke thistle, cardoon, desert artichoke, wild artichoke

Synonymous scientific names: none known

Closely related California natives: 0

Closely related California non-natives: *Cynara scolymus*

Listed: CalEPPC A-1; CDFA B

by Mike Kelly

HOW DO I RECOGNIZE IT?

Distinctive Features

Artichoke thistle (*Cynara cardunculus*) is a spiny thistle of the sunflower family, head high, occasionally taller, crowned by a cluster of showy, bright purple thistle flowerheads that are two to three inches in diameter from April through July. Rarely, patches of white-flowering plants are found (Munz 1963, Kelly pers. observation). One to several stout flower stalks rise from a bushy rosette up to five feet in diameter. A spray of basal leaves, each deeply lobed and gray-green, to four feet in length, arches gracefully up and out from the base. Smaller leaves, otherwise similar in appearance, grow from the flower stalk as it extends upwards. Stout spines on the leaves, stems, and bracts around the flowerheads make it easy to recognize.

Description

Compositae. Perennial herb. Stems: leafy, branched, stout, and generally erect unless heavy flowerheads cause bending, to 5 ft (1.5 m) in height. Leaves: basal until bolting, at which time cauline leaves appear off inflorescence stalk, alternate, 1-4 ft (3-12 dm), pinnately lobed to divided, very spiny. Gray-green upper surface loosely cobwebby, lower surface densely gray-tomentose. Rubbing leaf surface easily removes cobwebby material. Inflorescence: lower heads typical of aster family, discoid, large, with one to several per cyme in a loose cluster at top of stalk. Typically, the uppermost flower opens before lower or peripheral flowers on any given axis. Involucres: ovoid or hemispheric, 1-2.25 in (3-6 cm) in height,

1.5-2.75 in (4-7 cm) diameter, not including tips, tending to narrow above. Phyllaries surrounding flowerhead easy to identify, with stout spines at tips, overlapping in series, generally ovate, leathery, entire, glabrous, receptacle flat, fleshy, bristly. (Spiny tips on phyllaries on immature flowerhead easily differentiate this plant from the closely related *Cynara scolymus*, the common agricultural globe artichoke.) Flowers: striking, with purple, sometimes blue corollas, +/- 2 in (5 cm), tube very slender, throat widened abruptly, lobes linear; anther bases long-sagittate, tips oblong; style appendage long, cyclindric, minutely papillate, tip barely notched. Fruits cylindric to obconic, =/- 4-angled or +/- compressed, glabrous, attached at base; pappus of many stiff bristles, 1-1.5 in (2.5-4 cm), in several series, white or brownish, plumose below, fused and falling together (Hickman 1993). Corollas occasionally white (Munz 1963, Kelly pers. observation).

WHERE WOULD I FIND IT?

Artichoke thistle is found in disturbed places, to 1,650 feet (<500 m), throughout the state, except deserts (Hickman 1993). It is common in annual rangelands, especially with a coastal influence, but also is found inland in disturbed grasslands or abandoned agricultural fields and is associated with overgrazing (Thomsen *et al.* 1986). It was one of the worst pests on California rangelands by the 1930s, invading over 150,000 acres in thirty-one counties and requiring prodigious and expensive efforts to eradicate or control it. By the 1980s the worst concentrations of the plant were found in Orange, Solano, and Contra Costa counties, with locally dense populations elsewhere in the Coast Ranges, Central Valley, and Sierra Nevada foothills (Barbe 1990). Placement on the California Department of Food and Agriculture's B List reflects the fact that it became too widespread and difficult to eradicate in many areas, with the authorities opting for preventing its spread, and control when feasible.

Artichoke thistle has been observed colonizing riparian woodlands and natural openings in chaparral and coastal sage scrub, growing under willow, mulefat, and sycamore, as well as in native grasslands (Pepper and Kelly 1994). It does well in soils with a heavy clay content (Thomsen *et al.* 1986), which helps explain its invasion of grassland habitat occupied by endangered San Diego thornmint (*Acanthomintha ilicifolia*) (Kelly 1996).

WHERE DID IT COME FROM AND HOW DOES IT SPREAD?

Native to the Mediterranean (Hickman 1993), artichoke thistle became widespread on over 150,000 acres of California rangeland and also in Australia, New Zealand, and South America on grazing lands, especially the Argentine pampas (Thomsen *et al.* 1986). It is now recognized as the wild form of the cultivated globe artichoke, *Cynara scolymus* L. When grown from divisions of the perennial crown, globe artichoke will reliably produce the spineless, edible flowerhead and plant known to agriculture, but grown from seed it often reverts to a wild form, producing the inch-long spines around the flowerhead normally found on *C. cardunculus* (Thomsen *et al.* 1986).

Artichoke thistle was known as the "edible thistle" and mentioned in literary references dating back to several centuries before Christ. Cultivation of the edible wild form appears to have spread from Naples, Italy, into the broader Mediterranean during the fifteenth century (Warren 1996). Cultivated forms apparently were developed from artichoke thistle in monastery gardens during the Medieval period (Thomsen *et al.*1986). Emigrants from the Mediterranean region apparently carried one or more forms of the plant to other countries. During his voyage on the *Beagle*, Darwin (1989) found artichoke thistle had already reached the Argentine pampas and escaped cultivation: "Very many, probably several hundred square miles are covered by one mass of these prickly plants and are impenetrable by man or beast. Over the undulating plains where these great beds occur, nothing else can now live."

Botanical surveys in California from 1860 to 1864 reported the globe artichoke variety as having escaped cultivation. Artichoke thistle was already reported as having established itself outside cultivation in a pasture in San Diego County in 1897. Its appearance in California rangelands can be traced to its introduction for ornamental and culinary uses (Thomsen *et al.* 1986).

Although cultivated vegetatively from crown division, artichoke thistle in the wild spreads only by seed. The large seeds are dispersed by a variety of mechanisms. Bristles from the receptacle and the pappus (thistledown) remain attached to the seeds when they are released, aiding in dispersal. Seeds separate easily and quickly from the thistledown and, because of their large size and weight, they usually are not carried more than sixty-six feet (20 m) from the parent plant. Strong winds, however, may pile up great banks of thistledown against fence lines and road margins. However limited this dispersal mechanism may be, over a few years wind can significantly expand the patch size of a local infestation.

Most seeds seem to fall close to the parent plant. Birds feeding on the seed heads probably knock some seed to the ground and occasionally move it greater distances. Hillside patterns suggest that water and gravity carry seeds short distances on slopes. Seeds may attach to cattle and other mammals, and the plant is known to spread along game trails in coastal sage scrub in southern California. Its distribution along utility roads in some areas of San Diego County suggests that vehicle tires are transporting seeds.

WHAT PROBLEMS DOES IT CAUSE?

Artichoke thistle is an important rangeland problem because it reduces forage production and limits movement of livestock (Thomsen *et al.* 1986). The stout, upright yet spreading nature of the plant, its formidable spines, and high densities make wildlife movement through it difficult. The arching leaves shade a considerable area. Combined with its aggressive root system, artichoke thistle outcompetes native vegetation for light, water, and nutrients. At high densities it becomes a monoculture that excludes shrubs, herbaceous plants, and even annual grasses. For example, mature broom baccharis (*Baccharis sarothroides*) declines in vigor when in close proximity to artichoke thistle. The thistle also is a threat to the endangered San Diego thornmint.

There appears to be no alteration of soil chemistry or allelopathy, since other species, including those originally displaced, readily recolonize the site once artichoke thistle is dead. It is unlikely that the plant alters fire cycles, since it mainly displaces annual exotic grasses, which are highly combustible. However, one 4,000-acre infestation in San Diego was first colonized by this thistle after a grassland fire in the same area in the mid-1980s (Dumka 1997).

HOW DOES IT GROW AND REPRODUCE?

Artichoke thistle flowers as early as April in southern California and into July throughout its range. In a year of average rainfall in San Diego of nine inches (22.5 cm), a mature plant can produce more than a dozen flowerheads with as many as 200 seeds per head (Kelly 1996). In drier years most established plants sprout, flower, and produce seed, but these plants are usually smaller than plants growing during wetter years. Some established plants will sprout and then wither before flowering.

Bees, a common visitor to the flower, are probably the most important pollinator. Mature plants resprout repeatedly when cut or when herbicide is applied too early or in too small amounts (Pepper and Kelly 1994).

In southern California artichoke thistle germinates as early as December and as late as July. Seven years of control efforts with careful record keeping in Peñasquitos Canyon Preserve (San Diego) suggests that a sizable seedbank with an average duration of five years can be expected (Kelly 1991). In drier years seedling mortality is high. Growth is rapid during cool and wet winter months in coastal and southern California. Plants usually flower the first year if rainfall is equal to or greater than average, but flowering may be delayed until the second summer in drier years. Late-germinating plants barely reaching a foot in height are capable of producing flowers and setting seed. Plants form multi-stalked clumps after the first year and the clumps enlarge thereafter. A deep taproot develops during the first year (Parsons 1992). The roots can eventually reach eight feet in depth.

Rosette leaves usually die over the summer. The plant tends to produce single-species stands that can reach densities as high as 22,000 plants per acre (Thomsen *et al.* 1986). With a large, expanding underground taproot and tuber, a single plant quickly becomes a many-stemmed clump. Some populations become dense enough to restrict growth, resulting in tall, spindly plants rather than the robust, broad clumps typical of plants not under competitive pressures.

HOW CAN I GET RID OF IT?

Because of artichoke thistle's ability to resprout after chemical spraying and to build up a seedbank that lasts five years or more, yearly monitoring and repeat eradication are necessary. Eradication is most effective when mature plants are bolting, generally in early to mid-April in southern California and late April or May farther north. Chemical control efforts have proved successful on sizable populations in several open-space parks and on military lands managed for their natural resources in San Diego County. Seedbanks in some areas have been exhausted, and maintenance is minimal and routine (Kelly pers. observation).

Historically, artichoke thistle was controlled by hand grubbing, root-plowing by tractors equipped with a specially designed blade, displacement planting, and applying the herbidide 2,4-D from airplanes and helicopters. These methods often proved too labor-intensive or expensive to be used successfully in larger infestations. Costs of control efforts sometimes exceeded the value of the land (Thomsen *et al.*1986). Workers often wear chainsaw chaps as protection when working to control it.

Physical Control

Manual methods: Grubbing is practical when only scattered plants are present, but much of the taproot must be removed or new growth will develop from the cut surface. Cutting and removing seed heads can stop seed production in small populations where timely eradication of the plant is not possible (Pepper 1994).

Mechanical methods: Discing or plowing larger infestations in wildlands is impractical and not advised. Although it is theoretically possible to exhaust the carbohydrate reserves of the plant's tuberous roots, this would require many years of continued effort and several carefully timed passes each season because artichoke thistle can resprout repeatedly. Discing and plowing also disturbs the soil, opening it up to reinfestation by this species or other invaders. The deep root, reaching eight feet (3 m) in depth, also makes mechanical removal difficult.

Prescribed burning: This could be helpful in removing the above-ground biomass, making access for chemical control easier and more efficient. Burning would not be expected to kill the plant, given its perennial underground storage reserves, but might kill the surface layer of the seedbank. Applying herbicide to resprouts four to six weeks after they emerge would likely be most effective for eradicating this thistle from the burned area.

Biological Control

Insects and fungi: The plant is a food source for invertebrates, including earwigs, ants and their aphid consorts, harlequin beetles, and bees; the latter being found frequently on the flowers. No USDA approved biocontrol agents exist for this thistle in California. Given the close relationship of the wild artichoke to the cultivated crop artichoke, the purposeful introduction of a biocontrol fungus or insect is not likely. However, *Terellia fuscicornis* (artichoke fly), an exotic fly from the Mediterranean Basin, has been discovered, identified, and collected in 1994 from San Joaquin, Sonoma, and Madera counties. It has subsequently been identified in wild artichoke populations in a number of other counties. This exotic fly has been found on both *Cynara cardunculus* and *C. skolymus* and feeds on flowers and seed heads. Since it is a seed-eater, it does not currently threaten commercial artichoke production. Given the fly's widespread distribution and its potential as a biocontrol agent on artichoke thistle, a CDFA B rated weed, attempted extirpation is precluded under agricultural regulations, which is potentially good news for wildlands (Penrose 1994). By March 1997 the artichoke fly had been downgraded from a Q to a C listing by the California Department of Agriculture. The Q listing is a quarantine listing; it cannot be moved around. The C listing, is a conditional listing, allowing it to be moved if certain conditions are met with permits between counties (Darling 1997). Until the arrival of the artichoke fly, herbivory by insects or mammals, including cattle, offered little hope of controlling artichoke thistle.

Grazing: The spiny nature of the plant deters cattle and sheep from grazing heavy infestations, but hungry animals will eat the leaves (Parsons 1992). Several species of birds feed on the seeds, but not to any great extent. Herbivory by deer is not apparent, probably because of the spines. Rabbits occasionally eat it at the cotyledon stage, but not beyond. The occasional dead plant encountered in the wild appears to be the work of gophers, but their impact on the population is insignificant.

Chemical Control

Glyphosate is effective in killing artichoke thistle. Cut stump applications were reported as effective at any stage of growth in nature preserves in the Santa Monica Mountains (Pepper and Kelly 1994). Subsequent cut stump application on thousands of plants in Peñasquitos Canyon Preserve in San Diego have confirmed this method (Kelly pers. observation). The cut-stump method involves cutting the plants as close to the base as possible with a machete, loppers, or a brush cutter and applying a solution of 25 percent glyphosate (as Roundup®) to the stump. This method is useful for isolated plants or remote populations where spray equipment is impractical or when it is in close proximity to sensitive species and foliar spray is not advised. The cut-stump method has been successfully used to eliminate artichoke thistle from patches of the endangered San Diego thornmint with no harm to the latter.

A foliar spray of 2 percent glyphosate (as Roundup®) can achieve a kill of 95 to 98 percent on mature, bolting plants. Spraying plants that had gone to seed also achieved a similar high kill rate. Spraying plants in earlier stages of growth before the plant sends up its flower stalk kills the above-ground vegetative structures, but often does not kill all of the roots. In such cases the plant dies back, but up to 75 percent of sprayed mature plants resprout in the same season. Transport of fluids is generally up to the stems and leaves and less down to the roots in pre-bolting plants, probably preventing sufficient herbicide from reaching the roots. However, spraying seedlings with 2 percent glyphosate is effective. Cutting down dense patches with power tools or a tractor is a useful prelude to chemical treatment. It allows workers to penetrate patches with less damage from spines on standing dead plants and reduces the amount of herbicide needed.

A new herbicide, clopyrlid (as Transline®) appears to be effective when sprayed on this thistle at the rosette stage, but less effective on mature, bolting plants (Carrithers 1997).

Cytisus scoparius (L.) Link.

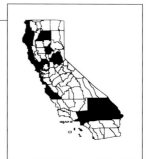

Common names: Scotch broom, English broom, common broom

Synonymous scientific names: *Sarothamnus scoparius, Spartium scoparius*

Closely related California natives: 0

Closely related California non-natives: 7

Listed: CalEPPC A-1; CDFA C

by Carla Bossard

HOW DO I RECOGNIZE IT?

Distinctive Features

Scotch broom (*Cytisus scoparius*) is a perennial shrub six to ten feet tall. Its sharply angled branches generally have five green ridges with hairs on them when young; as the branches mature the hairs fall off, and the branches become tan and lose the distinct ridges. Pods have hairs along the seams only. One or two golden yellow pea-like flowers cluster between the leaf base and stem. About half the photosynthetic (green) tissue is in the leaves and half is in twig tissue. Sometimes this species is confused with French broom (*Genista monspessulana*), which has pods with hairs all over them, stems that are not ridged or green, and more than eighty-five percent of its photosynthetic tissue in leaf tissue (Bossard and Rejmánek 1994).

Description

Fabaceae. Long-lived shrub. Stems: 5 angled, green and hairy when young, later glabrous. Leaves: on young branches there is usually one sessile leaf or three leaf-lets 0.3-0.7 in (5-18 mm) long; leaf/leaflets oblong and pointed on both ends, hairs may be flattened against the leaf or absent. Inflorescence: 1-2 flowers clustered in leaf axis; pedicels <0.5 in (<12 mm), without hairs. Flowers: shaped like pea flowers; calyx without hairs, <0.3 in (<6 mm); corolla golden yellow, banner 0.6-0.7 in (15-18 mm) may curl backward. Fruit: 1-2 in (25-50 mm) flat pod, brown or black at maturity, hairs present only along seams of pods (Hickman 1993). Seeds: 3-12 seeds per pod, greenish brown to black, shiny, round to oval, with a cream to yellow eliaosome (Hickman 1993).

WHERE WOULD I FIND IT?

Found along the California coast from Monterey north to Oregon border, Scotch broom is prevalent in interior mountains of northern California on lower slopes and very prevalent in Eldorado, Nevada, and Placer counties in the Sierra Nevada foothills. It is also reported from Los Angeles and San Bernardino counties. It is common in disturbed places, such as river banks, road cuts, and forest clearcuts, but can colonize undisturbed grassland, shrubland, and open canopy forest below 4,000 feet (<1300 m). Scotch broom prefers soil with pH less than 6.5; it is rare on limestone soils. It tolerates a wide range of soil moisture conditions and is competitive in low-fertility soils. Nitrogen-fixing bacteria found in small nodules on plant roots can fix nitrogen even at temperatures to 38 degrees F (4 degrees C) (Wheelor *et al.* 1988; Bossard 1991a; Partridge 1989).

WHERE DID IT COME FROM AND HOW DOES IT SPREAD?

Scotch broom is native to Europe and North Africa. Its natural range is broad, from Great Britain to the Ural Mountains and from Sweden to the Mediterranean. Introduced to California in the 1850s as an ornamental in the Sierra Nevada foothills, it was later used to prevent erosion and stabilize dunes (Geickey 1957, Schwendiman 1977). It spreads by prodigious seed production. One medium-sized shrub can produce over 12,000 seeds a year. After ballistic dispersal, seeds are further dispersed by ants, animals, or in mud clinging to road grading or maintenance machinery. Scotch broom is also readily dispersed by rain wash on slopes (Bossard 1991b). Plants can resprout from the root crown after cutting or freezing and sometimes after fire (Bossard and Rejmánek 1994).

WHAT PROBLEMS DOES IT CAUSE?

Scotch broom currently occupies more than 700,000 acres in central to northwest coastal and Sierra Nevada foothill regions of California (Barbe, pers. comm.). It displaces native plant and forage species and makes reforestation difficult. It is a strong competitor and can dominate a plant community, forming a dense monospecific stand. Scotch broom infestations can attain a biomass of over 44,000 to 50,000 kg/hectare in three to four years (Bossard and Rejmánek 1994, Wheelor et al.1988). Seeds are toxic to ungulates. Mature shoots are unpalatable and are not used for forage except by rabbits in the seedling stage (Bossard and Rejmánek 1994). Foliage causes digestive disorders in horses (Parsons 1992). Since Scotch broom can grow more rapidly than most trees used in forestry, it shades out tree seedlings in areas that are revegetated after tree harvest. Scotch broom burns readily and carries fire to the tree canopy, increasing both the frequency and intensity of fires (Parsons 1992). This species is difficult to control because of its substantial and long-lived seedbank.

HOW DOES IT GROW AND REPRODUCE?

Scotch broom becomes reproductive at two to three years on reaching a height of two to three feet (60-100 cm). It flowers in late March to April inland, April to June on the coast. Flowers appear before leaves. Long-lived seeds are copiously produced (to 12,000+ seeds/mature shrub) and mature in June and July. Seeds initially disperse ballistically from the pod, with an audible pop, and are further dispersed by ants and rain wash on the ground (Bossard 1990b, 1993). Seeds are known to survive at least five years in the soil (Bossard unpubl. data) and possibly as long as thirty years (Carson 1998). The seedbank can build to over 2,000 seeds/sq ft.

Seeds germinate from November to June inland, January to July along the coast (Bossard 1993) when provided with disturbance that creates open mineral soil. Germination may be enhanced by fire, but relies less heavily on water, wind, or animal distribution than do some other invasive plants, although its tough seed coat provides good protection from abrasion associated with water transportation (Carson 1998). Seedlings can tolerate even 90 percent shade. Approximately 35 percent of each seed crop becomes part of a rapidly developed seedbank. Plants can resprout from the root crown when cut, particularly during the rainy season (Bossard and Rejmánek 1994).

Scotch broom is host to nitrogen-fixing bacteria, which assists both its establishment on poor and disturbed sites and its ability to outcompete native species. It tends to acidify the soil (although not as strongly as does gorse, a relative). The period of most rapid vegetative growth is May to July, with some dieback occurring during seasonal periods of drought (Bossard and Rejmánek 1994). Most photosynthate is moving upward in the shrub toward branch tips during flowering, bud break, and seed set, which occur in late March to mid-April, April, and May, respectively). Photosynthate starts moving down toward roots after seeds are well grown but before seed release (Bossard, unpubl. data). On dry, hot sites Scotch broom will drop its leaves in late July or August. Its life span in California is longer than in its native range, with some individuals surviving up to seventeen years (Bossard 1990a).

Broom is considered to be primarily an early serial colonizer that will be shaded out once native species are established. There is, however, concern that its vigorous and prolific growth, along with acidification of the soil, inhibits establishment of other species.

HOW CAN I GET RID OF IT?

The best method for removal of a Scotch broom infestation depends on the climate and topography of the site, the age and size of the infestation, the relative importance of impact to nontarget species, and the type and quantity of resources available to remove and control broom at a given site. All methods require appropriate timing and follow-up monitoring.

Physical Control

Manual/mechanical removal: Pulling with weed wrenches is effective for broom removal. The wrench removes the entire mature shrub, eliminating resprouting. However, the resultant soil disturbance tends to increase the depth of the seedbank (Bossard 1991, Ussery and Krannitz 1998). Wrench removal is labor-intensive, but can be used in most kinds of terrain and allows targeting of broom plants with low impact on desirable species in the area. Golden Gate National Park has had success in using volunteers to remove broom with weed wrenches and then closely monitoring and removing broom seedlings for five to ten years.

Ussery and Krannitz (1998) found significantly more trampling of native species, and more soil disturbance and broom seedling regeneration, when adult broom plants were removed by pulling rather than cutting in British Columbia. Brush hogs, which twist off above-ground biomass, can be used for broom removal. They are less labor-intensive, but heavily impact non-target species and cannot be used on steep slopes. The twisting action is more destructive to tissues that initiate resprouting than is clean cutting. However, depending on the season of brush hog removal, resprouting can still be a serious problem. Brush hog removal has been used with limited success in Redwood National Park (Popenoe, pers. comm. 1997).

Saw cutting removes above-ground portions of shrubs, but depending on the time of cutting, may result in high rates of resprouting. In the Sierra Nevada foothills saw cutting undertaken at the end of the summer drought period (August to October) resulted in a resprouting rate of less than 7 percent, whereas cutting done at other times resulted in resprouting rates of 40 to 100 percent (Bossard and Rejmánek 1994). In British Columbia plants greater than one-quarter inch (3 mm) in diameter cut below two inches (5 cm) from the soil surface in July were found to have less than 1.5 percent resprout rate (Ussery and Krannitz 1998).

Prescribed burning: Burning uncut broom has been used with some success on Angel Island. Reburn of the removal site is usually necessary two and four years after the initial burn (Boyd, pers. comm. 1997). For prescribed burning of pretreated or cut broom see below under integrated methods.

Biological Control

Insects and fungi: Two USDA approved insects, a stem miner, *Leucoptera spartifoliella*, and a seed beetle, *Apion fusciostre*, were introduced in the 1960s as biocontrol agents, but have had limited success in California. New insect biocontrol agents are being tested in England and France for use on broom in Australia and New Zealand (Hoskings 1994). If proved safe and effective in California, these insects may ultimately become available for use as biocontrol agents in California.

Grazing: Heavy grazing by goats during the growing season for four to five years has been reported effective in New Zealand, and grazing by llamas has been tried at a few sites in California (Archbald, pers. comm. 1997). The disadvantage associated with using goats is that they are not selective, and native species that start to revegetate the area are also eaten.

Chemical Control

Foliar sprayed until wet, 2 percent glyphosate (as Roundup®) has been used to kill mature plants of Scotch broom. Adding surfactant improves effectiveness (Parsons 1992). The foliar spray impacts non-target species, and resprouting may occur. Triclopyr ester (25 percent) (as Garlon®) in Hasten®, Penevator®, or other seed press oil (75 percent) applied with a wick in low volume (2-3 drops) to basal bark has also proved effective (Bossard unpubl. data). This application technique does not affect non-target species, but it is more time-consuming and may be impractical for large infestations. Both of these chemical methods should be used during periods of active growth after flower formation. Chemical removal alone results in standing dead biomass that makes monitoring for and treatment of broom seedlings difficult. The standing dead biomass also presents a major fire hazard.

An Integrated Approach

The most effective removal treatment in a project in Eldorado Forest in the Sierra Nevada foothills was found to be cutting shrubs in September and October, allowing cut shrubs to dry on site, and then burning dried shrubs in late May and early June. This killed any resprouts and most of the seed within the top <1 in (2 cm) of the soil. Seeds within <1.6 in (4 cm) of the surface were scarified by heat, germinated within two weeks, and died during the summer drought period. This reduced the amount of seed in the soil by 97 percent. Although some seed remained below >2 in (>6 cm) in ant nests, the reduction in the seedbank significantly decreased the need for chemical or hand removal of new seedlings in succeeding years. Follow-up monitoring and treatment using this same combination of methods in a coastal area of Redwood National Park reduced the seedbank by only 52 percent and did not significantly reduce the time spent in follow-up control. The moister climate decreased the efficacy of this removal combination at the Redwood National Park site (Bossard 1991a, 1993, and unpubl. data).

Because of broom's seedbank, monitoring removal sites to locate and kill new seedlings is essential. Location and retreatment of resprouts is also imperative. If any single removal technique is used the site should be examined once a year, when seed germination ends in late spring, for five to ten years. Using the combined removal treatments, monitoring should occur late spring, yearly, for the first two years then again the fourth and sixth year after removal.

Cytisus striatus Rothm.

Common name: Portuguese broom

Synonymous scientific names: none

Closely related California native plants: 0

Closely related California non-native plants: 7

Listed: Cal EPPC A-1; CDFA nl

by Maria Alvarez

HOW DO I RECOGNIZE IT?

Distinctive Features

Portuguese broom (*Cytisus striatus*) is a shrub six to nine feet tall with many slender eight- to ten-angled stems that are silky-haired when young and become more or less smooth when mature. Stems are covered sparsely by small leaves consisting of one to three leaflets. Pale yellow, pea-like flowers arise from the leaf axils singly or in pairs. Mature fruit pods are densely white-hairy, and each contains several seeds. The main features that distinguish this species from Scotch broom (*Cytisus scoparius*) are the paler yellow flower color, the greater number of angles on the stem, the flat-hairy calyx (sepals), and the densely white-hairy fruit pods of Portuguese broom. The pods of *C. striatus* are generally larger than those of French broom (*Genista monspessulana*).

The inflorescences of French broom contain four to ten flowers per cluster with flower pedicels less than one-half inch, whereas Portugese broom pedicels are greater than one-third inch.

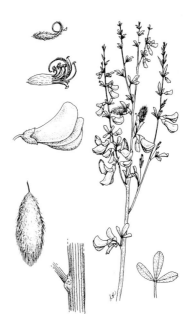

Description

Fabaceae. Long-lived shrub. Stems: young plants (1-3 years) have an upright form (erect) dominated by a leading apical stem. Plants become more branched and broaden with age, attaining a height of 7-10 ft (2-3 m). In windy locations growth is more compact and stunted, to 3-5 ft tall. Leaves: more abundant at outer ends of actively growing branches, giving the plant a broom- or antler-like appearance; leaflets 0.3-0.5 in (5-15 mm) long and obovate (pointed at both ends). Leaflet upper surface generally smooth (glabrous); lower leaflet surface silky-hairy. Leaflets on upper branches sessile, lower branches contain short petioled leaflets. Flowers:1-2 pea-like flowers cluster between leaf base and stem on pedicels 5-10 mm long. Flower calyx is appressed hairy and <0.3 in. (<8mm); corolla glabrous, banner 0.6-1 in (13-25 mm) long and not reflexed. Fruit: a pod, 0.7-1.75 in (15-40 mm), inflated (eliptical), and densely white-hairy. Number of seeds variable, ususally several per pod (Hickman 1993).

WHERE WOULD I FIND IT?

Portuguese broom is much less common than other broom species. It currently occupies sixty-five acres in the Marin Headlands, Marin County, where it forms dense cover, one mature shrub per two square meters. It is found occasionally in other parts of the Bay area, and has been reported in Mendocino and San Diego counties, with probable occurrence in central and south coastal counties. It is capable of invading and establishing dense populations in coastal prairie, coastal scrub, oak savannah, and open-canopy woodlands. In the San Francisco Bay Area it is particularly common on non-calcareous soils.

WHERE DID IT COME FROM AND HOW DOES IT SPREAD?

Portuguese broom is native to Portugal (Hickman 1993). Introduced to the Marin Headlands, Marin County, in the 1960s for landscaping and slope stabilization, it escaped and naturalized. Portuguese broom probably spreads like Scotch broom; that is, after ballistic dispersal, seeds may be further dispersed by ants, animals, by rain wash on open ground, or in mud clinging to road grading or maintenance machinery.

WHAT PROBLEMS DOES IT CAUSE?

Portuguese broom displaces native plant species, both herbs and other woody species. Its seeds are toxic to ungulates, and mature shoots are unpalatable. Like Scotch broom, it is fire prone and capable of carrying fire to the tree canopy layer.

HOW DOES IT GROW AND REPRODUCE?

Portuguese broom becomes reproductive at two to three years of age at a height of one and a half to three and a third feet (40-100 cm). It flowers in late March through May on the coast, resulting in copious production of long-lived seeds that mature in June and July. Seeds are released ballistically from the pod, then further dispersed by animals and water runoff along the ground. Seed germination is probably

similar to related broom species such as Scotch broom (*Cytisus scoparius*). Portuguese broom can resprout from its root crown when cut, particularly during the growing season.

Population growth is slower for Portuguese broom than for French broom, making containment a possibility. Drought conditions during summer cause growth to cease. Little dieback is observed during summer in the Marin population, probably because fog drip in late summer can be substantial. New growth resumes in winter and spring prior to flowering. Some large individuals can be killed by an unusually long freeze, perhaps limiting its geographic distribution (Alvarez pers. observation). It takes two to three years to reach reproductive maturity, but individuals can survive at least twelve years in California.

HOW CAN I GET RID OF IT?

Portuguese broom is difficult to remove because of the large size of individual plants, deep roots, and a long-lived seedbank. Removal should be followed by five years of monitoring and follow-up treatments to achieve control of this species. With limited resources, it is particularly important to determine the primary direction in which the population is expanding and start removal efforts there. Prevent seed dispersal into uninvaded areas by removing widely dispersed individuals from the main population center. Focus on preventing seed set and dispersal from all mature individuals each season so that no net increase in the seedbank can occur.

Physical Control

A combination of hand pulling, brush cutting, and mulching is the most effective way to remove mature plants and control reestablishment when chemicals, heavy equipment, and fire cannot be used.

Manual methods: Small plants can be hand pulled when the soil is moist. Larger plants can be pulled easily with a weed wrench if they are not too branched near the base of the plant. If densely branched, long-handled loppers or a pruning saw can be used to remove the lower limbs before pulling shrubs (1- to 3-inch-diameter trunks) with an appropriately sized weed wrench.

To maximize effectiveness of manual control for large broom plants, removal should be timed for late summer or early fall, when drought conditions exist. Plants can be cut at ground level, leaving the stump exposed. If warm conditions persist, it is likely that the plants will not resprout. Stems and stumps should be cut close to the ground to minimize resprouting. For additional insurance against resprouting from a stem, peel any bark back to the ground, or split the stump into shreds with a hand axe or the axe end of a pulaski. The GGNRA's Habitat Restoration Program has achieved 60 to 80 percent mortality by hand cutting alone during late summer and early fall in the Marin Headlands.

Mechanical methods: Brush cutters with four-pointed blades and chainsaws are also effective for removing Portuguese broom. However, it is difficult to brush cut broom flush with the ground on steep, uneven terrain, which may account for the resprouting of brush-cut broom in such areas.

Mowing has had mixed results on Portuguese broom. Marin State Parks and U.C. Extension have utilized various forms of mowing, and this method should be investigated further (Peterlee 1990, Nelson 1994). Mowing prior to flowering can be effective in preventing another seed crop. The overall length of the perimeter of the population should first be reduced by eliminating peninsulas or finger-like extensions so that seed dispersal and establishment are slowed. Then the infestation can be treated in a concentric fashion each season, reducing the size of the main infestation over time.

Mulching: If there is a large seedbank, three to four inches (7.5-10 cm) of straw (certified noxious weed free) should be applied. Mulching should be done during winter or spring before seedlings are over an inch tall. A controlled study by the Habitat Restoration Team demonstrated that mulching with rice straw was 99 percent effective in preventing French broom seedlings from emerging through straw for their entire germination period from December to April. Mulching also increased the mortality of brush-cut French broom in the same study when applied during winter (Alvarez unpubl. data).

Prescribed burning: Treatment should follow recommendations for Scotch broom.

Biological Control

Biological control agents have not been reported for Portuguese broom.

Chemical Control

Herbicides have not been used on Portuguese broom in the Marin Headlands, and there are no published accounts of their use elsewhere, but the treatments used on Scotch broom should be effective on Portuguese broom.

Delairea odorata Lemaire

Common names: cape ivy, German ivy

Synonymous scientific names: *Senecio mikanioides*

Closely related California natives: 36

Closely related California non-natives: 5

Listed: CalEPPC A-1; CDFA nl

by Carla C. Bossard

HOW DO I RECOGNIZE IT?

Distinctive Features

Cape ivy (*Delairea odorata*) is a perennial vine with shiny, five- to six-pointed leaves, usually with two small stipule-like lobes. There is one leaf at each node. Foliage is green to yellow-green and has a distinct odor. Plants have extensive waxy stolons running above and below ground. Below-ground stems are purple. Each flower is a yellow, round discoid head the size of a dime. Flowers are arranged in groups of twenty or more.

Description

Asteraceae. Long-lived, branching, glabrous perennial vine with shiny stolons covered by thick cuticle. Climbs over other vegetation and foliage; grows to 30 in (89 cm) deep on the ground. Underground stolons purple-mottled and root adventitiously at nodes. Leaves: evenly spaced, with shiny cuticle; blades 1.2-3.2 in (3-8 cm), +/- petiole (often with two stipule-like lobes at base) +/- round; blades sharply, palmately 5-9 lobed. Inflorescence: heads discoid, 20-40; main phyllaries +/- 8, 0.12-0.16 in (3-4 mm), tips green; flowers: 0 ray flowers, <40 disk flowers, bright yellow, head 0.2-0.5 in (5-12 mm) in length (Hickman 1993) Fruit: seeds glabrous, almost always sterile in California, often with pappus.

WHERE WOULD I FIND IT?

Cape ivy is invasive in Italy (Catalano *et al.* 1996), Australia (Fagg 1989), and the eastern United States, and currently occupies more than 500,000 acres in California (Grotkopf, pers. comm. 1998, Robison unpubl. data). It exists in many coastal forests the length of California, with populations found from Del Norte County in northern California to Canyons inland from San Diego. Typically found below 660 feet (<200 m) elevation, it prefers shady,

disturbed sites with year-round moisture, such as stream banks, coastal forests in a fog belt, or soils with a high water table (Chipping 1993). In recent years populations have appeared in grasslands, open oak forests, coastal scrublands, Monterey pine forest, coastal bluff communities, seasonal wetlands, and even a few serpentine soils. In habitats without year-round moisture sources this vine dies back in the dry season, then grows rapidly during the wet season. It can survive in the Central Valley if it is near a water source (Bossard and Grotkopf unpubl. data, Cudney and Hodel 1986).

WHERE DID IT COME FROM AND HOW DOES IT SPREAD?

Cape ivy is native to moist mountain forests of South Africa, where it has a limited natural range. Introduced in the 1850s as an ornamental in the eastern United States and to California by the 1950s (Elliot 1994), by the 1960s it had naturalized in Golden Gate Park, San Francisco, and Marin County (Archbald 1995, Howell 1970). It spreads vegetatively by stolons and fragmentation of stolons. Ninety-five percent of fragments of green stolons containing only one node establish, and drying stolon fragments in full sun for ten weeks does not stop them from rooting (Bossard unpubl. data).

WHAT PROBLEMS DOES IT CAUSE?

Cape ivy climbs over most other vegetation, forming a solid cover that blocks light and smothers other vegetation. The weight of the ivy mass sometimes causes trees to fall. Habitat for both plants and animals in protected natural reserves has been rendered worthless when large portions are occupied almost exclusively by cape ivy. Even in areas that have not become monospecific, native plant species richness can be reduced about 50 percent, with greater impact on annual than on woody perennial species (Alverez 1996). In the same study, native species seedling richness decreased 75 to 95 percent in areas containing cape ivy.

Cape ivy contains pyrrolizidine alkaloids (Stelljes *et al.* 1991) and xanthones (Catalano *et al.* 1996) that make it unsuitable forage for most fauna. Pyrrolizidine alkaloids such as retronecine, found in cape ivy foliage and flowers, are known to be toxic to mammals and to spiders. Initial experiments indicate that cape ivy foliage contains compounds that decrease fish survival (Bossard 1998).

Flood control function along streams is impacted by cape ivy infestations (Archbald 1995). Due to its shallow root system, cape ivy can contribute to serious soil erosion problems on hillsides (Cudney and Hodel 1986)

HOW DOES IT GROW AND REPRODUCE?

Cape ivy apparently reproduces only vegetatively in California. In Australia and in its native South Africa it reproduces by seed as well, which results from homogamous, radiate, self-incompatible flowers. Cape ivy flowers extensively in California, but evidently forms non-viable seeds. There have been occasional reports of "seedlings" from several populations of cape ivy in California. However, when thousands of seeds from twenty-six populations throughout California were examined by O'Connell and Bossard in 1994, not one embryo-bearing, germinable seed was found (Bossard unpubl. data). It is possible the naturalized cape ivy in California is derived from a single genetic stock and, since this species is self-incompatible, California populations do not produce viable seed.

The period of most rapid vegetative growth is February to June, with some dieback occurring during July to October in areas without a constant water source (Bossard and Benefield 1995). After flowering and a month or two of rapid vegetative growth, this vine starts storing sugars from photosynthesis as starch in underground storage organs. It flowers December to February. An experiment in Golden Gate Park (Alverez 1995) indicates that a single patch of cape ivy responds to resource changes more like an individual plant than a colony of individuals, allocating resources to developing regions or regions with decreasing resources from neighboring modules.

HOW CAN I GET RID OF IT?

Cape ivy is difficult to eliminate for two reasons: stolons and underground parts readily fragment while being removed, and plants will grow from almost any remaining fragment. The most effective control technique depends on the site topography, proximity of water, type of non-target vegetation on the site, age and size of the infestation, importance of impacts to non-target species, and type and quantity of resources available for control efforts. All methods require appropriate timing, and possibly supplemental revegetation by desirable species.

It is necessary to monitor removal sites every two months the first year and every four months the second year to locate and kill resprouts. Location and retreatment of resprouts is imperative or in six to eight months the ivy can reinfest the whole area from which it was removed. After the second year the site should be examined yearly to check for newly establishing populations. This is particularly important if a stream is present or if seasonal flooding occurs. Immediate removal of small new populations saves many hours of removal efforts a few months later and prevents further spread to other sites by fragmentation.

Physical Control

Manual removal has been attempted, sometimes successfully, in Volunteer Canyon, Marin County, Golden Gate National Recreation Area, San Francisco, and in parks in the Santa Cruz area. Manual removal requires clearing away native and invasive plant material to gain visual and physical access to locations with cape ivy stems emerging from the ground. Roots and stems must be teased out of the ground using a pointed or three-pronged mini-rake to loosen the soil. At some sites, where cape ivy is growing mat-like on the ground, it has been possible to roll up the entire infestation like a carpet using a potato hoe or rake (Archbald 1995). In Marin County volunteers were able to clear an average of 13.6 square meters of cape ivy per hour (Blumin and Peterson 1997). Removed cape ivy should be placed in or on plastic and, if feasible, removed from the area. Putting soda lime into cape ivy container bags will hasten the otherwise slow breakdown of this plant material. Manual control is sometimes followed with spot chemical treatment of resprouts.

Cape ivy tissues should not be put through a chipper or sent unbagged to a dump site. Both would likely result in spread of cape ivy. Returning at four- to eight-week intervals is necessary to locate and remove overlooked and resprouting plants. Manual removal is labor-intensive but can be accomplished where chemical applications cannot be used. The amount of disturbance to non-target species varies with the type of vegetative community infested, but it can result in increased erosion or in colonization by other invasive plants. Supplemental revegetation needs should be considered on a site-by-site basis.

Prescribed burning: This has not yet been attempted on cape ivy because of the high moisture content of its foliage.

Biological Control

Insects and fungi: An initial assessment of potential biocontrol agents for cape ivy conducted in South Africa suggests that there are seven promising insect candidates, including moth and beetle larvae and root-, stem-, and seed-feeding insects (Grobbelaar *et al.* 1999). Led by CalEPPC, an effort is underway to raise funds to continue to support the efforts of the USDA Albany lab in developing biocontrol agents for cape ivy. Some of the more promising insects are being tested in South Africa.

Chemical Control

A mixture of foliar-sprayed 0.5 percent glyphosate (as Roundup®) + 0.5 percent triclopyr (as Garlon 4®) + 0.1 percent silicone surfactant (as Silwit®) in water, applied as a foliar spray at 6.4 liters/ha proved effective in killing even long-established and extensive infestations of cape ivy in two applications, one year apart, in Golden Gate Park, San Francisco, and in a second test of this method near Morro Bay. Applications must be done in late spring when the plant is photosynthesizing actively but is past flowering, so the active ingredients move down with the sugars that are transported to underground storage organs. This mixture has a low concentration of active ingredients, which results in a slow, progressively deadly impact on dense cape ivy infestations with no measurable damage to non-target species (Bossard and Benefield 1995). Since the mixture contains a surfactant and triclopyr, it should be used cautiously and only within the guidelines specified on the label, especially where the water table is only a few inches below the surface or along pond or stream banks.

In Australia, clopyralid, (sold in California as Transline®), was used successfully in concentrations of 150 g/liter clopyralid at application rates of 6-8 liters/ha to remove dense infestations of cape ivy using the rope wick method of application. This was done in two applications a year apart. Clopyralid substantially damaged non-target species in the Asteraceae, Solonaceae, Urticaceae, and Bignoneaceae families, but no appreciable damage was found on non-target species of other plant families (Fagg 1989). This herbicide is more expensive but also more selective than most other herbicides.

Tests of glyphosate alone (as Rodeo®) and glyphosate + the surfactant R-11 at concentrations of 4 lbs/acre were carried out. While initial impact on above-ground foliage was high, extensive resprouting occurred from underground parts (Bossard and Benefield 1995). The rapid death of above-ground parts prevented translocation of the active ingredients to the roots and underground stems.

Digitalis purpurea L.

Common names: foxglove, purple foxglove

Synonymous scientific names: none known

Closely related California natives: 0

Closely related California non-natives: 0

Listed: CalEPPC Need More Information; CDFA nl

by Steven A. Harris

HOW DO I RECOGNIZE IT?

Distinctive Features

Foxglove (*Digitalis purpurea*) is an erect, knee-high to head-high herbaceous perennial with a basal rosette of leaves. In its second growing season it produces a leafy stock bearing a column of long, bell-shaped, nodding flowers on one side. Flowers are generally pinkish purple or white, with spots on the inside lower portion.

Description

Scrophulariaceae. Biennial or perennial herb. Stems: 1-5 ft (0.3-1.8 m) from taprooted caudex, branched above. Leaves: lanceolate to ovate, 4-12 in (10-30 cm) long, evenly spaced but reduced upward 2-8+ in (5-20+ cm), ovate, margins deeply dissected, upper surface green and soft-hairy, lower surface gray-tomentose, lower leaves soon deciduous. Inflorescence: pedicel 0.5-1 in (6-25 mm), spike 1-4 ft (30-130 cm), tomentose. Flowers: calyx lobes <0.7 in (<1.8 cm), lanceolate to ovate; corolla 1.6-2.4 in (4-6 cm), white to pinkish purple with darker spots on lower inside surface, lobes ciliate, sparsely hairy inside. Fruits: approximately 0.5 in (1.2 cm), ovoid with many 0.02 in (0.5 mm) seeds (Hickman 1993).

WHERE WOULD I FIND IT?

Foxglove is found along the California coast northward from Santa Barbara County, infesting moist meadows and roadsides. It is also reported from the northern Sierra Nevada foothills. A cultivated ornamental, it is often found escaping. It grows in full sun to part shade, in any well drained, fertile, acid soil in open woodland, pastures, roadsides, and disturbed places at less than 3,000 feet (1000 m) elevation. It thrives throughout the United States except in southern Florida and along the Gulf Coast, where it is suppressed by high humidity.

WHERE DID IT COME FROM AND HOW DOES IT SPREAD?

Native to Europe (especially western Europe), the Mediterranean, and northwest Africa, foxglove has been introduced to many areas as an ornamental and medicinal plant. By 1940 it was established in Humboldt and Mendocino counties (Robbins 1940). It escapes cultivation, and seeds are dispersed by wind and water.

WHAT PROBLEMS DOES IT CAUSE?

A source of the cardiac glycoside digitalis, a medically important heart stimulant, all parts of the plant are toxic. Foxglove is lethal to animals consuming small amounts of fresh or dried material (Scott 1997). It readily colonizes areas of soil disturbance, forming dense patches that displace natural vegetation.

HOW DOES IT GROW AND REPRODUCE?

Foxglove reproduces only by seed. In the spring of

the second year of growth, it rapidly produces stalks two to five feet (1-2 m) tall, lined with blossoms. Flowers mature in early summer, producing abundant seeds. Seeds are dispersed throughout the summer and remain viable in the soil at least five years (Scott 1997). Sprouts of the small seeds are not able to penetrate turf to any depth. Soil disturbance greatly increases establishment of seedlings. Stalks die back in winter.

HOW CAN I GET RID OF IT?

Sites will need to be monitored for five to ten years. Control efforts are required for at least five years.

Physical Control

Manual methods: Hand pulling of stalks is effective. In spring, while soils are moist, stalks and root masses are easily pulled from the ground. Pulled material must be removed from the site and destroyed (flower stalks left on site will continue to mature and release thousands of seeds). It is easy to strip flowers from the stalks, and little additional effort is needed to pull up the entire plant. If flower stalks are cut back before seeds ripen, the plant can bloom again in mid- to late summer. Therefore, above-ground treatments such as clipping and mowing may be counter-productive unless re- peated before resprouts have time to produce seed. Workers must protect themselves from extended contact with the poi- sonous leaves.

Prescribed burning: Fire as- sociated with other management programs is problematic, since stands of foxglove are not a good fuel source. Also, habitat in which foxglove typically be- comes established does not con- tain enough fuel to sustain a fire long enough to kill the plant. Smoke from burning leaves is toxic and has caused injury to workers on control projects (Scott 1997).

Biological Control

Foxglove is valuable com- mercially in horticulture, so bio- logical control has not been pur- sued.

Chemical Control

Herbicide trials were conducted in late summer and early spring by Scott (1997) on infestations of *Digitalis lanata* in Wilson county, Kansas. Metsulfuron methyl (as Escort®) at label strength and triclopyr (Garlon®) at 2 pts/acre showed some effect on the plants but did not kill all of them. Herbicides may work, but hand pulling is more efficient and effective with fewer effects on non-target plants.

Egeria densa Planchon

Common names: egeria, leafy elodea, dense waterweed, Brazilian waterweed, anacharis, Brazilian elodea

Synonymous scientific names: *Anacharis densa, Elodea canadensis gigantea, Elodea densa*

Closely related California natives: *Elodea canadensis, E. nuttallii*

Closely related California non-natives: none

Listed: CalEPPC A-2; CDFA nl

by Marc C. Hoshovsky and Lars Anderson

HOW DO I RECOGNIZE IT?

Distinctive Features

Egeria (*Egeria densa*) is a perennial freshwater aquatic herb in the waterweed family. It has stems up to fifteen feet long that are frequently branched. It is distinguished from related species by the absence of turions (shoots from underground stems) and tubers and by the presence of showy, white flowers that float on or just above the water. It is usually rooted in bottom mud, but may be found as a free-floating mat or fragments with its buoyant stems near the surface.

Description

Hydrocharitaceae. Aquatic perennial. Stems: Green to brown, slender 0.1 in (2-3 mm) thick, generally branched but long 10-16 ft (3-5 m). Leaves: opposite below, crowded and whorled above, 3 to 6 leaves per whorl, each narrowly oblong, 0.4-1.2 in long by 0.1-0.2 in wide (1-3 cm by 2-5 mm) and sessile with a finely serrate margin. Flowers: white, large, unisexual, exserted from sessile, +/- ovate, 2-toothed spathes formed in axils of upper leaves and carried to the surface on a thread-like extension of the perianth tube up to 3 in (8 cm) long. Female flowers solitary, with 3 spoon-shaped sepals, reflexed at maturity, and 3 broadly ovate but unequal sepals 0.2-0.3

in (6-8 mm) long and 0.2-0.3 in (5-8 mm) wide. Male flowers are in groups of 2-4 with 3 boat-shaped sepals and 3 showy petals 0.35-0.4 in (9-11 mm) long and 0.2-0.35 in (6-9 mm) wide. Fruits: cylindrical, 0.3 in (7-8 mm) long and 0.1 in (3 mm) diameter, developing within the female spathe; rare or absent in CA populations. Seeds: if present, spindle-shaped and 0.3 in (7-8 mm) long (Hickman 1993, Parsons 1992).

WHERE WOULD I FIND IT?

Egeria occurs in cool to warm freshwater ponds, lakes, reservoirs, and slowly flowing streams and sloughs. It can root up to seven meters below the water surface (Parsons 1992). In California, egeria occurs at less than 7,000 feet elevation in the Sierra Nevada, Central Valley, central coast San Francisco Bay, and San Jacinto Mountains (Hickman 1993).

WHERE DID IT COME FROM AND HOW DOES IT SPREAD?

Egeria is native to Argentina, Brazil, and Uruguay. It is has been distributed via the aquarium trade to many other parts of the world, including Chile, Mexico, the United States, England, New Zealand, and Australia, where it has escaped cultivation and become naturalized (Parsons 1992). The timing and location of its entry into California are unknown. Human dispersal via the aquarium trade is the most common means of egeria dispersal (Parsons 1992) and it can readily establish in natural water bodies after escaping from human settings. Once naturalized, egeria can spread along existing water courses into suitable new habitats without further human activity. Stem fragments at least two nodes long frequently break off and float away from the parent plant during active growth in spring (Parsons 1992). Fragments occur during all times of the year as a result of mechanical shearing of water flows, wave action, waterfowl activity, and boating.

WHAT PROBLEMS DOES IT CAUSE?

Egeria's dense underwater growth significantly retards water flow, interfering with irrigation projects, hydroelectric utilities, and urban water supplies. It may also slow water traffic and interfere with recreational and commercial activities such as boating, swimming, and fishing. Egeria reduces the abundance and diversity of native plant seeds in lake bottoms, and this is probably accentuated by increased sediment accumulation beneath the weed beds (deWinton and Clayton 1996).

HOW DOES IT GROW AND REPRODUCE

In California (and North America in general) reproduction and dispersal are via fragments of shoots and rhizomes, since only the male plant has become established. No seed formation has been documented (Anderson 1997, 1998, and pers. observation). Stem fragments can take root in bottom mud or may remain as free-floating mats. Roots range from fine and thread-like to long, thick, and robust adventitious roots, with many branches. Growth is most rapid during summer, as day length and temperature increase. Biomass in lakes reaches a maximum during late summer and fall. Thick mats form, consisting of long, intertwining, multi-branched stems below the water surface. No information is available on the rate of individual plant growth (Parsons 1992).

Egeria's ecological requirements are poorly investigated, although its nutrient and light requirements are similar to other members of the family. Its growth appears to be affected by nutrient status, light intensity, day length, temperature, and rate of water flow. It tolerates a wide range of nutrient levels, particularly phosphorus. Biomass increases with increased ammonium in stream water and with total nitrogen in sediments.

Egeria has a low light requirement. High light intensities cause discoloration and damage to the chlorophyll within about two weeks. Thus, turbid water is likely to favor rather than inhibit growth. Egeria thrives in red light spectra, which is more abundant near the water surface, and is killed or suffers under blue and green light spectra, which penetrate deeper below the surface. This may explain why the weed cannot establish itself more than twenty feet below the water surface (Feijoo *et al.* 1996).

HOW CAN I GET RID OF IT?

Several methods are useful in removing egeria, particularly in lakes and ponds where water movement is minimal.

Physical Control

Manual/mechanical methods: pulling, cutting, and digging with machines is costly, provide only temporary relief, and simultaneously encourage spread by fragmentation. Mechanical harvesting produces thousands of viable fragments per acre (Anderson, 1998).

Biological Control

Grazing: Two fish, the white amur or Chinese grass carp (*Ctenopharyngodon idella*) and the Congo tilapia (*Tilapia melanopleura*), have been introduced into water bodies to control egeria (Avault 1965). Currently, only the sterile (triploid) grass carp can be used in California and only in six southern California counties (Imperial, San Diego, Riverside, San Bernardino, Los Angeles, and Ventura). Permitted uses are authorized by the California Department of Fish and Game throughout the state with certain restrictions.

Chemical Control

Herbicides in aquatic systems must be handled carefully to avoid worsening the situation. A specialist in control of aquatic weeds should be consulted. At present the following herbicides can be used at label concentrations to control egeria in California: diquat (contact type); copper-containing products (contact type); acrolein (contact type and highly restricted uses where no fisheries are impacted); and fluridone (systemic type requiring 4 to 6 weeks of treatment at very low rates) (Anderson *et al.* 1996).

Ehrharta calycina, *Ehrharta erecta*, and *Ehrharta longiflora*

E. calycina Smith

Common name: perennial veldt grass

Synonymous scientific names: none known

Listed: CalEPPC A-2; CDFA nl

E. erecta Lam

Common names: ehrharta, panic veldt grass

Synonymous scientific names: none known

Listed: CalEPPC B; CDFA nl

E. longiflora Brey

Common name: annual veldt grass

Synonymous scientific names: none known

Listed: CalEPPC Need More Information

Closely related California natives: 0

Closely related California non-natives: no others

by Andrea J. Pickart

HOW DO I RECOGNIZE IT?

Distinctive Features

Perennial veldt grass (*Ehrharta calycina*) is a tussock-forming grass with numerous stems and flat, green to reddish purple-tinged, glaucous leaves three to eight inches long. The leaves are often wrinkled partway along the margins. Over time this grass can become a nearly continuous cover under shrubs, but individual tussocks can be distinguished. The inflorescence is a loose panicle above the leaves, four to six inches long, usually open but sometimes contracted.

Ehrharta (*Ehrharta erecta*) is another perennial grass, distinct from *E. calycina* in having a crabgrass-like habit with decumbent as well as ascending jointed stems. The sterile lemmas of *E. erecta* are without awns.

Annual veldt grass (*Ehrharta longiflora*) is similar to *E. calycina* in having wrinkled margin blades and purplish-tinged leaves. It is distinguished by the fact that it is annual and more sprawling, with stems growing out from the crown before becoming erect. The sterile lemmas of annual veldt grass are long-awned, unlike those of both perennial species of *Ehrharta*.

Description

Ehrharta calycina (perennial veldt grass). Poaceae. Erect perennial grass forming dense tufts (infrequently rhizomatous) 12-30 in (30-75 cm). Leaves: flat, glaucous leaf blades 3-8 in (7-20 cm) long, 0.08-0.3 in (2-7 mm) wide, often tinged reddish purple and wrinkled partway along the margin. Ligule a prominent membrane with several awn-like teeth at the apex. Inflorescence: 4-6 in (10-15 cm) long, contracted to open panicle. Bisexual compressed spikelets subsessile to stalked, 0.2-0.3 in (5-8 mm), falling as one unit. Glumes 0.2-0.3 in (5-7 mm), becoming purplish, about equal, longer than sterile florets. Three florets per spikelet, lower two sterile and without palea; upper floret fertile with palea. Sterile lemmas membranous (becoming hard), short-awned or pointed, and soft-hairy. Fertile lemma awnless with hairy veins (Smith 1993).

Ehrharta erecta (ehrharta). Poaceae. Perennial grass. Stems: culms erect or ascending from decumbent base, branching, 12-24 in (30-60 cm) tall. Leaves: flat leaf blades 2-5 in (5-12 cm)

Ehrharta calycina   *E. erecta*

Ehrharta calycina

long, 0.2-0.4 in (4-9 mm) wide. Inflorescence: 2-6 in (6-15 cm) long, contracted to open panicle. Laterally compressed sessile to subsessile spikelets, 0.1 in (3-3.5 mm), falling as one unit. Glumes 0.06-0.1 in (1.5-3 mm), about equal, longer than sterile florets. Three florets per spikelet, lower two sterile and without palea; upper floret fertile with palea. Sterile lemmas awnless, glabrous (Hickman 1993).

Ehrharta longiflora (annual veldt grass). Poaceae. Annual grass. Stems: erect. Leaves: flat leaf blades up to 8 in (20 cm) long, 0.4 in (10 mm) wide, wrinkled partway along the margin, collar sides purple. Ligule membranous and toothed. Inflorescence: spreading to contracted panicle, often fascicled. Compressed spikelets sessile to stalked, with branches and pedicels pubescent, sometimes purple, 0.3-0.4 in (8-10 mm). Glumes becoming purplish before fruit matures, the first 0.10-0.12 in (2.5-3.0 mm) and five-nerved, the second 0.16-0.18 in (4.0-4.5 mm) and seven-nerved, axis breaking above the glumes. Three florets, lower two sterile without palea, upper floret fertile with palea. Sterile lemmas membranous (becoming hard and white), seven-nerved, long-awned (about 0.3 in or 8 mm), and hairy at base of callus and base of keel. Fertile lemma seven-nerved, awnless (Brey 1996).

Ehrharta sp.

WHERE WOULD I FIND IT?

Ehrharta calycina (perennial veldt grass) is grass is characteristic of sandy soils and is known to occur in Santa Barbara and San Luis Obispo counties on the south central coast and in Sonoma County. It is prevalent in all of the upland habitats of the San Antonio Terrace (at Vandenberg Air Force Base) and throughout the Los Osos dunes southeast of Morro Bay. In recent years it has been increasing rapidly in dune scrub at the Guadalupe-Nipomo Dunes (Chesnut 1999) and on the maritime chaparral and coast live oak woodlands of the Nipomo Mesa.

Perennial veldt grass has also been reported from the Gaviota coastal grassland terrace, where it recently spread from a few roadside locations into a large disturbed area near the old Vista Del Mar school, threatening high-quality native grasslands and coastal sage scrub.

Although present at Bodega Bay (Hickman 1993), the species has increased significantly in extent in recent years. It is also widespread as a naturalized weed in Australia (Tothill 1962, Cade 1980). It appears to be in an explosive stage of invasion in California, and may already be present in other areas or may appear in the near future.

At Vandenberg Air Force Base, *E. calycina* was found to be more prevalent on hillsides than in the more densely vegetated swales (it is unknown whether this was a result of fewer openings in the swales or of some physiological tolerance of the grass). The species occurs on well drained soils and does not tolerate inundation (Tothill 1962, U.S. Air Force 1996). Tothill (1962) concludes that *E. calycina* will not survive in completely dry soil, but since its roots penetrate deeply, it is able to exploit dry, sandy soils. On heavy clay soils, it can be maintained only through removal of associated species, which compete aggressively with the delicate root system of *E. calycina* (Tothill 1962).

Ehrharta erecta (ehrharta) became established in northern California about 1930 (Stebbins 1985). It has been reported from the greater San Francisco Bay Area (Sigg 1996) and from La Jolla Shores in San Diego County (Brey 1996). Populations have been reported from the San Francisco Bay Area, as well as San Diego, Santa Barbara, Los Angeles, and Ventura counties (Smith 1993, Sigg 1996, Brey 1996). *E. erecta* is thought to be more invasive in the northern portion of its range, but more information is needed to verify this. In its native range it is widespread in winter-rainfall and tropical regions but does not extend into arid regions (Gibbs-Russell and Robinson 1983). Sigg (1996) observed it in a wide variety of habitats in both exposed and shady areas in the San Francisco Bay Area, including sand, heavy soils, and thin, rocky soils. Sigg reports that the species is still vigorously expanding its range, and that the degree of threat it poses may not yet be apparent.

Ehrharta longiflora (annual veldt grass) is an invasive annual that was recently discovered in the San Diego area (Brey 1996). It has also been introduced to Australia (Cade 1980). In southern Africa it has a wider distribution than most other species of *Ehrharta*, which may be an indicator of its potential for invasiveness on this continent.

WHERE DID IT COME FROM AND HOW DOES IT SPREAD?

All three species of *Ehrharta* present in California are native to southern Africa. *E. calycina* was first reported in California in 1929, imported as seed from Australia (Love 1948). *E. erecta* became established as an adventive near Berkeley around 1930 (Stebbins 1985). The former Soil Conservation Service (now Natural Resources Conservation Service) promoted *E. calycina* for forage improvement and erosion control during the 1950s and 1960s (Mulroy *et al.* 1992, U.S.

Air Force 1996), and it was planted on ranches and sowed on controlled burns in coastal San Luis Obispo and Santa Barbara counties. The U.S. Air Force stabilized sand dunes with *E. calycina* at Vandenberg Air Force Base in the late 1950s (U.S. Air Force 1996), where it has since spread invasively.

The three species of *Ehrharta* spread primarily by wind-borne seed. Invasions of *E. calycina* spread primarily in the direction of prevailing wind (Tothill 1962) and are enhanced by disturbance (Chipping, pers. comm.). *E. erecta* can also spread vegetatively (Sigg 1996).

WHAT PROBLEMS DOES IT CAUSE?

The invasion of *Ehrharta calycina* into native shrub communities causes a rapid shift toward grassland. The more open the original vegetation, the more rapidly invasion occurs (U.S. Air Force 1996). The species spreads readily into disturbed areas, such as roadsides, and from there into openings between shrubs. Once established, *E. calycina* inhibits or prevents germination and establishment of native dune scrub and chaparral species (U.S. Air Force 1996). Studies at Vandenberg Air Force Base have documented the dramatic and explosive increase in *E. calycina* cover between 1979 and 1996. The spread of this grass is now considered to be the most serious threat to sand dunes of the central coast of California (Chipping, pers. comm.).

Grass invasions have been shown to alter fire cycles, causing more frequent fires that favor the recovery of grasses over shrubs (D'Antonio and Vitousek 1992). In Australia, where *Ehrharta calycina* is also invasive, fire enhances invasiveness by reducing regrowth or establishment of native species and drastically increasing cover of *E. calycina* (Milberg and Lamont 1995). *E. calycina* creates a dense thatch during summer months, as the plant dries and the stems and leaves lean over. In addition to providing excellent fuel, this thatch may interfere with the germination and establishment of native plants during the wet fall and winter.

Because wildlife abundance and diversity are related to plant diversity, it is expected that the conversion of dune scrub to grassland will have negative impacts on wildlife. A small-scale study carried out by the U.S. Air Force (1996) at Vandenberg Air Force Base supports this premise.

There is no quantitative information available on the impacts of *Ehrharta erecta* and. *E. longiflora*. However, both have been observed to spread rapidly in wildland areas. Sigg (1996) reports that *E. erecta* is able to penetrate adjacent vegetation with decumbent stems and by going over it with ascending stems, as well as by spreading vegetatively. The dense turf that develops makes it difficult for seeds of other species to germinate. Noting the extremely rapid spread of *E. longiflora* in San Diego County, Brey (1996) calls its invasion a "red alert."

HOW DOES IT GROW AND REPRODUCE?

Ehrharta calycina (perennial veldt grass) spreads almost entirely by seed, although rhizomes are occasionally present. The species has been shown to accumulate persistent seedbanks (Pierce and Cowling 1991). *E. erecta* (ehrharta) reproduces both sexually and vegetatively by means of tillers (Sigg 1996). As an annual, *E. longiflora* (annual veldt grass) reproduces only by seed.

Ehrharta seeds germinate following winter rains. *E. calycina*, a perennial, grows and flowers throughout the rainy season (December to April) into early summer, when fog drip may prolong its growing season (U.S. Air Force 1996). Seeds are produced as early as January in wet years, although most fruits mature between March and June. Ripe seeds are shed progressively as the panicle matures, and flowering may occur for up to twenty-five weeks (Tothill 1962). Plants become dormant in summer in response to temperature, not lack of moisture (Tothill 1962). During summer months, when plants dry out, stems and leaves lean over, forming a dense thatch. The drop in soil temperatures in September and October releases dormancy, allowing plants to draw on deep soil moisture (Tothill 1962). Root growth is rapid and extensive after rains. *E. calycina* grows in dense tufts, which can survive some burial by sand. As sand accumulates over the base of the plant, buried shoots sometimes form lateral bunches, giving the appearance of rhizomes. True rhizomes occur rarely. *E. erecta* can create a continuous turf in moist areas, with plants spreading both vegetatively and by seed.

HOW CAN I GET RID OF IT?

Nearly all documented attempts to control *Ehrharta* species have been limited to *E. calycina*, and the following discussion centers on this species. It is likely, however, that techniques used on *E. calycina* would be effective on the other two species, and Brey (1996) reports good success with glyphosate used on *E. erecta*. As an annual, *E. longiflora* control efforts would need to be appropriately timed to its reproductive and growth periods.

Ehrharta control efforts are still new, and more information is needed. Large-scale experiments with herbicide have only recently been initiated. Research to identify potentially competitive native species should also be undertaken. Regardless of the method used, more than one year of treatment will be necessary, due to its extensive seedbank, the persistence of which is unknown. Unless a biological control method is found, treated areas are likely to require perpetual management to prevent reestablishment.

The invasion of *E. calycina* is an established phenomenon along the central coast, and where the species has become common or dominant it is unlikely to be totally eradicated. Regular monitoring and treatment will be needed to detect and control regrowth or new infestations. Prevention is the preferred strategy, and if *E. calycina* is detected as a new invader low in abundance, the highest priority should be given to its control.

Physical Control

Manual methods: manual removal of *Ehrharta calycina* (perennial veldt grass) has been undertaken at Los Osos Dunes with mixed success (Chipping, pers. comm.). Care must be taken to remove the buried base of the plant, or resprouting will occur. Removal by hand is labor-intensive and probably stimulates germination from the seedbank. Extremely high densities of emerging seedlings have been observed following manual removal (Cicero, pers. comm.). Manual removal must be repeated as plants emerge from the seedbank. In Morro Bay, areas treated two years in a row continued to support *E. calycina*, although densities declined.

Prescribed burning: Although fire is sometimes used as a control method for grasses, it is inappropriate for *Ehrharta* species, as studies have shown that fire increases the invasiveness of this species (Milberg and Lomont 1995).

Biological Control

Insects and fungi: No biological control efforts have been attempted for *Ehrharta* species in California. Doidge (1948) reported on results with the fungal pathogen *Uredo ehrhartae-calycinae* on *E. calycina* in South Africa.

Grazing: Sources have documented that *Ehrharta calycina* is not a suitable forage species because it is easily stressed by grazing, especially during flowering. Rossiter (1947) reports that *E. calycina* was unable to withstand continuous or even rotational grazing at normal stocking rates with sheep. A South African research project (Van der Westhuizen and Joubert 1983), which involved mowing of *E. calycina* during anthesis, demonstrated that cut plants recovered to a large extent, but were lower in total available carbohydrates (46 percent of control plants). The researchers, who were not attempting to control *E. calycina*, cautioned against severe defoliation during anthesis in the year of establishment. These observations point to the possibility of using sheep grazing as a potential method of control.

A ten-acre (4 ha) rotational grazing experiment was initiated in 1999 at the Guadalupe-Nipomo Dunes (Chesnut 1999); the results are pending.

Chemical Control

Glyphosate (as Roundup®) applied as a foliar spray at 2 percent concentration with added surfactant was shown to be effective against *Ehrharta calycina* under a wide variety of conditions at the Vandenberg Air Force Base dunes (Mulroy *et al.* 1992, U.S. Air Force 1996). Plants of different ages, with and without supplemental watering and/or mowing, were killed after one application. Spraying typically is carried out when the grass is actively growing and green. The use of glyphosate is believed by some to be most appropriate when *E. calycina* is growing as a near-monospecific stand, since it will cause damage to associated native plants. However, some managers have found that careful treatment of *E. calycina* bunches with a backpack sprayer can reduce or eliminate impacts to other native species. Under these circumstances it may be necessary to return and treat bunches of *E. calycina* that did not receive sufficient coverage with the first application.

The grass-specific herbicides sethoxydim and fluazifop-p (Fusilade®) have also been tested for use on *Ehrharta calycina* at Vandenberg Air Force Base, but results were inconclusive (U.S. Air Force 1996). Currently, efforts are underway to test fluazifop-p at Morro Bay dunes using a concentration of 0.75 oz herbicide and 0.5 oz surfactant to 1 gallon of water (Chipping, pers. comm.).

A study conducted by the U.S. Air Force (1996) indicates that the use of glyphosate (as Roundup®) in conjunction with seeding of native species shows potential for restoration of native dune scrub invaded by *Ehrharta calycina*. The study was undertaken in an area that had only recently become dominated by the species. First application was in December, with five follow-up treatments over the next two years. Seeds of eleven native dune scrub species were sown a week after the initial spraying. Most of the established *E. calycina* plants were killed in the initial application. Damage to adjacent native plants was minimal. Native cover showed a slight but significant increase, and open cover (also a constituent of the native vegetation type) increased substantially. Use of transplants or nursery propagated plants might enhance restoration success.

The Land Conservancy of San Luis Obispo (1999) recently carried out a demonstration project in which they combined the use of mowing, glyphosate, woodchip mulch, outplanting of native container stock, and irrigation. Long-term results will be forthcoming.

Eichhornia crassipes (C. Martius) Solms-Laubach

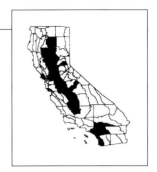

Common names: water hyacinth, common water hyacinth

Synonymous scientific names: *Eichhornia speciosa, Piaropus crassipes, Pontederia crassipes*

Closely related California natives: 0

Closely related California non-natives: 0

Listed: CalEPPC Listed A-2, CDFA nl

by Kris Godfrey

HOW DO I RECOGNIZE IT?

Distinctive Features

Water hyacinth (*Eichhornia crassipes*) is a floating aquatic plant with bright green, waxy leaves and attractive, violet flowers that have yellow stripes on the banner petals. These plants tend to form mats on the water surface. Sometimes water hyacinth can be found growing in muddy soils near the edge of an aquatic system. The leaves are arranged in a rosette. The leaf stem usually is somewhat to completely swollen and filled with spongy tissue and thus acts as a float. In plants anchored in mud, the leaf stem tends not to be swollen. The blade of the leaf is oval to round and usually much smaller than the leaf stem.

Description

Pontederiaceae. Perennial, floating aquatic plant. Stems: Stout, erect; may be connected by stolons; may be greater than 12 in (30 cm) in length. Leaves: Blade is generally oval to round and usually not greater than 4 in (10 cm) in width; leaf stem is somewhat to completely swollen, filled with spongy parenchyma tissue, 1-1.5 in (2-3 cm) diameter, generally longer than the blade (McClintock 1993). Inflorescence: funnel-shaped flowers borne on 2-6 in (5-15 cm) spikes.

Flowers: perianth varies from pale to deep lavender to blue or white; banner petal has a diamond-shaped yellow-orange spot surrounded by a wide pale purple border (Penfound and Earle 1948, McClintock 1993). Fruit: a capsule, usually surrounded by the remains of dried flowers; +/- 50 seeds per capsule; seeds longitudinally ribbed (Penfound and Earle 1948, Hickman 1993).

WHERE WOULD I FIND IT?

Water hyacinth can be found in both natural and man-made freshwater systems (ponds, sloughs, rivers). It will not tolerate brackish or saline water with salinity levels above 1.8 percent (Penfound and Earle 1948). In California water hyacinth typically is found

below 660 feet (200 m) elevation in the Central Valley, San Francisco Bay Area, and South Coast. The Sacramento-San Joaquin Delta and several of the rivers drained by this delta are heavily infested (Thomas and Anderson 1984).

WHERE DID IT COME FROM AND HOW DOES IT SPREAD?

Native to the Amazon River basin of tropical South America, water hyacinth has now spread to all tropical and subtropical

countries and is universally regarded as one of the most serious of the world's weeds (Parsons 1992). It was introduced into the United States in 1884 as an ornamental plant for water gardens. The plant quickly spread throughout the country, becoming a major weed in southern states from Florida to California. By 1897 it had clogged many waterways and was interfering with shipping (Parsons 1992). It was found in California in 1904 (Thomas and Anderson 1984).

Water hyacinth spreads through fragmentation of established plants and may resprout from rhizomes or germinate from seeds (Penfound and Earle 1948). Dispersal also occurs by water-borne seeds and by seeds that stick to the feet of birds. Migratory birds may be important in long-distance dispersal (Parsons 1992).

The major means of dispersal, and the most difficult to control, is active transport by people who, ignorant of its impacts, seek to propagate it in other ponds and lakes. In Australia almost every new infestation has come from deliberate planting or the disposal of surplus material from deliberate planting. Humans also contribute to its spread in some areas by using the plant as a packing material and as cushions in boats (Parsons 1992).

Water hyacinth has many functional properties, and expanded use of the plant may contribute to its spread. For example, it is attracting attention in the United States for use in sewage and industrial waste treatment. A 0.5 ha lagoon of the plant can purify the daily sewage waste of 1,000 people. It is being used in some countries to treat effluent from paper mills, tanneries, and factories producing explosives, rubber, photographic chemicals, and palm oil. It can remove heavy metals and pesticides from contaminated water and is being considered as a means of removing radioactive contaminants from nuclear power plant wastewater (Parsons 1992).

WHAT PROBLEMS DOES IT CAUSE?

Water hyacinth can quickly dominate a waterway or aquatic system because of rapid leaf production, fragmentation of daughter plants, and copious seed production and germination. It degrades habitat for waterfowl by reducing areas of open water used for resting, and when decomposing it makes water unfit for drinking. It displaces native aquatic plants used for food or shelter by other wildlife species.

Water hyacinth causes problems for humans by obstructing navigable waterways, impeding drainage, fouling hydroelectric generators and water pumps, and blocking irrigation channels. By 1975 nearly 700,000 hectares of waterways in Louisiana alone were infested, and rivers were blocked for distances of forty kilometers. Agricultural production in California's Central Valley was threatened at one point because of an 80 percent reduction in the efficiency of irrigation channels and pumping equipment. From one district alone, 20,000 truckloads of the weed were removed from waterways in 1981-82. The problem has diminished markedly in recent years as a result of control efforts (Parsons 1992).

The protected water within mats of water hyacinth makes ideal breeding sites for mosquitoes and other vectors, which, in tropical countries, increases the danger of malaria, schistosomiasis, and other diseases (Parsons 1992).

Water hyacinth increases water losses from lakes and rivers because of the plant's high transpiration rate, calculated to be almost eight times the evaporation rate of open water surfaces (Parsons 1992). It changes water quality beneath the mats by lowering pH, dissolved oxygen, and light levels, and increasing CO_2 tension and turbidity (Penfound and Earle 1948; Center and Spencer 1981). This affects the health of fish, while decaying plants make water unfit for drinking by humans, livestock, and wildlife.

The weed is a major concern in other countries as well. It has resulted in tremendous losses annually in fish and paddy rice production in India. In the Sudan it had infested over 3,000 kilometers of rivers by 1979, resulting in an estimated 10 percent loss in the normal flow of the Nile River and costing more than $3 million per year in control efforts (Parsons 1992).

HOW DOES IT GROW AND REPRODUCE?

Water hyacinth can reproduce either sexually or vegetatively. Flowering (i.e., sexual reproduction) occurs in mid-summer and early fall. The flower stalks bend back into the water once they are pollinated (it is thought to be self-pollinated) and release the seeds. Seeds sink to the bottom and can remain viable in sediments for several years.

Water hyacinth seeds require warm, shallow water and high light intensity for germination. Submerged seedlings root in the substrate and then form four or five linear leaves 0.2-0.6 in (5-15 mm) in length. The sixth and later leaves contain large amounts of spongy parenchyma tissue, which adds buoyancy to the seedlings. When seedlings are sufficiently buoyant, they break off from the roots and float to the surface. Once on the surface, seedlings form numerous fibrous and adventitious roots beneath the stem. The plants continue to grow by adding leaves in a whorl-like fashion, with the oldest leaves on the outside of the plant (Center and Spencer 1981). When daughter plants are of sufficient size, they break off the parent plant.

Vegetative reproduction occurs from late spring through fall. Parts of the stem may break off at the water surface to form independent plants called daughter plants (Penfound and Earle 1948). These daughter plants are capable of producing additional reproductive stem segments within weeks.

Water hyacinth grows rapidly. Growth of more than one ton of dry matter per day per hectare is not uncommon. One plant may be able to produce enough growth to cover 600 square meters in one year. Infestations break up into "rafts" that drift wherever the winds and currents take them, rapidly infesting entire river systems (Parsons 1992).

HOW CAN I GET RID OF IT?

The best method of controlling water hyacinth is to prevent it from being introduced into a freshwater system. This can be done by educating the public about the problems that occur from disposal of unwanted water garden or aquarium plants into freshwater systems or by not properly cleaning boats, trailers, other water sports equipment, bait buckets, or fishing equipment to remove all plant material before moving the equipment to another freshwater system.

Physical Control

Manual/mechanical methods: For small ponds or lakes infested with water hyacinth, harvesting and removal of plant material from the water can be attempted. Care must be taken to remove all plant material, including small fragments. Harvesting and removal of plant material is labor-intensive and expensive. A less expensive method of containing water hyacinth is the use of floating barriers that can contain the weed in a small area. Dredges, which drag plants onto river banks, are effective if the material is allowed to dry and is then burned. These are costly efforts, and they have been replaced in most areas by chemical control (Parsons 1992).

Biological Control

Insects and fungi: Biological control has been successful in many, but not all, areas. Three insects and a fungus have been extensively studied and subsequently released by the USDA to control water hyacinth. The insects include two weevils, *Neochetina eichhorniae* Warner, and *N. bruchi* Hustache (Coleoptera: Curculionidae), and a moth, *Sameodes albiguttalis* Warren (Lepidoptera: Pyralidae). The fungus is *Cercospora rodmanii* Conway (Fungi Imperfecti: Moniliales), which was first found in Florida in 1976 (Conway 1976).

In the southern United States the weevils have been most effective in reducing water hyacinth populations (Center *et al.* 1989). In Florida the weevils combined with the fungus have also produced good results. In California all three species of insects have been released. However, only *Neochetina eichhorniae* has established, and its impact on density of water hyacinth is slight. The fungus is currently unavailable for use in California.

Grazing: Most animals, except rabbits, do not readily eat the plant, possibly because its leaves are 95 percent water and have a high tannin content.

Chemical Control

Water hyacinth can be controlled using glyphosate as a foliar spray (formulated as Rodeo®) and copper complexes used only as a foliar spray. Herbicide use is more highly regulated in aquatic systems than in terrestrial systems. A current label for the herbicide must be obtained to determine suitability for a given system and amount of active ingredient to be applied. Both suitability and the amount of active ingredient may change from one year, habitat type, and/or jurisdiction to another. Consult your county agricultural agent or a certified herbicide applicator.

Elaeagnus angustifolia L.

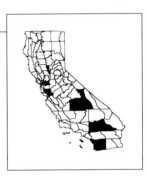

Common names: Russian olive, oleaster

Synonymous scientific names: none known

Closely related California natives: 0

Closely related California non-natives: 0

Listed: CalEPPC A-2; CDFA nl

by Laurie Deiter

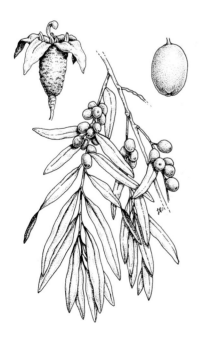

HOW DO I RECOGNIZE IT?

Distinctive Features

Russian olive (*Elaeagnus angustifolia*) is a perennial deciduous tree up to twenty-three feet tall. It has flexible, often spiny branches with dark, smooth bark, small, narrow, silvery leaves, and small clusters of aromatic yellow flowers that produce a hard olive-shaped fruit.

Description

Eleagnaceae. Tree to 23 ft (6.9 m) tall, especially in cultivation. Branches: flexible, often spiny, reddish brown, coated with gray and scaly pubescence, becoming glabrous with age. Leaves: alternate, blades simple, narrowly ovate to lanceolate, 1-3.5 in (2.5-9 cm) long, tip pointed to blunt, margins entire and commonly undulate, upper surface light green and covered with stellate pubescence, lower surface silvery white and densely covered with scales

with edges that fray to appear like stellate pubescence, petiolate 1/4-1/2 in (6-13 mm). Flowers: yellowish green sepals, sepals 4, fused below, lobes spreading or reflexed to erect, silvery-scaly outside, yellowish green inside, petals none, fragrant. Fruit: drupe-like, oval 0.5-0.6 in (13-15 mm) long, mealy, greenish yellow to brown, densely covered with mealy scales. Seed: brown, oblong, 1/4-1/2 in (6-13 mm) (Stubbendieck *et al.* 1994).

WHERE WOULD I FIND IT?

In California, Russian olive is found in disturbed, seasonally moist places, generally below 5,000 feet (1500 m) elevation. It exists in the San Joaquin Valley, San Francisco Bay Area, southern Sierra Nevada, San Diego County, and parts of the Mojave Desert near springs (Hickman 1993). It is common in riparian zones and floodplain forests and sub-irrigated pastures and irrigation ditches, but is also found in drier sites such as railroad beds, fence lines, along highways, and in grasslands.

Russian olive tolerates a wide range of soil and moisture conditions, from sand to heavy clay, and can withstand flooding and silting. It grows best in deep sandy or loamy soils with only slight salt and alkali content. There are dense, healthy stands in river bottoms where the water table is seldom more than two feet (0.6 m) below the surface, but it survives considerable drought (Borell 1971). Russian olive is tolerant of considerable salinity or alkalinity, although it prefers slightly alkaline soils with low to medium concentrations (100-3,500 ppm) of soluble salts (Olson and Knopf 1986). It can withstand temperatures ranging from -50 degrees F (-45 degrees C) to 115 degrees F (46 degrees C). It occurs from sea level to at least 8,000 feet elevation in several western states. Russian olive is somewhat shade tolerant and can withstand competition from other shrubs and trees (Borell 1971).

WHERE DID IT COME FROM AND HOW DOES IT SPREAD?

Native to southern Europe and western Asia, Russian olive was introduced into North America as an ornamental in the early 1900s (Christensen 1963). A popular landscaping tree, it escapes cultivation readily. Russian olive was actively promoted for many years for windbreaks, soil stabilization, and wildlife habitat improvement. It spreads by seed, primarily ingested by birds and deposited elsewhere. Small mammals also spread seeds by gathering and stockpiling them (Olson and Knopf 1986).

WHAT PROBLEMS DOES IT CAUSE?

Russian olive has escaped cultivation and naturalized in seventeen western states, colonizing edges of riparian zones first (Knopf and Olson 1984). It is able to regenerate under a wide variety of floodplain conditions with little or no mortality after seedling development. Native cottonwoods and willows, having narrow germination and establishment requirements and intolerant of shade, are unable to regenerate under advancing populations of Russian olive (Shafroth *et al.* 1994). Once established, Russian olive increases stream bank stabilization and reduces river stage levels, creating a relatively dry upland site with Russian olive as the climax species. Russian olive stands provide lower-quality habitat for terrestrial vertebrates compared to native riparian woodlands (Howe and Knopf 1991). Bird species richness is lower in Russian olive stands than in native riparian forest. By displacing native trees, monospecific stands of Russian olive reduce the availability of nesting cavities and insect food for many native birds (Shafroth *et al.* 1994).

HOW DOES IT GROW AND REPRODUCE?

Russian olive reproduces by seed, which is usually produced after trees are four to five years old. It generally flowers from May through June (Stephens 1973, Vines 1960). The fruits mature from August to October and remain on the tree throughout the winter or until the crop is consumed (Olson 1974, Borell 1971). Seeds are ingested with the fruit by birds and small mammals and dispersed in their droppings. The outer layer of the seed is impermeable to digestive juices. Seeds can remain viable for up to three years and are capable of germinating over a broad range of soil types (Knopf and Olson 1984). Germination is enhanced by stratification in moist sand for ninety days at 41 degrees F (5 degrees C) (Vines 1960). Spring moisture and slightly alkaline soil tend to favor seedling growth (Olson 1974).

Russian olive is a deep-rooted tree with a medium to rapid growth rate. It can grow up to six feet (1.8 m) per year (Williams 1991) and creates a well developed lateral root system (Borell 1971). It can resprout from from the root crown and sends up root suckers (Williams and Hanks 1976, Bovey 1965). It tolerates a wide variety of growing conditions, making it appealing to consumers for landscaping.

HOW CAN I GET RID OF IT?

Control methods vary with tree size, habitat, and use of the area (e.g., agricultural field vs. sensitive wetland). Removal should be undertaken before seeds are fully developed to prevent further spread of seeds. Control is difficult once trees mature, so early detection and control are important.

Physical Control

Manual/mechanical methods: Russian olives with small diameters (3.5 in) can be pulled out with a weed wrench when soils are moist. In certain situations larger trees can be removed using a tractor/chain. Any remaining exposed roots should be cut off below ground level and buried. Girdling and cutting are not effective controls by themselves. The tree may resprout below the girdled or cut area or along root line.

Fire control: Stump burning has been shown to be successful, but it is time-consuming compared to other control techniques (Knopf and Olson 1984).

Biological Control

Research on biocontrol agents has not been undertaken for this species.

Chemical Control

Most translocating herbicides (e.g., glyphosate) are effective at label strength when applied during the growing season. Some dormant-season herbicides (e.g., imazipyr as Chopper RTU®) are labeled for Russian olive control. Foliar spraying has been successful, as has injecting herbicide capsules around base of trunk. When injecting herbicides into the cambium of a standing tree, monitoring should occur the same year to ensure that the entire tree is affected. Cut-stump treatments can be effective when combined with burying the stump or painting the cut surface with chemicals. Cuts should be made as close to the ground as possible and immediately be followed by 5-10 cc of glyphosate (as Roundup®) applied at full strength to the cambium. For trees that do not have to be removed or immediately taken down, exposing more than 50 percent of the cambium by cutting into the bark with a saw or ax close to ground level and introducing herbicides into the exposed areas is effective. A syringe (size 14 needle) works well for both of these combinations. Brushing also works, but requires a larger amount of herbicide. Burying a stump after cutting can also prevent regrowth from the stump, but exposed roots should be monitored for resprouting. Monitoring regrowth of cut stumps or roots should be done one year after treatment.

Erechtites glomerata   (Poiret) DC.

Common names: cutleaf fireweed, cutleaf burnweed, New Zealand fireweed, Australian burnweed

Synonymous scientific names: *Erechtites arguta, Senecio glomeratus, Senecio arguta*

Listed: CalEPPC B; CDFA nl

Erechtites minima   (Poiret) DC.

Common names: Australian fireweed, little fireweed, coastal burnweed, Australian burnweed

Synonymous scientific names: *Erechtites prenanthoides, Senecio minimus, Senecio prenanthoides*

Listed: CalEPPC B; CDFA nl

Closely related California natives: 0

Closely related California non-natives: 2

by Gavin Hoban and Marc C. Hoshovsky

Erechtites glomerata (left); *Erechtites minima* (right)

HOW DO I RECOGNIZE IT?

Distinctive Features

Both cutleaf fireweed (*Erechtites glomerata*) and Australian fireweed (*E. minima*) are annual or short-lived perennials, four to eight feet in height. Flowers are dull yellow, arranged in cylindrical or oval-like groups to one foot across. Cutleaf fireweed has lance-shaped leaves that are sharply and unevenly toothed compared to the deeply pinnately lobed leaves of Australian fireweed (Hickman 1993, Robbins *et al.* 1941).

Description

Erechtites glomerata

Asteraceae. Annual herb or weak perennial. Stems: 0.3-6.6 ft (10-200 cm) tall, erect from a deep, often branching taproot. Branches ascending, thinly villous to tomentulose. Leaves: alternate, oblong-ovate to lanceolate in outline, light gray-hairy, becoming somewhat smooth; 2.8-5.9 in (7-15 cm); lower leaves petioled, deeply pinnately lobed or pinnatifid, villous to tomentulose, becoming somewhat glabrous; upper leaves sessile, reduced, and pinnately to irregularly toothed. Inflorescence: heads in terminal, somewhat corymbose clusters or panicles; radial (discoid and salverform), involucre 0.2-0.3 in (5-8 mm) long, cylindrical, stalked. Phyllaries in 2 unequal series, oblong, glabrous to thinly tomentulose, outer much shorter than inner, apices acute. Flowers: two forms, outer pistillate, inner bisexual, corollas tubular, pale or dull yellow. Fruit: achenes 0.04-0.08 in (1-2 mm) long, cylindrical, ribbed; pappus 0.2

Erechtites minima

in (5 mm) long, white, composed of fine capillary bristles. Flowers in California from April to October (Robbins *et al.* 1941).

Erechtites minima

Asteraceae. Annual herb or weak perennial. Habit and size similar to *Erechtites glomerata*, but subglabrous to obscurely puberulent. Leaves: arachnoid (cobwebby) beneath, linear-lanceolate, evenly and finely dentate, 2.8-7.9 in (7-20 cm), not lobed or pinnately cleft. Inflorescence: large, to 30 cm broad. Flowers: pistillate, pappus 6-7 mm long (Robbins *et al.* 1941).

E. minima

Erechtites sp.

WHERE WOULD I FIND IT?

Both species occur occasionally along the coast at low elevations (<500 m) from central Oregon to Santa Barbara County and the Channel Islands. They are found primarily in disturbed areas, especially roadsides, stream banks, pastures, and as post-burn opportunists. They tend to prefer grasslands, woodlands, and coastal scrub habitats (Hickman 1993). Australian fireweed has been reported from redwood forests on the North Coast (Robbins *et al.* 1941). Although ubiquitous, *Erechtites* species do not now represent a major ecological threat to native plant communities on California's north coast (Popenoe 1999). They invade areas following clearcut logging operations, sometimes dominating overall plant cover, but other shrubs and trees typically replace them in about five to ten years (Muldavin *et al.* 1981). Martin and Popenoe (1984) found buried seed of *Erechtites* to be more abundant than seed of any other species in old-growth redwood forests of Redwood National Park. Combined average seed density for the two species was 522 per square meter in old-growth blocks in the vicinity of recently harvested clearcut second growth.

Erechtites species can grow in a variety of climates. Muldavin *et al.* (1981) found *E. minima* in clearcuts with compaction and topsoil removal. They apparently are well suited to exploit fertile, freshly disturbed ground, but are demanding of nutrients and are weak competitors. *Erechtites* plants are uncommon in north coast vegetation types until competing vegetation is removed, either mechanically or by fire. The plants are also uncommon along roads where soil is compacted or topsoil has been removed.

WHERE DID IT COME FROM AND HOW DOES IT SPREAD?

Both species are native to Australia and New Zealand. Australian fireweed was naturalized in Humboldt County by 1918 (Robbins 1940). It has become naturalized on the West Coast only in southern Oregon and California. Cutleaf fireweed was reported before 1941 in redwood forests from Mendocino to Del Norte County (Robbins *et al.* 1941). Plants spread by wind dispersal of seeds.

Erechtites species quickly dominate grasslands and fields, and they are among the most serious plant pests in Channel Islands National Park (Halvorson 1992).

HOW DOES IT GROW AND REPRODUCE?

Little information is available on growth and reproduction of either species. A related species, *Erechtites hieracifolia*, is better known and may have similar characteristics.

In germination studies of *Erecthtites hieracifolia*, a close relative of Australian fireweed, Baskin and Baskin (1996) found that if conditions were not favorable for germination during the late summer and early fall, about half of the seeds at maturity in September became dormant. These seeds would not germinate under any test conditions, whereas the other half of the seeds would germinate under only a narrow range of test conditions. Australian fireweed flowers in July-September; cutleaf fireweed flowers in June-August (Allen 1997).

In germination studies of *Erechtites hieracifolia*, Baskin and Baskin (1996) found that half the seeds at maturity in September would not germinate under any test condition, whereas others would germinate under only a narrow range of conditions. Seeds appear to need minimal temperatures to germinate, such as conditions in spring. Since 89 percent of seeds were viable after eight years of burial, it appears that, although seeds of this species are wind-dispersed, they also have the potential to form a long-lived seedbank. Thus, soil disturbance at any time from May to September could result in establishment of plants from seeds. Australian fireweed has been reported to be facultatively mycorrhizal, which may explain its rapid invasion capabilities in relatively arid grasslands on San Miguel and other coastal California islands (Allen 1999).

On the Channel Islands cutleaf fireweed apparently does not need disturbance to become established (Halvorson 1992). Elliott *et al.* (1997) describe *Erechtites* species as early successional and shade-intolerant. Optimum conditions for *E. hieracifolia* growth occurred at soil pH 5.3 to 5.5 (Stephenson and Recheigl 1991). At Point Reyes National Seashore, Australian fireweed commonly sends roots into the soil or into rotting wood from any stem that has been laid on the ground either by heavy rain or the weight of its long (to six feet) stems. Rooting stems continue to grow either vertically or horizontally. Many branch further to produce what looks like an entire flowering plant connected to its neighbor by one stem (Allen 1999).

Erechtites species tend to exhibit moderate to rapid rates of infestation, particularly after fire. In less than one year of initial establishment on San Miguel Island, Santa Barbara County, in 1984, cutleaf fireweed spread to cover 173 acres (70 ha) with a maximum density of 3,237 plants per acre (8,800/ha). Within a year of the Vision fire in October 1995 at Point Reyes National Seashore, thick stands of Australian fireweed dominated the burned areas where native bishop pine seedlings were reestablishing. Over 1.2 million plants of Australian fireweed were removed in 1996 (Allen 1997).

HOW CAN I GET RID OF IT?

Monitoring three times a year is suggested because of the large seedbank. Disturbance caused by removal efforts could exacerbate the infestation.

Physical Control

Manual methods: Channel Islands National Park uses volunteers to manually remove cutleaf fireweed (Halvorson 1992).

Biological Control

Insects and fungi: No biological control agents have been approved by the USDA for use on *Erechtites* species.

Plant competition: Popenoe (1999) suggests that control experiments might try to examine nutrient requirements and ability to compete. Success might be possible by decreasing nitrogen availability by mulching with sawdust, and sowing or planting natives to increase competition. Natives might have a better chance of successful competition if they are innoculated with mycorrhizae to increase their hold on available nutrient pools.

Chemical Control

No information on chemical control of these species is available. Herbicide effects may be similar to those for *Erechtites hieracifolia*. Olney (1971) reported that atrazine gave excellent control of *E. hieracifolia*, whereas diuron was ineffective. Herbicides were applied as directed sprays to *E. hieracifolia* in Hawaii. Diuron at 4 lb/acre gave excellent control for twenty weeks in both trials. Linuron at 4 lb/acre gave comparable results (Higaki 1973). In Indonesian tea plantations, three liters of glyphosate in 700 liters of water/ha completely controlled *Erechtites* species (Sukasman 1979). Check with a certified herbicide applicator to assess current chemical methods registered for the habitat type at any site at which removal of *E. minima* or *E. glomerata* is desired.

Eucalyptus globulus Labill.

Common names: blue gum, Tasmanian blue gum, common eucalyptus

Synonymous scientific names: none known

Closely related California natives: none

Closely related California non-natives: 8

Listed: CalEPPC A-1, CDFA nl

by David Boyd

HOW DO I RECOGNIZE IT?

Distinctive Features

Blue gum (*Eucalyptus globulus*) is a tall (150-180 foot), aromatic, straight-growing tree, with bark that sheds in long strips, leaving contrasting smooth surface areas. Adult leaves are waxy blue, sickle-shaped, and hang vertically. Juvenile leaves are oval, bluish green, and have square stems. Fruits are blue-gray, woody, and ribbed. Trees produce abundant fruit drop and leaf and bark litter. Blue gum is distinguished by tall growth habit, smooth bark, long leaves, and large, solitary, waxy buds and fruits (Chippendale 1988).

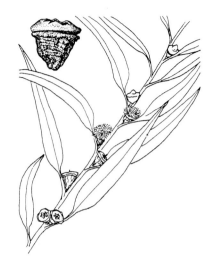

Description

Myrtaceae. Tall, long-lived tree. Bark: usually rough, grayish or brownish at tree base, peeling off above in long strips, leaving a smooth yellowish or grayish surface. Leaves: juvenile: opposite, sessile, elliptic-ovate, 4-6 x 2-4 in (11-15 x 6-10 cm) firm, uniform green color. Flowers: simple, axillary, usually 1-flowered, occasionally 3-flowered, peduncle sometimes absent or very short and stout, pedicels usually absent, buds with 4 (occasionally more) distinct ribs, extremely glaucous. Fruit: sessile, sub-spherical to more or less hemispherical, 0.4-0.8 x 0.5-0.9 in (1-2.1 x 1.4-2.4 cm), with 4 (occasionally more) distinct ribs, glaucous on hypanthia, disk broad, more or less level or ascending with slight lobes, valves 4 or 5. Rim of fruits has a distinct, concave calycine ring (Boland 1984).

WHERE WOULD I FIND IT?

Blue gum has been planted extensively worldwide because of its rapid growth and adaptability to a wide variety of site conditions. It does especially well in Mediterranean climate regions, characterized by cool, wet winters and dry, warm summers, such as portions of California, Chile, Portugal, Spain, and South Africa (Skolmen 1983). In California it is most widely planted in central coast locations, but found below 1,000 feet elevation of the north, central, and south coasts, as well as inland throughout the Central Valley. It is most frequently found growing in small groves or windbreaks within grassland habitats where initial plantings took place. Large specimen trees are found in urban and rural settings.

Blue gum grows well on a wide range of soils, but requires good drainage, low salinity, and a soil depth of two feet (0.6 m) or more. In California it grows best on deep alluvial soils because of the greater moisture supply (Skolmen and Ledig 1990). Hawaiian soils supporting blue gum eucalyptus are about three feet (0.9 m) deep. They are usually acidic, moderately well drained, silty clay loams (Skolmen 1983). Blue gum does well with only twenty-one inches (530 mm) of annual rainfall accompanied by a pronounced dry season, primarily because frequent fogs compensate for lack of rain (Skolmen and Ledig 1990).

WHERE DOES IT COME FROM AND HOW DOES IT SPREAD?

Native to Australia, where it occurs mainly along the east coast of Tasmania, blue gum was first cultivated in California in 1853 as an ornamental. Widespread commercial planting occurred after 1870, primarily for timber and fuel. A second planting boom took place in the early 1900s (Groenendaal 1983). By the 1930s planting in California had lost popularity because of unsuitable characteristics of the wood for lumber production and a decrease in demand for fuel wood. Blue gum aggressively invades neighboring plant communities from original plantings if adequate moisture is available for propagation by seed. Invasive in coastal locations, blue gum is rarely invasive in the Central Valley or in dry southern California locations. It is most invasive on sites subject to summer fog drip.

WHAT PROBLEMS DOES IT CAUSE?

Within groves, biological diversity is lost due to displacement of native plant communities and corresponding wildlife habitat. Abundance and diversity of understory vegetation is dependent on stand density. Under-

story establishment is inhibited by the production of allelopathic chemicals and by the physical barrier formed by high volumes of forest debris consisting of bark strips, limbs, and branches. The fuel complex formed by this debris is extremely flammable, and under severe weather conditions could produce drifting burning material with the potential to ignite numerous spot fires. Because stringy bark is carried away while burning, eucalyptus forests are considered the worst in the world for spreading spot fires. The Oakland hills firestorm was both intense and difficult to control because of the many stands of eucalyptus. Individual trees growing near structures or in public use areas are hazardous because of the potential for branch failure. Stature and growth form are distinctive and unlike native tree species, which compromises the visual quality of natural landscapes.

HOW DOES IT GROW AND REPRODUCE?

Blue gum reproduces by seed and by resprouting. In California flowering occurs from November to April. Flowers are pollinated by insects and hummingbirds. Fruit ripens from October to March, about eleven months after flowering. Seed set begins at approximately four to five years of age. Good seed crops are produced in most locations at three- to five-year intervals. Seeds are small and abundant. Capsules open immediately on ripening, and seed is dispersed by wind within one to two months. Dispersal distance from one 131-foot (40 m) tall tree, with winds of six miles per hour (10 km/h), was sixty-six feet (20 m). Newly released seeds germinate within a few weeks under suitable conditions. California eucalypts have highly variable germination rates, ranging from 2 to 80 percent within a thirty-day germination period. Seedlings often survive in sufficient abundance to significantly invade neighboring plant communities. Establishment of eucalyptus saplings within groves is inhibited by forest litter and duff, but can be significant following disturbances such as fire or harvesting operations (Skolmen and Ledig 1990, Krugman 1974).

Blue gum sprouts readily from the main trunk, from stumps of all sizes and ages, from the lignotuber, and from roots. Large masses of foliage are produced by sprouting stumps after tree felling. Numerous clusters of shoots later thin to one stem per cluster. A number of small-diameter stems can continue to thrive on each stump, resulting in bush-like growth. Production of lignotubers, which may live for many years in soil, may account for sprouting that sometimes occurs away from the main stump of cut trees (Skolmen 1983).

Blue gum typically grows in dense monospecific stands. Rapid growth is characteristic of this species. Most height growth occurs within the first five to ten years, and 60 to 70 percent of total height growth is achieved in about ten years. Growth is dependent on site quality, but trees can reach heights of eighty feet (24 meters) in ten years (Skolmen 1983).

Blue gum generally does not form a taproot. It produces roots throughout the soil profile, rooting several feet deep in some soils. Blue gum is shade intolerant, and failure to regenerate within forests in the absence of fire is related to low light intensities. It is drought tolerant and somewhat frost hardy. Frost resistance increases with maturity (Skolmen and Ledig 1990).

HOW CAN I GET RID OF IT?

Physical Control

Manual/mechanical methods: Removing trees is a difficult task and can be expensive if individual trees are felled. It is also unlikely that this cost can be offset because of the low value of the wood as fuel. An effective method to control stump resprouting is absolutely necessary. Stump grinding can eliminate sprouting, as well as remove all evidence of trees. Where there are few stumps and the terrain is gentle, this may be a preferred method. It is expensive to treat many stumps this way, even if a powerful and efficient self-propelled grinder is used. Care must be taken to grind all underground portions of stumps to a depth of approximately two feet. Provision must be made to fill resulting craters with soil.

Manual removal of eucalyptus sprouts from stumps results in eventual control as food resources are exhausted. This method is expensive and impractical if a large number of stumps are to be treated. Manual removal should be limited to situations where close attention can be given to a few stumps.

Prescribed burning: This method can reduce fuels in blue gum stands, but the species is fire tolerant. Only seedlings can be killed by fire. Fuel replenishment is rapid.

Biological Control

Because blue gum is valued as an ornamental tree in many settings, biological control cannot be considered as a control option. However, pests have inadvertently been introduced, including the long-horned borer, which may increasingly affect the health of these trees.

Chemical Control

The most effective control of sprouting is achieved through application of triclopyr or glyphosate directly to the outer portion of the stump's cut surface at the time of tree felling. Triclopyr (as Garlon 4® and Garlon 3A®) should be applied at the rate of 80 percent in an oil carrier. Imazapyr (as Arsenal or Stalker) can be used as an alternate to Garlon. Glyphosate (as Roundup® or Rodeo®) should be applied at 100 percent. Stumps should be cut as low to the ground as practical and brushed clean of sawdust to maximize absorption of the herbicide. For best results, herbicides should be applied to the freshly cut surface as soon after cutting as possible. Maximum success is achieved if cutting occurs in fall (Carrithers, pers. comm.). Complete control of sprouting on every stump will not always be achieved. Any resprouts, when three to five feet tall, should be treated with a foliar application of 2 percent of triclopyr or glyphosate.

Triclopyr (as Garlon 4®) offered the best results of the herbicides currently available in California for a 1996 eucalyptus removal project at Angel Island State Park in Marin County. A high concentration was used (80 percent Garlon 4®, 20 percent oil carrier; an alternative is 100 percent Garlon 3A). Glyphosate (as Roundup®) was used in 1990 on a similar eucalyptus removal project on Angel Island, but with less consistent results. When sprouting occurred following the 1990 eucalyptus removal project on Angel Island, excellent follow-up control was achieved by applying triclopyr as Garlon 4® (80 percent Garlon 4®, 20 percent oil carrier) to overlapping frill cuts. These cuts were made on portions of the vertical surfaces of stumps with live cambium.

Application of these herbicides to the foliage or stems of sprouts is less effective. Several years of foliar applications of triclopyr at Annadel State Park, Sonoma County, following a major eucalyptus removal project produced only incremental results. The visual impact of tall, herbicide-killed sprouts must also be considered.

At The Nature Conservancy's Jepson Prairie preserve near Rio Vista, Solano County, over 1,200 eucalyptus stumps were killed by repeated foliar herbicide treatments over one summer period (Serpa, pers. comm.). The stumps were not treated at the time of felling, and sprouts were allowed to grow to a height of about ten feet. The sprouts were then cut and the resulting resprouts were treated with glyphosate as Roundup® (5 percent solution). By the third herbicide application all of the stumps were dead. It is possible that the dry climate of this site contributed to the success of this method.

Euphorbia esula L.

Common names: leafy spurge, wolf's milk

Synonymous scientific names: *Tithymalus esula, Galarrhoeus esula*

Closely related California natives: 25

Closely related California non-natives: 10

Listed: CalEPPC A-2; CDFA A

by L. Butch Kreps

HOW DO I RECOGNIZE IT?

Distinctive Features

   Leafy spurge (*Euphorbia esula*) is a perennial, rhizomatous, erect herb to three feet tall with distinctive bluish green leaves. It reproduces vegetatively from vigorous rootstocks and by seed. The entire plant contains a milky sap and has forked branching. Leaves are alternate, narrow, and long. Many stems may arise from a single rootstock and often appear clustered. The flowers are yellow-green, inconspicuous, and arranged in numerous small clusters, each cluster enclosed by paired, heart-shaped, yellow-green bracts. The massed flower clusters and accompanying bracts of dense infestations are conspicuous from a distance. The roots are brown and bear numerous pink buds, which may produce new shoots or roots. Seeds are oblong, grayish to yellow-brown, contained in a three-celled capsule, each cell containing a single seed (Whitson *et al.* 1991).

Description

   Euphorbiaceae. Perennial herb. Plant has milky sap throughout and extensive rootstocks. Roots may extend to depths of 15 ft (4.6 m) or more, spreading and most abundant in the upper foot of soil, woody, brown with many pink buds that can produce new shoots if upper portion destroyed. Stems erect, glabrous, branched at the top, to 3 ft (0.9 m) tall. Leaves sessile (no petiole), alternate, broadly linear to narrowly oblong-lanceolate or inverted lanceolate, 1-3 in (2-8 cm) long, bluish green. Flowers borne in distinctive clusters (cyathia), generally with 1 female flower at the center surrounded by 11-21 male flowers, clusters surrounded by 5 bracts fused into a bell-shaped involucre 1.5-2.5 cm, topped by crescent-shaped, 2-horned, bright yellow-green glands, 1.5-2 mm. Flowers small, male flowers stalked, with 5 small sepals, no petals; female flowers stalked, style divided half the length, ovary exserted from involucre bearing large 3-lobed

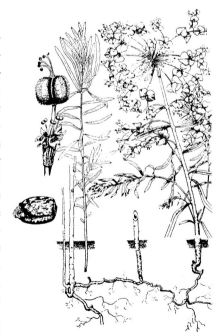

pistil that matures into a capsule containing up to 3 seeds if pollinated. Seeds elliptic-oval, about 0.08 in (2 mm), oblong, smooth, light gray to yellow-brown, with yellow (or white) emarginate caruncle (description after Whitson *et al.*1991, Hickman 1993).

WHERE WOULD I FIND IT?

Leafy spurge may be found at scattered locations in northern California, particularly in the far northern portion of the state. Over 500 acres were infested in Siskiyou County in the late 1960s, but, after twenty years of eradication efforts, only ten to twenty acres remain infested. Small infestations were found and eradicated in Lassen and Modoc counties. Populations have been found in Sonoma County and as far south as Los Angeles County. Leafy spurge is widespread in the western United States and extremely troublesome in prairies, pastures, rangelands, and other grasslands from western Minnesota and the Dakotas to northern Idaho and northeastern Oregon and south to Colorado. It also invades pine savannahs, riparian areas, cultivated fields (grains, alfalfa), and roadsides throughout the northern plains and northern Rockies. It is usually found in patches or large infestations rather than as single plants. Patches and infestations are easiest to spot when spurge is in full bloom, generally for several weeks between late May and late July.

Leafy spurge grows in a wide range of habitats. It is most aggressive in semi-arid areas, but it can be found in xeric to subhumid and subtropic to subarctic habitats. It will tolerate flooding for more than four months. It is only slightly limited by shade. Leafy spurge occurs most commonly on untilled, non-crop areas such as pastureland, rangeland, woodland, prairies, roadsides, stream and ditch banks, and waste sites. It grows on all kinds of soils, but it is most abundant on coarse-textured soils and least abundant on clay soils. Root growth and vegetative reproduction are highest on coarse-textured soils. Sexual reproduction, germination, and seedling establishment are highest on clay soils (Butterfield and Stubbendieck 1999).

WHERE DID IT COME FROM AND HOW DOES IT SPREAD?

Leafy spurge is native to Eurasia, where it is known from Spain, Italy, and Germany to central Russia. It was first recorded in North America in Massachusetts in 1827 and was probably introduced to the United States repeatedly in contaminated grain, particularly in the northern plains. It was established in California after 1900, being found in Modoc and Siskiyou counties by 1917 (Robbins 1940). The invasive leafy spurge found in North America may be a hybrid or series of hybrids of two or even three *Euphorbia* species (including *E. virgata*) that were native to Eurasia and had the chance to interbreed following introduction here.

Leafy spurge spreads by seed, by vegetative growth, and by root fragments, which may be cut up by plowing and carried on road maintenance or farm machinery. Pieces of root as small as 0.5 inches (1.3 cm) long and 0.125 inches (0.3 cm) in diameter can produce shoots that grow rapidly. Animals, birds, insects, equipment, seed, hay, grain, and the natural dehiscence of the capsule all assist in the dispersal of leafy spurge (Butterfield and Stubbendieck 1999).

WHAT PROBLEMS DOES IT CAUSE?

Leafy spurge can invade and dominate a variety of vegetation types, including prairies, grasslands, and pine savannahs, crowding out native plant species. At present it infests nearly 2.5 million acres in North America. Stem densities of 1,000 plants per square yard are not uncom-

mon in infested areas. This results in almost complete exclusion of native forbs and grasses and other desirable vegetation. Exclusion of other plants may result in part from the allelopathic chemicals that have been found in leafy spurge (Butterfield and Stubbendieck 1999).

Leafy spurge is unpalatable and often toxic to most native ungulates, including deer, elk, and antelope, as well as to cattle and horses. It has been reported to cause severe irritation to the mouth and digestive tract in cattle and can result in death (Whitson *et al.* 1991). Sheep and goats can be induced to feed on spurge, and in some cases will acquire a taste for it and help to reduce its cover. Leafy spurge has an extensive root system with nutrient reserves that, once plants are established, make it extremely difficult to remove and control.

HOW DOES IT GROW AND REPRODUCE?

Although a successful seed producer, leafy spurge spreads primarily through its extensive lateral root system. Vegetative reproduction occurs from both crown and root buds. Most plants in the center of a patch are the result of crown buds, while plants growing on the edge of a patch are primarily from lateral root buds. Crown buds are the first to form, developing seven to ten days after seedling emergence. Lateral roots and buds begin to develop as plants mature. Roots may be either long or short. Long roots can produce shoots and may reach nearly seventeen feet (5 m) laterally and about 15 feet (4.6 meters) in depth. Up to 300 buds have been counted on a single long root. Because of the large numbers of buds, any tillage technique may quickly spread the plant. An experiment showed that rototilling increased the density of leafy spurge to 316 shoots/m^2 compared with 134 shoots/m^2 in an untilled area. Root fragments only 0.66 inches (1.5 cm) in length produced new shoots (Butterfield and Stubbendieck 1999).

Leafy spurge is one of the earliest plants to emerge in spring, usually in mid-March to late April. Once the stem emerges, elongation occurs rapidly. Initiation of the inflorescence occurs within one to two weeks of stem emergence. Yellowish bracts form in May, making leafy spurge conspicuous from late May through June. Flowering in the terminal inflorescence generally ends in late June to mid-July. If conditions are favorable, leafy spurge may continue flowering throughout summer and into fall. Plants may produce seed until frost. Pollination of leafy spurge is entirely by insects. Over sixty species of insects have been found on leafy spurge flowers (Butterfield and Stubbendieck 1999).

Seeds mature about thirty days after pollination. Each plant produces from ten to fifty capsules, with a seed yield range of 200 to 250 seeds per plant. Seeds can be propelled up to fifteen feet (4.5 m) from parent plants. Sixty to 80 percent of fresh seeds are viable. Seeds can remain viable in the soil for five to eight years. However, annual viability in the soil decreases by about 13 percent each year. Ninety-nine percent of viable seeds will germinate in the first two years. Temperature is the most important requirement for germination. Temperatures between 68 and 85 degrees F (20-30 degrees C) are optimal. Alternating freezing and thawing, wet and dry periods, and shortened photoperiod promote germination. Peak germination is from late May to early June. If adequate moisture is present, germination can occur throughout the growing season (Butterfield and Stubbendieck 1999).

Much early seedling growth is devoted to establishing a root system rather than shoot elongation (Robbins 1970). Leafy spurge emerges from the soil from root buds or from germinating seeds after soil begins to warm in early spring. Shoots from root buds grow rapidly, reaching maximum height between late May and mid-July when they flower. Shoots set seed three to six weeks after flowering. Leafy spurge shoots remain green until the first heavy frost in fall when foliage dies.

HOW CAN I GET RID OF IT?

Leafy spurge is extremely difficult to control, and the best approach is to detect and eliminate or contain new infestations as quickly as possible. An integrated management approach using chemical, biological, cultural, and grazing control methods will usually yield the best results with established populations. Regardless of the method of control used, all formerly infested areas should be monitored for new spurge plants for at least ten years. Leafy spurge seeds may remain dormant in the soil for several years before germinating and reestablishing an infestation. Spurge roots may also lie dormant for more than five years before producing new shoots, particularly after they have been treated with herbicides. Mapping infested areas will make reinfestations easier to detect, and this will make eradication or containment more efficient.

Physical Control

Manual/mechanical methods: Opinions differ on the effectiveness of mechanical control methods. Butterfield and Stubbendieck (1999) do not generally recommend cultivation for leafy spurge control, because of the plant's ability to sprout from buds. Other researchers report that properly timed cultivation and/or planting of competitive species can be effective. Cropped areas infested with leafy spurge must be cultivated twice in fall or every two weeks during the entire growing season to ensure reasonable control. Cultivation should be continued for three growing seasons to prevent reinfestation. Mowing can reduce above-ground stands, but it stimulates underground shoot development (Butterfield and Stubbendieck 1999).

Prescribed burning: Burning is not effective in controlling leafy spurge (Butterfield and Stubbendieck 1999).

Biological Control

Insects: At least thirteen USDA approved biological agents have been released to control leafy spurge in North America. Insects that have shown the most promise to date include several species of flea beetles: *Aphthona nigriscutis*, *A. czwalinae*, *A. lacertosa*, *A. flava*, and *A. cyparissiae*. The larvae of these beetles feed on leafy spurge roots, which may explain why they have been successful in controlling some infestations that herbicides failed to kill. Different species of insects are known to be more effective in various climates or soil types or against different populations of spurge, but it is still difficult to explain why this is so. Thus far few, if any, of these beetles have been released in California, since leafy spurge infestations here are currently small. They have been released in southern Oregon, however, where there are also scattered leafy spurge infestations, and they may help prevent these populations from spreading southward (Lym and Zollinger1995).

When a new release is made, leafy spurge outside the release site is sometimes treated with herbicides to prevent the infestation from increasing during the time it will take for the insect population to grow. As the insects become established, herbicide applications are reduced. A gall midge, *Spurgea esulae*, and a stem-boring beetle, *Oberea erythocephala*, were also released for leafy spurge control, but they have had little or no success in controlling spurge populations (Lym and Zollinger1995)

Grazing: Sheep and goats have been useful in reducing stands of leafy spurge. Goats have been most effective because they tend to graze spurge regardless of plant density, while sheep consume less spurge as its density in the vegetation declines. Sheep also may take two to three weeks before beginning to feed on spurge, while goats begin immediately. Goats, however, are harder to manage and less profitable (Lym and Zollinger 1995). Unfortunately, both sheep and goats can pass viable leafy spurge seeds through their digestive systems, so animals that have been feeding in areas with flowering or fruiting leafy spurge should be confined for four to five days before being allowed into spurge-free areas.

Plant competition: Few, if any, smother crops can eliminate this plant. However, sowing perennial grasses after tilling can reduce leafy spurge populations as much as 80 percent. Local university or government sources can advise on the best choice of competitive grasses in the area. Good management of existing vegetative cover can often reduce the likelihood of a new invasion.

Chemical Control

Picloram or picloram + 2,4-D are the most widely used herbicides for leafy spurge control, but picloram is not currently registered for use in California. Compounds registered for use in California, including 2,4-D alone, glyphosate, and triclopyr, have not been as effective for leafy spurge control in experiments and field trials.

Applying 2,4-D in spring and summer and glyphosate in fall have been the most commonly used and effective methods for eliminating leafy spurge in California. Costs can be minimized by ensuring that herbicide rates and times of application control against leafy spurge while leaving surrounding vegetation to compete with spurge regrowth.

Sometimes chemical treatments should not be used because of local environmental factors. A well timed combination of cultivation, planting competitive grasses, and the use of herbicides is probably the most effective method for controlling leafy spurge.

Ficus carica L.

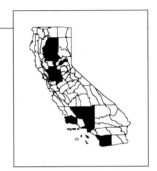

Common names: edible fig, common fig

Synonymous scientific names: none known

Closely related California natives: 0

Closely related California non-natives: 0

Listed: CalEPPC A-2; CDFA nl

by John M. Randall

HOW DO I RECOGNIZE IT?

Distinctive Features

Edible fig (*Ficus carica*) is the familiar fig tree that produces edible fruits sold fresh, dried, and as paste used as filling for cookies (fig newtons) and other sweets. Mature trees often have multiple trunks and may grow to thirty feet tall. The heavy trunk and branches are covered with a smooth, light gray, flaky bark. The sap is thick, sticky, and slightly milky. The leaves are rough to the touch, bright green, with three to five lobes, the classic fig-leaf shape. They are two to eight inches wide and two to ten inches long with a two- to four-inch-long petiole (stem). Edible figs are deciduous, dropping their leaves in mid- to late autumn and leafing out again in early spring (March-April). The fruits are shaped like small sacks full of sand, two to four inches long and nearly as wide. Edible fig has been widely cultivated, and the color of ripe fruits of different cultivars ranges from dark purple-black in the 'Mission' fig to pale greenish yellow in the 'Kadota' and 'Genoa' figs to white with a purple blush in 'Conadria'

figs. Cultivated trees produce large numbers of fruits, which can become a nuisance when they ripen and drop, but those growing wild in canyons and riparian (riverside) forests around California's Central Valley usually produce only a few greenish yellow fruits that ripen in late summer or fall.

Description

Moraceae. Tree up to 30 ft (10 m) tall, winter-deciduous, leaves dropping in mid- to late autumn and emerging again in early spring (March-April) in California. Leaves with petioles <4 in (<10 cm) long, blades with 3-5 palmate lobes usually halfway to the midrib, blades broadly ovate to round in overall outline, 2-8 in (5-20 cm)

wide and 2-10 in (5-25 cm) long, bright green to dark green, scabrous on upper surface, hairy on lower surface. Plant is gynodioecious (two sexual forms), one, the caprifig, with staminate (male) flowers and short styled pistillate (female) flowers; the other, the fig, with only long-styled pistillate flowers. Both forms have a complex inflorescence called a synconium. Clusters of flowers are enclosed inside the pear-shaped synconium, which has a small, scale-

covered opening (ostiole) at its distal end. Flowers visible only if the synconium is cut open, and are unisexual, small, and more or less radial with 4 sepals and no petals, male flowers with 4 stamens, female flowers with superior ovary and simple style. Synconium develops into a multiple fruit, consisting of fleshy receptacle surrounding the drupelets that form from each pistillate flower. Synconium 1.5-3 in (4-8 cm) long, color dependent on variety (on plants outside cultivation generally bright green, turning pale green to pale yellow as they ripen).

The multiple fruits of caprifigs are dry and chaffy and regarded as unpalatable, while those of figs are fleshy and savored by many birds and mammals, including humans. Botanically speaking, the fleshy outer tissue of the fig is actually the receptacle, while the true fruits, enclosed within the receptacle, are single-seeded drupelets with hard, thin outer walls that give figs their characteristic crunchy texture (description from Hickman 1993, Lisci and Pacini 1994).

WHERE WOULD I FIND IT?

Edible figs invade and dominate riparian forests, streamside habitats, levees, and canal banks in and around California's Central Valley, surrounding foothills, the south coast, and the Channel Islands (Hickman 1993). They are also widely cultivated for fruit and ornament in areas below 2,500 feet (800 m) elevation. California is one of the world's largest producers of fig fruits (Ferguson *et al.* 1990).

Edible fig is most likely to escape where soils stay moist throughout the summer. It has invaded many nature preserves and parks in California. Plants form dense thickets covering roughly twenty-five acres along a seven-mile-long section of Dye Creek at the Dye Creek Preserve northeast of Chico and have begun to invade the riparian forest at Woodson Bridge State Park along the Sacramento River to the west. Several rapidly expanding fig thickets were found in the most pristine valley oak riparian forest on the Cosumnes River Preserve south of Sacramento. These thickets were repeatedly cut and the stumps treated with herbicide, but they were difficult to eliminate. Edible figs have also invaded parts of the riparian forest in Caswell State Park near Stockton. They are found on the Santa Cruz Island Preserve in disturbed sites, and scattered along coastal flats and in coastal scrub (Junak *et al.* 1995).

WHERE DID IT COME FROM AND HOW DOES IT SPREAD?

Edible fig is probably native to the fertile region of southern Arabia (Ferguson *et al.* 1990). It was probably first domesticated outside its native region, in Mesopotamia, in the valley of the Tigris and Euphrates rivers, in what is today Iraq. It was among the earliest known fruits to be cultivated, with records dating back to the Sumerian era, roughly 4,900 years ago (Ferguson *et al.* 1990). The area where figs were cultivated gradually grew to encompass all nations of southwestern Asia and the Mediterranean Basin. Edible fig was first introduced to the New World in the West Indies by Spanish and Portuguese missionaries in the 1520s and then to what is now the East Coast of the United States in 1575 (Ferguson *et al.* 1990). Edible fig trees were introduced to California by Spanish missionaries beginning in 1769; hence this variety is known as 'Mission' or 'Franciscan' fig. Many other varieties were introduced to California for food and ornamental plantings after 1850. Commercial production of figs was hampered until the fig wasp, *Blastophaga psenes*, was successfully introduced to the state in 1899 (Ferguson *et al.* 1990). Currently, California ranks third in the world in fig production, behind Turkey and Greece, although figs rank only twenty-second in value and eighteenth in acreage among crops produced in California (Michailides *et al.* 1996).

It is not clear how edible fig spreads into preserves and other wild areas. It grows quickly and can spread vegetatively by root sprouts, soon forming dense thickets that exclude most other plants. Limbs that have been cut or broken and fallen to the ground can take root, and it is thought that branches broken off during storms or floods may wash up and root at downstream sites. Many birds eat the fruits and may spread the seeds. Hujik (pers. comm.) reports that deer also feed on the fruits. Seeds germinate only if they are removed from the fleshy synconium during passage through an animal's gut or by mechanical means such as heavy rainfall (Lisci and Pacini 1994).

WHAT PROBLEMS DOES IT CAUSE?

If not controlled, edible fig trees could crowd out native trees and understory shrubs characteristic of California's riparian forests. Riparian forests are already rare in California, especially in the Central Valley, where over 95 percent have been converted to cropland, pasture, or developed areas in the past 150 years. No published or unpublished reports are available with quantitative information on the impacts of edible figs invading natural vegetation in California or elsewhere.

The leaves of edible fig contain at least two furocoumarin compounds that are activated on exposure to a certain waveband of light and can then cause a skin rash in humans (Damjanic and Akacic 1974, Evans and Schmidt 1980). The activated furocoumarins are primary irritants, meaning they chemically or mechanically irritate the skin rather than causing an allergic response. The mode of action of these compounds is not known, but they may photobind to DNA and/or ribosomal RNA in epidermal cells following exposure to ultraviolet light in the 320-370 nm waveband (Evans and Schmidt 1980).

HOW DOES IT GROW AND REPRODUCE?

Most edible fig fruits with viable seeds are produced in late summer and in autumn. Studies in Europe indicated that, once freed from the fleshy synconium, edible fig seeds may germinate in autumn or in spring, depending on climatic conditions (Lisci and Pacinia 1994), and the same presumably is true in California. Seeds germinate at temperatures between 50 and 85 degrees F (10 - 30 degrees C), but only if humidity remains high or if they are in contact with soil that is continuously wet (Lisci and Pacini 1994).

Edible fig grows quickly in soils with enough moisture and with exposure to high light levels. It is winter-deciduous, and the timing of leaf-out and leaf drop varies with the cultivar and with climatic conditions.

Edible fig reproduces by seed and by vegetative growth. Most of the world's *Ficus* species depend on a species-specific agaonid wasp (family Agaonidae, Hymenoptera) for pollination. *Ficus carica* depends on the wasp *Blastophaga psenes*. The wasps are in turn dependent on *F. carica* because they breed only inside its fruits (Kjellberg *et al.* 1987). Fertilized female *B. psenes* wasps squeeze through the scale-covered ostiole in the end of the synconia of caprifigs and lay one egg in each of several of the short-styled female flowers. Each wasp larvae destroys a female flower as it feeds and grows. Female flowers that escape the egg-laying may each produce a single viable seed if pollinated. Adult male wasps emerge first and quickly cut into flowers containing female wasp larvae and mate with them. These female wasps emerge two to three weeks later and, as they make their way out of the synconium, they pick up pollen from the male flowers clustered near the opening. They then search for another, younger synconium and squeeze through the narrow opening to reach the flowers inside. Galil and Newman (1977) found that the opening is so small that it is likely all pollen on the body surface of the wasp is scraped off as it squeezes through. They speculate that the wasps carry pollen in cavities on their bodies (e.g., between abdominal segments) and that these pollen grains fall out and onto fig flowers when the wasp's body swells and twists as it attempts to deposit eggs. The wasps insert their ovipositors down the style tube to deposit their eggs, but the styles of the female flowers in fig synconia are so long that they prevent the wasps from successfully depositing eggs. During their exertions, however, the wasps deposit pollen and fertilize these flowers. The undamaged, pollinated long-styled female flowers develop into tiny seeds within the synconia. Seeds from these flowers can produce both types of tree, caprifigs and figs (Beck and Lord 1988). Once a female *B. psenes* wasp has entered a synconium to lay eggs, it cannot leave, so these individuals are destined to die without producing offspring (Kjellberg *et al.* 1987).

In California caprifigs produce three crops of fruit, one each in winter, spring, and summer. If climate allows, figs may produce two crops, a small one initiated in spring and maturing in June or July and a main crop initiated between May and July and ripening between August and December (Ferguson *et al.* 1990). Staggering of fruit crops enables the fig wasp to survive throughout the year and ensures pollination of flowers in fig synconia. The synconia of the main crop of Smyrna varieties of edible fig, including the 'Calimyrna' cultivar, will abort and fail to develop if flowers are not fertilized. Synconia of common figs, including the 'Mission,' 'Adriatic,' and 'Kadota' cultivars, can develop fully even if the flowers are not pollinated and no viable seeds are produced (Ferguson *et al.* 1990).

Because the wasps can carry fungal disease spores, California fig growers often grow caprifigs apart from fig trees and distribute ripe caprifig synconia with emerging wasps around their orchards only after checking them for disease (Michailides *et al.* 1996). It is not known whether the edible fig trees that have invaded natural areas in California are all of one variety, nor is it known whether the individuals are caprifigs (with male and short-styled female flowers) or figs (with only long-styled female flowers).

The fleshy tissue of the synconium apparently contains inhibitors and/or creates a microenvironment with high osmotic pressure that prevent seeds from germinating (Lisci and Pacini 1994). Birds and mammals feed on and pick apart the fig fruit and then excrete or drop seeds, releasing them from this inhibition and usually distributing them as well. Seeds may also be washed free of the synconium by hard rains after they fall to the ground and split open.

Edible fig may begin to produce fruit (synconia) within one year if propagated by cuttings or within two to three years if propagated by seed under favorable conditions in orchards. Numbers of fruit are small the first few years, but orchard plantings usually bear harvestable crops by their fifth year.

HOW CAN I GET RID OF IT?

An efficient control method for edible fig has not yet been developed. The trees resprout vigorously after cutting and are difficult to control without herbicides.

Physical Control

Manual/mechanical: Edible figs are shallow-rooted in heavy, wet soils typical of riparian forests and can be pulled up fairly easily when young. They often root-sprout, however, so that what looks like one small sapling may be one of many sprouts from a large network of roots. A small or medium-sized weed wrench may help remove some of the mid-sized specimens. Repeated cutting of resprouts may eventually exhaust the root reserves of a tree or small thicket if the interval between cuttings is short enough, but this has not yet been demonstrated.

Biological Control

Insects and fungi: No biological control species are approved by the USDA for this species. However, figs are subject to damage from nematodes, tree borers, and rust.

Chemical Control

At the Cosumnes River Preserve all trunks and sucker shoots in a thicket were cut six to eighteen inches above the ground and the cut stumps treated with a 100 percent solution of an amine formulation of triclopyr (sold under the names Garlon3A® and Brush-B-Gone®). This was successful, although some thickets had to be retreated at least once because there was some resprouting. The retreatments were carried out at yearly intervals, but shorter intervals (two to six months) might have improved their impact by giving the plants less time to replenish root reserves. Managers at the Cosumnes River Preserve recently have been using a hack-and-squirt method, applying 100 percent triclopyr amine formulation to the wounds, but it is too early to tell if this will be as effective as the cut-stump treatments. This method was also tried at the Dye Creek Preserve, but was not effective there.

Herbicide may be applied in an eight- to twelve-inch-wide band around the uncut trunks of trees with trunk diameters up to two or three inches and perhaps greater. This is known as basal bark application, and it has been shown to be highly effective for a variety of trees and shrubs. Other herbicides, including glyphosate (marketed under a variety of names, including Rodeo® and Roundup®) and imazapyr (as Chopper® and Arsenal®) may be at least as effective as triclopyr against edible fig, but studies of this have yet to be conducted.

Foeniculum vulgare Miller

Common names: fennel, anise, sweet fennel, aniseed, sweet anise, sweet fennel

Synonymous scientific names: *Anethum foeniculum, Foeniculum officinale*

Closely related California natives: 0

Closely related California non-natives: 0

Listed: CalEPPC A-1; CDFA nl

By Rob Klinger

HOW DO I RECOGNIZE IT?

Distinctive Features

Fennel (*Foeniculum vulgare*) is an erect perennial herb, four to ten feet tall, with finely dissected, almost feathery leaves and characterized by a strong anise scent originating from stems and leaves. The flowers are yellow and small (one-quarter inch across), and are clustered in large, rounded, umbrella-like groups (compound umbels), roughly four inches across, that are conspicuous from April through July. During the growing season plants usually include a mixture of living and dead hollow stems (canes). Branches arise from the stems at conspicuously jointed nodes, and leaves arise both from the root crown and from the stems. Leaves sheath the stems where they meet. Seeds of wild fennel look like the fennel seed commonly used as a flavoring in foods: they are oblong, dorsally compressed, and ribbed.

Description

Apiaceae. Perennial herb, 3.3-12 ft (1-3.5 m) high with a characteristic anise or licorice scent. Roots: mature plants have a thick, deep taproot from which erect, solid glaucous-green stems arise. Stems: 10-20 stems originate from a basal cluster in late winter, then die back the following September-November. Leaves: petioles 2.8-5.6 in (7-14 cm) long sheaths hug the stem; leaf blades triangular-ovate in outline and 0.3-2 ft (10-60 cm) long, 1-1.3 ft (30-40 cm) wide, finely dissected into nearly thread-like segments. Inflorescence: compound umbel with 15-40 spreading-ascending rays, each 0.4-1.6 in (1-4 cm) long. Flowers: no sepals, yellow petals with narrowing tips, 5 small stamens, inferior ovary topped by two short styles. Fruits: 0.1-0.2 in (2.5-4.5 mm), oblong-ovate, dorsally compressed, with thick, prominent ridges.

WHERE WOULD I FIND IT?

In California fennel is found in mesic locations with a Mediterranean climate from sea level to 2,000 feet. It usually colonizes disturbed areas, especially weedy sites adjacent to fresh or brackish water, and pastures, abandoned lots, and roadsides. Common in open habitats such as grasslands, coastal scrub, savannas, and the banks of creeks, estuaries, and bays. Dense local populations have been reported from Santa Cruz Island, in fields around the San Francisco Bay region, Palos Verdes Peninsula (Los Angeles County), and Camp Pendleton (San Diego County). It is widely scattered in fields and ditches throughout the Sacramento, Salinas, and San Joaquin valleys and foothills, and in hillside pastures of most coastal counties from Mendocino south to San Diego. Fennel is particularly aggressive in areas subjected to plowing or medium-heavy grazing and recently abandoned (Beatty 1991).

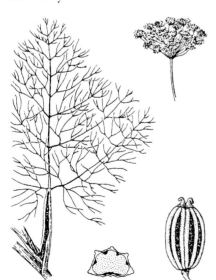

Fennel occurs in soils with pH ranging from 4.8 to 8.3, but appears to prefer more acidic than alkaline conditions. The preferred soil type appears to be well drained, sandy soils, but it has been observed to thrive in sites with a high clay content (pers. observation). Fennel forms dense stands in localized areas (Beatty 1991, Beatty and Licari 1992, R. Klinger, unpubl. data).

WHERE DID IT COME FROM AND HOW DOES IT SPREAD?

Fennel is native to southern Europe and the Mediterranean region, where it has been used for centuries as a spice and for medicinal purposes (Garland 1979). Although details about its introduction are unknown, it has occurred in California for at least 120 years and is presumed to have escaped from cultivation repeatedly (Robbins *et al.* 1941).

Fennel will reproduce from both root crown and seed. Seeds are dispersed by water and on vehicles and clothing. Birds and rodents eat the seeds and may disperse them as well.

WHAT PROBLEMS DOES IT CAUSE?

Fennel will invade areas where the soil has been disturbed and can exclude or prevent reestablishment of native plant species. It can drastically alter the composition and structure of many plant communities, including grasslands, coastal scrub, riparian, and wetland communities. It appears to do this by outcompeting native species for light, nutrients, and water and perhaps by exuding allelopathic substances that inhibit growth of other plants (Granath 1992, Colvin 1996, Dash and Gliessman 1994). It develops dense, uniform stands. On Santa Cruz Island fennel can achieve 50 to 90 percent absolute cover and reach heights of ten feet (Brenton and Klinger 1994).

Once established, fennel is tenacious and difficult to control. Because of its prolific seed production and seed viability, a long-lived seedbank can build up rapidly.

Most impact assessment for fennel has focused on native plants, but fennel's value to animals is unknown. Grazers will feed on early-season regrowth, and feral pigs will seek out and eat the roots, but mature stems are generally not used as food. Birds and rodents eat the seeds.

Fennel stand development and successional patterns are poorly understood, especially with regard to persistence. It is unclear whether fennel stands are an edaphic climax, or whether another plant community will replace them after several decades. In parks and preserves where fennel removal is part of a restoration program, transitional communities will occur after fennel is removed, but these may be dominated by other non-native species (Brenton and Klinger 1994). Klinger and Brenton (in prep. and in review) found there was a significant increase in native herbaceous species shortly after removal of fennel, but the areas quickly became dominated by non-native grasses.

HOW DOES IT GROW AND REPRODUCE?

Fennel reproduces from both root crowns and seeds. Flower production generally begins when individuals are eighteen to twenty-four months old. Flowering stems begin to be produced in late winter to early spring, and flowers appear by early May. Seed production is prolific and can begin as early as May and continue through early November. Generally, seed production peaks in August and September. Seeds are dispersed by water, by animals, and by humans by clinging to clothing or mud on vehicles.

Seeds may persist in soil for several years without germinating. Germination can occur almost any time of the year. Vegetative growth begins in mid-winter and peaks in July to August. Initial growth during winter and spring is slow, then becomes rapid in early summer. Flowering stems die during late fall and early winter, although some remain alive and begin to produce new leaves with the onset of rains. Plants have a thick taproot.

There is little quantitative data on the population biology of fennel. Data on germination rates, seed production, survival, and longevity, density, and viability of the seedbank would be useful for developing management programs.

HOW CAN I GET RID OF IT?

Little published information is available on controlling fennel (Brenton and Klinger 1994, Dash and Gliessman 1994). Management plans should include a survey of where fennel occurs, the current land use, land use in adjacent areas, anticipated changes in land use, and primary dispersal mechanisms.

In areas where fennel stands are already well established, management will require a long-term commitment of time and resources. Management efforts should focus on preventing or reducing disturbance favorable to further spread (soil disturbance, moderate to heavy grazing) and reducing fennel density within dense stands. Dash and Gliessman (1994) reported that non-native species dominated all areas following fennel control regardless of the technique used. For these reasons, fennel removal should be considered only a first step in a larger restoration process that will require other actions to favor recolonization by native species (Dash and Gliessman 1994). It is probably impossible to completely eradicate fennel from wildlands, but reducing stand density and disturbance will minimize its impacts.

Physical Control

Manual/mechanical methods: Manual methods are most effective when infestations are light and locally restricted (Dash and Gliessman 1994). Digging out individual plants by hand is preferred to plowing or bulldozing because it minimizes soil disturbance, but it is labor-intensive.

Cutting, mowing, and chopping temporarily reduce the height of fennel plants within a stand, but they are ineffective as methods of removal and minimally impact the spread of fennel stands (Dash and Gliessman 1994, Colvin 1996). These techniques leave the roots intact, alive, and ready to support regrowth of shoots. Repeated cuts may have more impact by helping to exhaust the resources of the taproot over time. However, intervals between cuts must be short, because fennel recovers rapidly from cutting and begins to replenish its root energy supplies (Brenton and Klinger in review, Dash and Gliessman 1994). Cutting while plants are producing seed will promote dispersal.

Prescribed burning: Experiments on Santa Cruz Island indicate that burning is not an effective control method by itself (Klinger and Brenton in prep.). However, fall burns (November-December) followed by herbicide sprays the following two springs can reduce fennel cover 95 to 100 percent (Klinger and Brenton in prep.). For reducing fennel in large areas with dense stands, this method is effective but costly.

Biological Control

Insects and fungi: No biological controls agents for fennel are known.

Grazing: Use of livestock to control fennel will probably be ineffective except where stands are small, not very dense, and young. In older and/or dense stands grazing will spread fennel further (Brenton and Klinger 1994). Since fennel can reproduce by roots as well as seed, removal of above-ground shoots will slow, but not prevent, vegetative spreading. If livestock are in pastures when fennel is producing seed, they will spread the seed to new areas. Most heavily infested areas of Santa Cruz Island were formerly used as cattle pastures (Beatty 1991).

Chemical Control

Brenton and Klinger (1994 and in review) found that 95 to 100 percent kill was achieved when amine and ester formulations of triclopyr (Garlon 3A® and Garlon4®, respectively) were applied to fennel in early spring at rates of 6 lbs/100 gallons water (1 lb active ingredient/acre) on Santa Cruz Island. Lower concentrations (3.0 and 4.5 lbs/100 gallons) of both the amine and ester formulations were less effective, and all treatments were less effective when administered in late summer rather than early spring. Cutting fennel and treating the cut stems did not increase the effectiveness of the herbicide.

Dash and Gliessman (1994) reported that glyphosate (as Roundup®) sprayed in spring at the manufacturer's recommended rate reduced fennel cover 75 to 80 percent. Cutting prior to spraying did not increase the effectiveness of the treatments.

Genista monspessulana (L.) L. Johnson

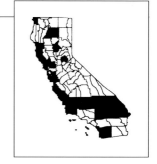

Common names: French broom, soft broom, canary broom, Montpellier broom

Synonymous scientific names: *Cytisus monspessulanus, C. racemosus, C. canariensis, Genista monspessulanus, Teline monspessulana*

Closely related California natives: 0

Closely related California non-natives: 4

Listed: CalEPPC A-1; CDFA C

by Carla Bossard

HOW DO I RECOGNIZE IT?

Distinctive Features

French broom (*Genista monspessulana*) is an upright, evergreen shrub, commonly to ten feet tall. The round stems are covered with silvery, silky hair, and the small leaves are ususally arranged in groups of three. About eighty-five percent of the photosynthetic tissue of French broom is in leaf tissue. The small (less than half-inch) yellow flowers are pea-like and clustered in groups of four to ten. The mostly inch-long pods are covered with hairs.

This species sometimes is confused with Scotch broom (*Cytisus scoparius*), which has pods with hairs only at the seam, green stems that are five-angled and ridged, flowers that are golden yellow and larger than half an inch, and only about fifty-five percent of total green tissue as leaves (Bossard and Rejmánek 1994).

Description

Fabaceae. Shrub, usually <10 ft (3 m), but occasionally to 16 ft (5 m). Stems: twigs silvery silky-hairy. Leaves: alternate; stipules <0.1 in (2 mm); deciduous; leaflets of trifoliate leaves

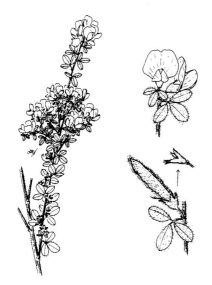

0.3-0.5 in (10-15mm), oblanceolate to obovate with length about twice width, upper surface glabrous, lower surface with appressed or spreading hairs hairiness; petioles <2 in. (5 mm). Inflorescence: 4-10 flowers in subcapitate racemes (on axillary short shoots); terminal or central flower usually opening last; pedicels <0.1 in. (1-3 mm). Flowers: shaped like pea flowers; calyx silky-hairy, 0.2-0.3 in (5-7 mm); banner 0.3-0.5 in (10-15 mm), corolla yellow to light yellow. Fruit: a pod, 0.5-1 in (15-25 mm), covered in dense silky hairs, dark brown or black at maturity. Empty seed pods curled. Seeds: 3-8 seeds per pod, brown to black, shiny, round to oval, with a cream to yellow eliaosome (description from Hickman 1993 and pers. observation).

WHERE WOULD I FIND IT?

French broom is found primarily in central coastal counties from Monterey County north to Mendocino County and inland in Lake, Solano, and Contra Costa counties. It is also known from Del Norte County, northern Sierra Nevada foothill counties to 800 meters, and in Kern, San Bernardino, and San Diego counties.

This broom is common on coastal plains, mountain slopes, and in disturbed places such as river banks, road cuts, and forest clearcuts, but it can colonize grassland and open canopy forest. It is found growing in varied soil moisture conditions, but prefers siliceous soils. Unlike other broom species in California, it grows reasonably well on alkaline soils with pH 8. It is competitive in low-fertility soils because of mutualistic relationships with nitrogen-fixing bacteria found in small nodules on roots. While Scotch broom is a problem species in many parts of the world, French broom is especially problematic in California and Australia (Partridge 1989, Parsons 1992). French broom seedlings are less tolerant of frost than are those of Scotch broom and consequently are less often found at higher elevations.

WHERE DID IT COME FROM AND HOW DOES IT SPREAD?

Native to countries surrounding the Mediterranean and in the Azores, French broom is thought to have been introduced to the San Francisco Bay Area in the mid-1800s as an ornamental. It spreads via prodigious seed production. A medium-sized shrub can produce over 8,000 seeds a year (Bossard unpubl. data). After pods open explosively, flinging seeds up to 4 m, the seeds are further dispersed by ants, birds, and animals and in river water and rain wash (McClintock, pers. observation), in mud, and on road grading or maintenance machinery (Parsons 1992). It resprouts readily from the root crown after cutting, freezing, and sometimes after fire (Bossard *et al.* 1995).

WHAT PROBLEMS DOES IT CAUSE?

French broom currently occupies approximately 100,000 acres in California (D. Barbe, pers. comm.). It displaces native plant and forage species, and makes reforestation difficult. It is a strong competitor and can dominate a plant community, forming dense monospecific stands. In an experiment in New Zealand French broom had a higher growth rate than any other broom species found in California, reaching an average height of more than 4.5 feet (141 cm) in two growing seasons. Since it can grow more rapidly than most trees used in forestry, it shades out tree seedlings in areas that are revegetated after harvest.

French broom foliage and seeds are toxic, containing a variety of quinolizidine alkaloids, especially in young leaves (Montlor *et al.* 1990). In some livestock, ingestion of plant parts can cause staggering followed by paralysis (McClintock 1985). Foliage can cause digestive disorders in horses (Parsons 1992). Infestations of broom degrade the quality of habitat for wildlife by displacing native forage species and changing microclimate conditions at soil levels. French broom is believed to be responsible for reducing arthropod populations by one-third in Golden Gate National Recreation Area (Lanford and Nelson 1992). It burns readily and carries fire to the tree canopy layer, increasing both the frequency and intensity of fires. French broom along roadsides obstructs views, requiring expensive ongoing road maintenance. This species establishes a dense, long-lived seedbank, making it difficult to eradicate.

HOW DOES IT GROW AND REPRODUCE?

French broom becomes reproductive at two to three years of age, on reaching a height of one and a half to two feet (45-60 cm). It flowers in late March-May inland, March-July on the coast. Flowers appear just prior to new leaves. Long-lived seeds are copiously produced (Hoshovsky 1995) and mature in June-July. Seeds are known to survive at least five years in soil (Bossard unpubl. data). French broom seedbanks have been found to contain 465 to 6,733 seeds per square meter (Hoskings 1994, Parker and Kershner 1989). Seeds germinate December-July (Bossard unpubl. data). Cheng (in press) reports that heat treating seeds with temperatures of 65 degrees C improved germination of seed in some populations but not in others. Seedlings can tolerate up to 80 percent shade (Bossard unpubl. data). Plants can resprout from the root crown after cutting. Once seedlings are taller than approximately eight inches (20 cm), their rate of resprouting after cutting can be over 90 percent, particularly if cut in the rainy season (Bossard unpubl. data).

The period of most rapid vegetative growth is April-July. As in other brooms, most photosynthate is moving up in the shrub toward branch tips during flowering, bud break, and seed set. Photosynthate starts moving down toward roots of this broom after seeds are well grown but before seed release (Bossard *et al.* 1995). French broom retains much of its foliage in coastal areas, and is more deciduous in inland areas. Its life span is typically ten to fifteen years (Waloff, pers. comm.).

HOW CAN I GET RID OF IT?

As with other broom species, the best method for removal of a French broom infestation depends on climate and topography, age and size of the infestation, importance of impact to non-target species, and type, quantity, and duration of resources available to remove and control broom at the site. All methods require appropriate timing and follow-up monitoring. Because of the seedbank, monitoring removal sites to locate and kill new seedlings is essential. Location and retreatment of resprouts is also necessary. Sites should be examined once a year, when the seed germination period ends in late spring, for five to ten years and every two years thereafter.

Physical Control

Manual/mechanical removal: In general, when using hand removal or mechanical methods it is best to start in areas with small infestations and many desirable species that will reseed naturally. Desirable species should be given some assistance by hand weeding of French broom. Next work on areas with an intermediate degree of infestation (Fuller and Barbe 1985). Finally, tackle larger areas and dense concentrations of French broom using other techniques (fire, chemicals) to augment or replace hand pulling.

Pulling with weed wrenches is effective for broom removal in small infestations or where an inexpensive, long-duration labor source is dedicated to broom removal. The weed wrench removes the entire mature shrub, eliminating resprouting. However, the resultant soil disturbance tends to increase depth of the seedbank and prolong the need for monitoring. Wrench removal is labor-intensive, but can be used on slopes. It also allows targeting of broom plants while minimizing impact on neighboring species. Golden Gate National Recreation Area has had success in using volunteers to remove broom with weed wrenches and closely monitoring and removing broom seedlings for five to ten subsequent years.

Brush hogs, which twist off above-ground plant material, can be used for broom removal. Although less labor-intensive than weed wrenches, they damage neighboring species and cannot be used on steep slopes. The twisting action is more destructive to tissues that initiate resprouting than is clean cutting. However, depending on the season of brush hog removal, resprouting can still be a problem. Brush hog removal has been used with limited success in Redwood National Park (Popenoe, pers. comm.).

Saw cutting: Archbald (1996) reported success with saw and/or brush cutter following four steps: (1) cut shrubs at or below ground level in late July or August, after broom has gone to seed and soil moisture is at a seasonal low; (2) move cut broom plants to sites appropriate for disposal or burn in spring after plants dry (use tarps to avoid spreading mature seed to uninfested areas while moving sawed broom); (3) the following summer, after grasses are dry and have dispersed their seed, destroy new French broom seedlings by mowing as low to the ground as possible with a heavy-duty brush cutter with a four-pointed metal blade; and (4) repeat for the next five or six seasons or until the seedbank is exhausted. Timing and height of cutting are critical in using this technique. Cutting French broom in June in Mendocino County at 5-8 cm above soil surface resulted in extensive resprouting (Bossard *et al.* 1995).

Mulching: A 10 cm deep wood bark mulch significantly decreased seedling emergence of French broom in experiments conducted by Cheng (in press) in the San Francisco Bay Area. This suggests that mulching could be used to suppress regrowth from the seedbank after removal of mature shrubs.

Prescribed burning: Using fire to remove uncut French broom in late spring or early summer has had some success at Mt. Tamalpais State Park in Marin County (Boyd 1994). Reburning of the removal site is usually necessary two and four years after the initial burn (Boyd, pers. comm.). Reburnings are most effective in killing resprouts and seedlings if there are either naturally occurring or reseeded grasses to carry the fire.

Ken Moore (pers. comm. 1999) reports that California State Parks has been very successful (100 percent mortality) using a propane torch to remove French broom seedlings up to 20 cm in height that emerge from the seedbank after removal of adult brooms. The torch is set so it is hot but not flaming and it is passed over the French broom seedlings. The heat does not cause the seedling to burn but within a day the seedling is wilted and dead. This is done at the end of the rainy season when seedlings are up but there is no fire danger. Flame throwers have been used to spot-treat road edges or small areas with seedlings emerged from the seedbank after removal of mature brooms in Redwood National Park and in New Zealand (Popenoe, pers. comm., Johnson 1982). For prescribed burning of pretreated or cut broom see below under integrated methods.

Biological Control

There are no USDA approved biocontrol agents for French broom. The distribution and effects of the native pyralid moth, *Uresiphita reversalis*, on French broom were investigated by Montllor *et al.* (1990). While this insect may defoliate some French broom shrubs, plants grow new leaves after the larval stages undergo metamorphosis. Other insect biocontrol agents are being tested in England and France for use on Scotch broom in Australia and New Zealand (Paynter 1997). Some of these agents may use French broom as a host as well. However, the insects known to feed on and impact mature Scotch and French brooms (some *Sitona* sp.) are likely to feed on *Lupinus* species as well and consequently would not be appropriate for release in California (Paynter 1997).

Grazing: Heavy grazing by goats for four or five years during the growing season has been reported as effective in New Zealand and has been tried at a few sites in Marin County in California (Archbald, pers. comm.). The disadvantage is that goats are not selective, and native species that may start to revegetate the area are also eaten.

Chemical Control

A solution of 3 percent glyphosate sprayed on foliage until wet has been used to treat mature French broom shrubs. Adding surfactant improved effectiveness (Parsons 1992). However, the foliar spray impacts non-target species, and resprouting often occurs. Triclopyr ester (25 percent), in Hasten® or Penevator® oil (75 percent) in one spot, low-volume basal bark application with a wick has proved effective in killing French broom (Bossard *et al.* 1995). Dye should be added to the herbicide solution to help avoid missing stems. It was necessary to spot only the main stem with 2 or 3 drops of herbicide, within 8 cm of the ground surface, to obtain a 99 percent kill of the eight-year-old French broom plants in this experiment conducted in Mendocino

County. Soil analyses showed no contamination by the triclopyr, even in plots that were later burned. However, killing the mature shrubs was not sufficient to remove the infestation of French broom because of its well developed seedbank (Bossard *et al.* 1995). This application technique does not impact non-target species, but it is time-consuming if the site is large. Both of these chemical methods should be used during periods of active growth after flower formation and seed set but before seed dehisces.

The herbicide 2,4-D, alone or with additives such as diquat, picloram, dicamba, and sodium chlorate, has been used to control French broom. Not all of these herbicides are registered for use in California. French broom seedlings are least resistant to auxin-minimizing herbicides such as 2,4-D at the four- to six-inch (10-15 cm) size. Chemical removal alone results in standing dead biomass, which makes monitoring for and treatment of broom seedlings difficult. Standing dead biomass also presents a fire hazard.

An Integrated Approach

The most effective removal treatments in a project in Jackson State Demonstration Forest conducted by the CalEPPC broom committee (Bossard *et al.* 1995) was a combination of treatments that began in early July with low-volume basal bark application of triclopyr ester (as Garlon®) (25 percent) in Hasten or Penevator oil (75 percent) and a purple dye in a low-volume basal bark application (2-3 drops in one spot <8 cm from the soil surface) with a squirt bottle on mature dense stands of French broom. After four weeks all broom shrubs were dead and were cut down, left on site, and burned. This flushed the seed from the seedbank by increasing germination rate with the next rains. French broom seedbanks in burned plots were reduced to less than 5 percent of their original size three years after prescribed burns. Seedbanks of unburned plots otherwise treated the same were reduced to 15.5 percent of their original size, and control plots exhibited no significant decrease in seedbank size. For the next two years, in July, seedlings in plots were treated with either 2 percent glyphosate (as Roundup®, label-recommended strength) or cut with a four-blade gasoline powered brush cutter. Glyphosate was applied with a backpack sprayer, and non-target vegetation was avoided. Brush-cut plots had 1.6 resprouts per square meter, whereas plots in which glyphosate was applied to seedlings had 0.2 resprouts per square meter. Mean percent cover by French broom was reduced from 87 percent to less than 0.2 percent in plots treated with basal bark triclopyr, cut, burned, and seedlings treated with glyphosate.

Halogeton glomeratus (M. Bieb) C. Meyer

Common name: halogeton

Synonymous scientific name: *Anabasis glomerata*

Closely related California natives: 0

Closely related California non-natives: 0

Listed: CalEPPC Red Alert; CDFA A

by Steven A. Dewey

HOW DO I RECOGNIZE IT?

Distinctive Features

Halogeton is an annual herbaceous plant, typically six to twelve inches tall, with short, fleshy, sausage-like leaves less than half an inch long. One of its most distinctive features is the conspicuous, soft, slender spine at the bluntly rounded tip of each leaf. Plants are often bluish green in spring and early summer, turning yellow, salmon, pink, purplish, or even reddish by late summer or early fall. Stems often turn pink or red while leaves are still blue-green. Plants can resemble Russian thistle in early stages of growth, but are distinguished easily by the unique leaf tips and the presence of tiny, cotton-like hairs in the leaf axils.

Description

Chenopodiaceae. Annual semi-succulent herb ranging from 2-18 in (5-45 cm) tall, and 2 to 18 in (5-45 cm) wide. Stems: main stems branch from the base, spreading at first, then bending upward to become erect, with numerous short lateral branches. Leaves: sausage-like, smooth, fleshy, 0.1 to 0.5 in (2-12 mm) long, positioned alternately on stems. Inflorescence: numerous inconspicuous flowers in compact clusters in leaf axils of upper stems. Flowers: two types, larger with 5 pale yellow or greenish yellow, fan-like membranous sepals (no petals), each about 0.08-0.12 in (2-3 mm) wide; smaller with tooth-like sepals (no petals). Both flower types generally have 2-5 stamens and 2 stigmas per flower. Halogeton also has two types of seeds. Fruits: wingless "brown" (light tan-colored) seeds are produced in early summer, while winged "black" (dark chocolate brown) seeds are produced in late summer (Welsh *et al.* 1987).

WHERE WOULD I FIND IT?

In 1980 halogeton was reported in Inyo, Kern, Lassen, Los Angeles, Modoc, Mono, and Nevada counties in California. Today it can be found throughout southern California and in all counties bordering Nevada. It has also been reported from Siskiyou and San Diego counties. Halogeton is widely distributed over millions of acres throughout at least eleven western states from about 2,500 to 7,000 feet elevation. It is highly suited to the alkaline and saline soils of the region's semi-arid high-desert environments, but it also may be found on heavy clays, clay loams, sandy loams, and loamy sands. The abundance of halogeton depends upon year-to-year precipitation, so outbreaks may appear sporadically. Annual precipitation at most halogeton sites is from five to thirteen inches (127-330 mm) (Cooke 1965).

Halogeton is found mainly on disturbed arid sites in saltgrass, salt desert shrub, mixed desert shrub, or pinyon-juniper plant communities. Annual weeds typically associated with halogeton include cheat grass (*Bromus tectorum*) and Russian thistle (*Salsola tragus*). It is especially common along roadsides, on the edges of alkaline flats, in livestock bedding or feeding areas, in abandoned dryland farms and townsites, and around desert watering sites (Pemberton 1986, Bellue 1951).

WHERE DID IT COME FROM AND HOW DOES IT SPREAD?

Halogeton is native to Eurasia or Siberia and was introduced into the United States in northern Nevada in the early 1930s, possibly for use in grazing experiments. The first herbarium specimen was collected in 1934 in Nevada. It spread quickly into desert lands throughout Nevada, Utah, California, and adjacent states (Cook and Stoddart 1953). Halogeton can be dispersed by wind, water, animals, and human activity. Local dispersal by wind is aided by winged bracts on the black seeds. Entire plants can become tumbleweeds, driven by wind or catching on the undercarriage of vehicles, distributing seeds for great distances. Whirlwinds or dust-devils can transport dry stems with seeds up to two miles. Animals can spread large amounts of seed great distances because the seeds are resistant to digestion (Whitson *et al*. 1991, Dayton 1951, Stoddart *et al*. 1951). Halogeton seeds are rapidly spread along roads by equipment, especially road graders (Cronin 1965).

WHAT PROBLEMS DOES IT CAUSE?

Halogeton is not an extremely competitive plant, but it can quickly invade disturbed or overgrazed sites, and it can prevent reestablishment of desirable species. It is poisonous to livestock.

Sheep are especially prone to poisoning, although cattle also can be affected (James *et al.* 1980). Leaves and stems are rich in a toxic substance called sodium oxalate. Halogeton is readily consumed by hungry or thirsty livestock and is responsible for thousands of livestock deaths. Signs of poisoning include depression, weakness, reluctance to move, rapid and shallow respiration, drooling, recumbency, coma, and death within two hours to several days after ingestion (Bohmont *et al.* 1955). The best defense against poisoning is to keep livestock away from infested sites, especially after drinking or following a rain or snow storm (Burge 1950, Cook and Stoddart 1953).

HOW DOES IT GROW AND REPRODUCE?

Reproduction of halogeton is exclusively by seed. Flowering typically begins in June. Fruiting and seed production are generally from July through October. Halogeton can produce seventy-five seeds per inch (35 seeds/cm) of stem, or 200 to 400 pounds of seeds per acre (222-449 kg/ha). More than 100,000 seeds can be produced on a single large plant. Two types of seeds are produced, which is important to the plant's spread and persistence. Black seeds germinate during the first growing season after production; brown seeds can remain viable but dormant in the soil for ten years or more (Cronin and Williams 1966).

Halogeton takes advantage of infrequent desert precipitation by emerging quickly when soil moisture becomes available from February to mid-August. Plants grow rapidly and produce an abundant seed crop before frost ends the growing season in fall. Halogeton usually establishes first in disturbed sites, such as along road shoulders or livestock trails and bedding areas (Erickson *et al.* 1951).

HOW CAN I GET RID OF IT?

If eradication of small, isolated patches of halogeton is the goal, it will be necessary to monitor infestations at least once or twice each year to make sure no seedlings survive to produce seed. This will have to be done for at least ten consecutive years, because halogeton seeds can remain dormant but viable in the soil for that length of time. Controlling extensive infestations will require elimination or reduction of the disturbance that allowed halogeton to invade and simultaneous reestablishment of competitive perennial grass, forb, and/or brush species (Fenley 1952).

Physical Control

Manual/mechanical methods: Because halogeton is a simple shallow-rooted annual, it can be controlled effectively by tillage or pulling. Plants are easiest to control as seedlings or in early vegetative growth. Plants not controlled until after flowering begins may contain seeds and should be removed and destroyed to prevent reseeding. Periodic mowing close to the soil surface can significantly reduce but not completely prevent seed production. Surviving branches below the reach of mower blades will continue to produce viable seeds.

Prescribed burning: Halogeton is not controlled effectively by burning, and it is one of the first plants to reestablish following wildfire on infested rangeland.

Biological Control

Insects and fungi: A stem-boring moth, *Coleophora parthenica*, was introduced by the USDA into the United States for possible control of halogeton, but it failed to establish. Other potential biological control agents have been identified in Central Asia, but they have not yet been developed and tested.

Grazing is not a control option because of the toxicity of the plant.

Chemical Control

Glyphosate (as Roundup®; 2 percent solution) in a spot treatment on small infestations will kill emerged halogeton plants if applied before the bloom stage. Repeat treatments will be necessary to control any flushes emerging later in the season. Herbicides containing the active ingredient metsulfuron (in Escort® or Ally®) effectively control more extensive infestations of halogeton in pastures and rangeland without causing injury to desirable grasses. Colorado State University studies comparing various rates of metsulfuron, dicamba (as Banvel®), and picloram (as Tordon®) reported good initial control from each herbicide when applied in spring to small, one- to three-inch (2.5-7.5 cm) tall halogeton plants. However, effects of many treatments were temporary. Five months after application, control from metsulfuron remained good to excellent (73-94 percent). However, dicamba provided poor to good control (48-78 percent), and picloram treatments were rated as poor (19-53 percent control) (Sebastian and Beck 1993).

Hedera helix L.

Common name: English ivy

Synonymous scientific name: none known

Closely related California natives: 0

Closely related California non-natives: 0

Listed: CalEPPC A-1 list, CDFA nl

by Sarah Reichard

HOW DO I RECOGNIZE IT?

Distinctive Features

English ivy (*Hedera helix*) is the familiar vining plant often allowed to grow up building walls. It has two forms: an evergreen woody vine and an evergreen shrub. Both forms have deep green, glossy, leathery leaves. Vining plants do not produce flowers or fruits, and their leaves have lighter-colored veins and three to five lobes. Upright shrubby plants may produce flowers and fruits, and their leaves are ovate rather than lobed. In both forms the leaves may have a strong odor when crushed. The white flowers are in clusters on the ends of stems produced in fall, and the fruits are dark blue or purplish drupes. English ivy may be distinguished from grape vines (*Vitis* sp.) and *Ampelopsis* species by its evergreen leaves, which are not hairy or fuzzy (pubescent), and by vines that have no tendrils. It differs from cape ivy (*Delairea odorata*) in having leaves that are evergreen in all climates, with a deep cleft at the leaf base that makes the lower lobes appear larger than the others. Cape ivy may also be distinguished from this species by its small, yellow composite flowers.

Description

Araliaceae. Perennial, evergreen woody vine to 99 ft (30 m) (juvenile plant) or shrub (adult plant). Stems: creeping juvenile stems have roots at leaf nodes with adventitious rootlets that allow the plant to climb vertical surfaces by adhering to, but not penetrating, bark and brick. Adult flowering stems erect and non-climbing. Leaves: leathery, simple, and alternate. On juvenile plants leaves have 3-5 lobes and are 1.6-4 in (4-10 cm) long and about as wide. Terminal lobe about as broad as long; the two basal lobes may be reduced or absent. Lobes often more pronounced on leaves of climbing stems. Leaf base cordate and veins markedly lighter in color. Leaves on flowering stems mostly unlobed, ovate to rhombic, base shallowly cordate to cuneate, and veins slightly lighter in color. Petioles on both forms about as long as the leaf. Young shoots, leaves, and peduncles covered with stellate hairs and scales; older shoots and leaves glabrous. Inflorescence: a raceme that appears umbellate. Flowers: bisexual, radial, usually 0.2-0.3 in (5-7 mm) across, with 5 sepals fused at the base and persistent but small. There are 5 separate white to yellowish green petals; stamens usually 5 and alternate with petals; 1 style with 5-lobed stigma; ovary inferior. Fruits: berry-like drupe about 0.24-0.36 in (6-9 mm) in diameter containing 4-5 seeds; drupe usually dark blue to black, lighter on some cultivated varieties. English ivy flowers in fall, and fruits are produced the following spring in April and May (Putz and Mooney 1991).

WHERE WOULD I FIND IT?

English ivy is found in northern California forests south to at least Santa Cruz. It has also been observed in Shasta and Butte counties and along the south coast from Santa Barbara County to San Diego. It is a serious problem in the coastal Pacific Northwest from central Oregon into British Columbia. On the eastern seaboard it also spreads into woods, particularly from Virginia north to New York. English ivy is generally found in open forests, especially those with a deciduous component, from sea level to 3,300 feet (1,000 m) elevation. It is especially common in forests near urban areas. It climbs up tree trunks and along branches into the canopy and may also cover the ground. English ivy will invade riparian zones where flooding has disturbed the soil, but it does not grow well in areas where the water table is high and soil is waterlogged (Thomas 1980). It grows well in acid and basic soils.

WHERE DID IT COME FROM AND HOW DOES IT SPREAD?

English ivy is native to England, Ireland, the Mediterranean region, and northern Europe west to the Caucasus Mountains. In its native range English ivy is widespread and usually found in woods and along rocky areas. It is often considered a weedy pest in its native range (Wyman 1954). It was introduced into North America in early colonial times as an ornamental (Wyman 1969). English ivy has been planted to control soil erosion in many parts of the United States because of its habit of rooting at the leaf nodes along the stem. It is perhaps the best known of all

evergreen vines in cultivation. Birds disseminate the seeds. Once established, it spreads quickly by vegetative means.

WHAT PROBLEMS DOES IT CAUSE?

English ivy can alter natural succession patterns in forests. It forms "ivy deserts" of vigorous vines in forests where nothing else seems able to compete. It inhibits regeneration of understory plants, including forest wildflowers and new trees

and shrubs (Thomas 1980). By blocking regeneration in forests, it jeopardizes their long-term persistence. English ivy also kills trees in the understory and overstory by shading them out (Thomas 1980). It tends to grow up tree trunks into branches, especially those of deciduous trees. The increased winter light under deciduous trees apparently allows this evergreen vine to grow rapidly upward in winter (Thomas 1980). Once in the canopy, English ivy can shade out deciduous foliage during summer months, suppressing the growth of the tree that supports it. As the tree dies back, its increasingly open crown allows the vine to grow even more (Thomas 1980). In addition to shading, the additional weight of water and/or ice on the evergreen ivy leaves may increase storm damage to trees, especially in the presence of high winds. This effect has been observed in trees infested with *Vitis* sp. in Connecticut (Siccama *et al.* 1976).

English ivy may replace species used by native wildlife. Its leaf litter adds nitrogen to the soil, which may disadvantage native species that compete best under lower nutrient levels (Tremolieres *et al.* 1988). The sap can cause dermatitis in some people, and both berries and leaves are toxic (Hickman 1993).

HOW DOES IT GROW AND REPRODUCE?

While vegetative reproduction is a key to the success of English ivy, the plant also reproduces prolifically by seed. English ivy flowers in fall, and fruits are produced the following spring in April and May. The juvenile period is long, often ten years or more, but when it becomes reproductive, it produces large numbers of bisexual flowers in fall that are attractive to pollinating bees. Seeds ripen the following year, and on average about 70 percent are viable (Dirr and Heuser 1987). English ivy seed has a hard coat that must be scarified before it can germinate, a condition easily met as the seed passes through the digestive systems of birds that disperse the fruits. The fruits are eaten by several species of birds. English ivy reproduces vegetatively by adventitious roots along the stem and may regenerate from stem fragments if they remain in contact with the soil. The vines can persist a long time; there have been reports of a vine that was 433 years old (Putz and Mooney 1991).

English ivy grows as a woody vine when young, becoming increasingly shrubby as the plant becomes reproductive. Because its leaves are evergreen, it can photosynthesize year-round and will grow rapidly if unchecked. Its ability to climb structures using adventitious roots suggests that it is well adapted to establishment in late successional forests (Carter and Teramura 1988). The root system is shallow. Growth in the adult shrub form is slower. English ivy tolerates shade, but its growth is stimulated by light. Thomas (1980) found that in heavy shade (4 to 7 percent of full sunlight) English ivy survived but began to slowly decline, while under 65 to 68 percent of full sunlight the plant flourished.

HOW CAN I GET RID OF IT?

Control of English ivy has not received sufficient attention or research. Research in the past has focused on establishing new cultivars rather than on controlling or eliminating the plant.

Physical Control

Manual/mechanical: The best method for controlling English ivy may be hand removal of vines using pruners to cut the vines and then pulling the plants up from the forest floor and down from the trees. Removing and killing vines that spread up into trees is especially important because the fertile branches grow primarily on upright portions of the vine. If vines are cut at the base of the tree the upper portions will die quickly but may persist on the tree for some time; vines on the ground around the tree should also be removed to prevent regrowth up the tree. Care should be taken to minimize disturbance during removal. If the forest floor becomes disrupted, appropriate native species should be planted on the site to inhibit reinfestation by English ivy or another invader (Humphries *et al.* 1991).

Prescribed burning: An extreme method that has been used with some success is to burn ivy plants and resprouts with a blow torch at regular intervals; the energy used by the plant to regrow will eventually be depleted. Obviously, this approach requires considerable caution. No other attempts to use fire to control English ivy have been reported.

Biological Control

Insects and fungi: There have been no attempts to introduce biological control agents, and it is extremely unlikely that such agents will ever be used. English ivy is an important landscape plant and has strong support from the horticultural community, including a society dedicated to its study and promotion (the American Ivy Society).

Grazing: The palatability of English ivy to grazing animals is unrecorded.

Chemical Control

English ivy is tolerant of preemergence herbicides (Derr 1992). Its waxy leaves make effective application of post-emergent herbicides difficult, even when a surfactant is added. Glyphosate (as Round-up®) applied at a rate of 2.7 lb/acre effectively controlled young plants, especially in early spring (Neal and Skroch 1985), but tests on more mature plants suggested that adult upright English ivy is tolerant of this herbicide. This was true even when surfactants, high application rates (4 lb/acre), and second applications were used (Derr 1993), although growth may be retarded as much as 60 percent.

Workers at the Washington Park Arboretum in Seattle have had some success using a string trimmer to remove most of the leaves and young stems and then immediately spraying triclopyr (as Garlon 4®) at a rate of 6.5 oz/gal plus a surfactant. Two years after application the treated plants were dead, although the area was being reinfested from surrounding populations. In smaller infestations the herbicide can be brushed onto cut stems. Check with a certified herbicide applicator for concentrations of these herbicides currently registered for use in California.

Helichrysum petiolare Hilliard & Burtt

Common name: helichrysum, licorice plant

Synonymous scientific names: *Helichrysum petiolatum, Gnaphalium lanatum*

Closely related California natives: 0

Closely related California non-natives: 0

Listed: CalEPPC Red Alert; CDFA nl

by Jake Sigg

HOW DO I RECOGNIZE IT?

Distinctive Features

Helichrysum is a white-woolly, sprawling, waist-high shrub that is capable of climbing. It has aromatic leaves the size of a nickel or a quarter, densely matted on both surfaces with soft white hairs, making the plant appear white. This whiteness makes it easy to spot infestations from a distance. Its cream-colored flowers are in densely clustered heads that resemble those of native everlastings and especially cudweeds (*Gnaphalium* spp.), although native cudweeds are much smaller than this plant.

Description

Compositae. Perennial shrub to 3 ft (1 m) high. Stems: to 3 ft (1 m) high. Leaves: aromatic, roundish-triangular 0.7-1 in (2-2.5 cm) with upper and lower surfaces covered in woolly hairs. Inflorescence: flowers densely clustered in 1-2 in (2.5-5 cm) heads that are corymbs. Flowers: cream-colored, discoid.

WHERE WOULD I FIND IT?

Helichrysum has been reported outside cultivation in four locations in California, all within a few miles of the ocean in Monterey, Marin, and Mendocino counties. The Monterey infestation is in the Del Monte Forest, the two Marin populations are within the Golden Gate National Recreation Area, and the Mendocino population is near Gualala (Howell 1969; Smith and Wheeler 1990). The plant invades coastal scrub communities, but has the potential to invade coastal grasslands. It is sun-loving, and it is unlikely to tolerate much shade.

The species' known tolerances (mild winters and summers) indicate that it has the potential to invade along the California coast to a few miles inland, especially the south and central coast. It thrives in either sandy or heavy soil. Because it is a garden plant, special efforts should be made to locate incipient infestations near human habitations in the coastal belt.

WHERE DID IT COME FROM AND HOW DOES IT SPREAD?

Helichrysum petiolare is indigenous to the coastal belt at the extreme southern tip of Africa. It was offered by specialist nurseries in the 1960s, and present infestations are undoubtedly garden escapes. Vegetative propagation may be an important means of spreading because branches can root if in contact with the ground. Seeds are wind-dispersed, but they generally do not carry long distances.

WHAT PROBLEMS DOES IT CAUSE?

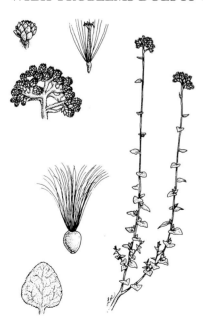

The Stinson Beach infestation in Golden Gate National Recreation Area demonstrates that *Helichrysum petiolare* can invade undisturbed sites, since it is penetrating an otherwise apparently healthy, dense California sagebrush-coyote brush scrub. It forms a closed canopy in places and displaces native plant species.

HOW DOES IT GROW AND REPRODUCE?

Helichrysum petiolare spreads vegetatively and by seed. Flowering is in mid-summer, and abundant seed is produced by early autumn; seed longevity is not known. Its branches are lax and sprawling, and those touching ground will take root, hence vegetative propagation may be an important means of spread in California.

Seed germination, as with most Mediterranean-climate plants, is in fall after the first rains. Plants grow actively in spring and early summer; they flower in mid-

summer followed by seed set, after which they are semi-dormant (Sunset 1988).

HOW CAN I GET RID OF IT?

There have been no reported efforts to control *Helichrysum petiolare*. Small infestations are easily removed by hand pulling. For large patches, it will be necessary to experiment with an herbicidal application. The dense woolly hairs protecting the leaves may inhibit absorption of many herbicides. A certified herbicide applicator should be able to assist in selection of an effective surfactant. It is not known whether this species is adapted to regenerate from seed or root crowns following fire.

Hydrilla verticillata (L.f.) Caspary

Common names: hydrilla, water thyme, Florida elodea

Synonymous scientific names: *Hottonia serrata, Hydrilla angustifolia, H. dentata, H. ovalifolia, H. wightii, Leptanthes verticillatus, Serpicula verticillata, Vallisneria verticillata*

Closely related California natives: 0

Closely related California non-natives: 0

Listed: CalEPPC Red Alert; CDFA A

by Kris Godfrey

HOW DO I RECOGNIZE IT?

Distinctive Features

Hydrilla (*Hydrilla verticillata*) is a perennial, submersed aquatic plant, consisting of a series of individual green stems that bear tightly packed whorls of two to eight triangular leaves at each node. It has small leaf scales at the base of each leaf that can be seen with 10x magnification. The spear-shaped leaves are about five to seven times as long as wide, with serrated leaf margins and small spines on the lower surface of the leaf midrib. Hydrilla produces distinctive subterranean vegetative propagules (tubers) and swollen shoots (turions) in its leaf axils (Anderson 1987).

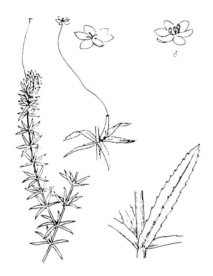

Description

Hydrocharitaceae. Perennial aquatic. Stems: erect and elongate from rhizomes; may branch extensively; produces tuber-like rhizomes that spread horizontally. Leaves: small, oblong or spear-shaped, 0.38-0.75 in (1-2 cm) long and 0.06-0.08 in (1.5-2 mm) wide, arranged in whorls with generally 2-8 leaves per whorl; leaf margin serrate; midrib may have small spines or tooth-like conic bumps. Inflorescence: flowers solitary and inconspicuous. Flowers: there are two biotypes of hydrilla: dioecious (staminate and pistillate flowers on different plants) and monoecious (staminate and pistillate flowers on same plant); both occur in California. Staminate flowers (male) are deciduous, free-floating, perianth in 2 whorls, each with 3 parts; 3 stamens present, 0.12-0.2 in (3-5 mm). The pistillate flowers are persistent, floating, perianth in 2 whorls each with 3 parts, 3 stigmas. Fruit: cylindrical; 0.2-0.5 in (5-13 mm) long, <0.25 in (3-6 mm) in diameter (Hickman 1993).

WHERE WOULD I FIND IT?

Hydrilla is capable of infesting any freshwater aquatic system in California. It has been observed in the Mojave and Colorado deserts, south and central coasts, San Francisco Bay Area, and Central Valley. Currently, isolated infestations of hydrilla are found in Shasta, Yuba, Lake, Calaveras, Madera, Mariposa, and Imperial counties. Typically, it is found in shallow (<11.5-16.5 ft or 3-5 m) water, but if the water is clear enough it may be found growing to depths of forty-eight feet (15 m) (Langeland 1990). It can tolerate some salinity and is sometimes found in upper estuaries. It grows better on mud than on sand. Growth is enhanced in water with agricultural runoff that raises nutrient levels (Parsons 1992).

WHERE DID IT COME FROM AND HOW DOES IT SPREAD?

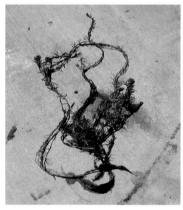

Hydrilla is thought to have been brought into the United States from Eurasia, its native region, as an aquarium plant. It is endemic to Asia, southern Europe, and Africa and has naturalized in the South Pacific and Australia. It escaped from aquaria in the 1950s and 1960s (Balcuinas 1985) and was first found in California in Yuba County in fall 1976 (California Department of Food and Agriculture 1991). Hydrilla spreads mainly by stem fragmentation and sprouting from tubers and turions that break free of parent plants. Monoecious hydrilla, present but rare in the United States, can produce seed (Sutton and Van 1992).

WHAT PROBLEMS DOES IT CAUSE?

Hydrilla forms large mats that fill the water column and can block or severely restrict water flow. Physical blockage reduces recreational quality (e.g., swimming and boating) of infested water systems, crowds out native plants, decreases habitat for fish and other wildlife, degrades water quality, and slows water flow in canals, thereby increasing sedimentation rates and impairing irrigation and drainage (Anderson 1987, Langeland 1990). Hydrilla is easily spread by people (e.g., by fragments on boats or fishing equipment) or by wildlife. Once established, it produces a bank of tubers and turions in the soil that may remain viable for three to five years (Van and Steward 1990, Anderson *et al.* 1992).

HOW DOES IT GROW AND REPRODUCE?

Hydrilla reproduces primarily by vegetative means in the United States. It can reproduce by fragmentation of stems, rhizomes (underground stems), and root crowns, and by the production of tubers and turions. A single viable node can produce stems and rhizomes, leading to production of an independent plant. Tubers are produced from stem tissue beneath the surface of the sediments. Turions are produced on the terminal portion of stems containing leaves (Anderson 1987). Monoecious hydrilla is also capable of producing seed, although seedlings have never been observed in a natural setting in the United States (Anderson 1987).

Although it is a perennial, hydrilla acts like an annual. Dieback of above-ground portions of the plant usually occurs in late fall and winter. In spring, when water temperatures exceed 59 degrees F (15 degrees C), hydrilla begins to grow, producing large amounts of biomass by late summer and early fall (Anderson 1987). Tubers and turions also begin to sprout, forming new plants when sediment temperatures rise (Haller *et al.* 1976). Dioecious and monoecious hydrilla have different growth patterns. Dioecious hydrilla stems elongate rapidly to form a dense canopy in the water column; once the canopy is formed, plants spread horizontally by producing rhizomes. Monoecious hydrilla spreads horizontally across the sediments by producing rhizomes and new root crowns; stems then elongate and form a canopy in the water column (Anderson 1987).

HOW CAN I GET RID OF IT?

Hydrilla is a CDFA A-rated noxious weed and is targeted for eradication whenever it is found in California. By law hydrilla eradication efforts are the responsibility of state and county governments. Suspected infestations of hydrilla should be reported to the local county agricultural commissioner's office. Once its presence has been confirmed, eradication efforts will be coordinated by the CDFA Integrated Pest Control Branch. The first action is to quarantine the infested aquatic system to minimize risk of further spread. Where possible, this includes temporarily closing the area to all traffic (e.g. boaters, campers, hikers). When it is not possible to close the area, an intensive education and inspection effort is undertaken to minimize spread of hydrilla fragments, tubers, or turions.

Methods used to control and eradicate hydrilla vary with the size and condition of the infestation and the environmental sensitivity of the area. If the infestation consists of just a few plants, divers may be used to harvest and dispose of above-ground and below-ground plant parts. Herbicides may be applied to larger infestations or, where possible, the entire system may be drained and dredged. Dredged spoils and the sediments remaining in the system are then fumigated with a soil sterilant. Sterile, triploid grass carp (*Ctenopharyngodon idella*), a hydrilla-eating fish, were used along with harvesting, herbicides, and dredging to eradicate a large infestation in the Imperial Irrigation District. Grass carp are not available for general use in California because of concerns that they could displace native fish. Thus far they have been approved for use only in the Imperial Irrigation District.

Lepidium latifolium L.

Common names: perennial pepperweed; tall white top, broadleaved pepperweed

Synonymous scientific names: none known

Closely related California natives: 15

Closely related non-natives: 4

Listed: CalEPPC A-1; CDFA B

by Ann Howald

HOW DO I RECOGNIZE IT?

Distinctive Features

Perennial pepperweed (*Lepidium latifolium*) is a multi-stemmed herb that grows three to eight feet tall with a heavy, sometimes woody, crown and a spreading underground root system. Stems and leaves are dull gray-green and waxy, sometimes with reddish spots. The tiny white flowers are borne in dense clusters at the tops of the stems. Flowering from May to July, plants produce many small, roundish, light brown fruits. Perennial pepperweed is somewhat similar to whitetop (*Cardaria draba*), but perennial pepperweed is much taller. The upper leaves do not clasp the stem as do those of whitetop.

Description

Brassicaceae. Perennial herb. Roots: deep and spreading. Stems: 1 to many above-ground stems 3-8 ft (1-2.7 m) tall. Stems and leaves glabrous, with gray waxy coating, appearing dull gray-green, sometimes with small reddish spots. Leaves: young plants have petioled leaves 1-2 in (2.5-5 cm) wide and 4-12 in (10-30 cm) long, arising near base of stem; older stems have alternate, sessile leaves, reduced in size upward. Leaf margins smooth or with rounded shallow teeth. Inflorescence: a panicle, 5-6 in (25-27.5 cm) wide. Flowers: white, 0.1 in (3 mm) wide; 4 sepals, white, oval, <1 mm long; 4 petals, white, spatulate, 0.06 in (1.5 mm) long; 6 stamens. Fruit: a silicle, round-ovate, about 2 mm long, with 2 flattened ellipsoid seeds (Robbins 1951).

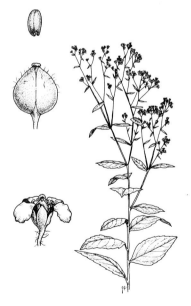

WHERE WOULD I FIND IT?

Perennial pepperweed invades brackish to saline or alkaline wetlands throughout California, from the coast to the interior and north and eastward into the Great Basin and Columbia Basin. It is also found in native (unplanted) hay meadows and as a weed in agricultural fields where the soil is slightly alkaline or saline. It has

been found in all counties in California except Del Norte, Humboldt, and Imperial (Young and Turner, 1995) and is well established in marshes of the San Francisco Bay and Delta (May 1995), including at Benicia State Park. Large infestations occur in Suisun Marsh, especially at Grizzly Island Wildlife Area. It is also common in the Sacramento and San Joaquin valleys, including the Sacramento National Wildlife Refuge, Gray Lodge Wildlife Area, Kesterson National Wildlife Refuge, and Los Banos Wildlife Area. Small infestations occur along roadsides in the Sierra Nevada, such as along State Highway 50 south of Kyburz and at Echo Summit (Howald, pers. observation). It is also found east of the Sierra Nevada in native hay meadows and managed alkaline wetlands, such as at Honey Lake, and in the Owens Valley near Bishop. Its range in southern California is not well documented.

According to observations of wildlife area managers and others, within the last fifteen years perennial pepperweed populations in California have expanded, and the plant has significantly increased its overall range. In California the plant typically grows in full sun in heavy, moist soils that are often saline or alkaline, but it also grows in drier sites and on other soil types. Its precise tolerance limits for aridity, alkalinity, and salinity are unknown. In Wyoming perennial pepperweed is found on soils of high alkalinity (pH 9.2), and it appears to tolerate, but not require, saline conditions. It is found in all western states, and there are large infestations in Nevada.

WHERE DID IT COME FROM AND HOW DOES IT SPREAD?

Perennial pepperweed is native to Eurasia, and it is now found from North Africa north through Europe to Norway and east to the western Himalaya (May, 1995). It has been introduced to Australia, Mexico, and throughout much of the United States. In its native range it grows in a wide variety of habitats, including fresh, brackish, and saltwater wetlands, in and around agricultural fields, in waste places, and even on stony slopes, from sea level to above 10,000 feet (3,049 m) elevation (May, 1995).

The first published record of perennial pepperweed in California is from 1936, when it was collected on a ranch north of Oakdale in Stanislaus County (Bellue 1936). Robbins *et al.* (1951) state that it may have been introduced to California as a contaminant of sugar beet seed, although no evidence is presented to support this. Recent localized infestations along State Highway 50 may have been initiated from seed or plant fragments that were contaminants in rice straw bales, since these infestations are found in areas of recent construction where straw bales were used for erosion control (Howald pers. observation).

Perennial pepperweed can be distributed by seeds or by pieces of the underground stems. The small seeds have no special adaptations for long-distance dispersal. They are capable of being transported by wind, water, and possibly waterfowl.

WHAT PROBLEMS DOES IT CAUSE?

Perennial pepperweed is an aggressive invader of coastal and interior wetlands throughout

California. It forms dense monospecific stands that exclude other plants, including natives (Corliss 1993, Trumbo 1994). At Grizzly Island Wildlife Area in Suisun Marsh it is encroaching on several rare plant populations, including soft bird's-beak (*Cordylanthus mollis* ssp. *mollis*), Suisun Marsh thistle (*Cirsium hydrophilum* var. *hydrophilum*), and Suisun Marsh aster (*Aster lentus*) (Skinner and Pavlik 1994). In most areas it prefers habitat slightly higher than that dominated by pickleweed (*Salicornia* spp.), but it has invaded *Salicornia*-dominated marshes in the Alviso Slough area (May, 1995), and thus poses a threat to the habitat of the endangered salt marsh harvest mouse, California black rail, and California clapper rail. In waterfowl

nesting areas it outcompetes grasses that provide food for waterfowl. It is also an aggressive invader of some agricultural lands in the Central Valley and east of the Sierra Nevada. In Lassen County it has become widely established in native (unplanted) hay meadows, reducing the value of the hay crop (Young, pers. comm.).

HOW DOES IT GROW AND REPRODUCE?

Perennial pepperweed reproduces from seed, as well as vegetatively from intact root systems or from pieces of rootstock. Flowering time varies from May to July in different parts of California. Peak bloom lasts for several weeks. Seeds mature by June or July. Each mature plant has the capacity to produce thousands of seeds each year. Seeds typically germinate in spring in wet sand or mud. Germination studies (Robbins *et al.* 1951, Miller *et al.* 1986) have shown high germination rates (64-100 percent) under a variety of conditions. Miller *et al.* (1986) found the highest rates occurred under alternating temperature regimes yielding germination rates as high as 64 to 96 percent in laboratory studies.

Seedlings grow rapidly and can produce flowering stems the first year. In fall and winter aerial stems die back to the ground, creating a thick thatch of dead stems in heavily infested areas (Young *et al.* 1995). In early spring new shoots begin to form from the rootstocks. A single intact root crown can produce several flowering stems. New plants readily grow from pieces of root-stock less than one-third of an inch (0.8 cm) in diameter and less than one inch (2.5 cm) long (Wotring *et al.* 1997).

HOW CAN I GET RID OF IT?
Physical Control

Manual/mechanical methods: Mechanical methods are unlikely to control perennial pepperweed because new plants quickly regenerate from pieces of rootstock left in the soil (Young *et al.* 1995). Segments much shorter than one inch (2.5 cm) are capable of resprouting. Disking of perennial pepperweed at Grizzly Island Wildlife Area resulted in a significant increase in distribution (Feliz, pers. comm.). Young *et al.* (in press) attempted to control *Lepidium latifolium* in native hay meadows near Honey Lake, Lassen County, in tillage experiments conducted from 1991 to 1992, using monthly disking throughout the growing season. They concluded that this treatment resulted in no permanent reduction in perennial pepperweed cover; the year following disking perennial pepperweed reestablished approximately 100 percent cover.

Prescribed burning: Experiments at Malheur National Wildlife Refuge in southern Oregon indicate that fire alone is unlikely to be effective in controlling *L. latifolium*, in part because typical fuel loads in infestations of this plant are inadequate to sustain burns.

Inundation: Perennial pepperweed may be intolerant of prolonged inundation. At West Navy Marsh in Contra Costa County, perennial pepperweed distribution and abundance were significantly reduced after a diked marsh was returned to tidal action, increasing inundation time (May, 1995). Young and Turner (1995) report that perennial pepperweed does not appear to survive lengthy periods of flooding during the growing season.

Biological Control

Insects and fungi: Development of a biological control program seems unlikely because of risks to many important crop plants that are members of the mustard family (Brassicaceae) (Young *et al.* 1995, Birdsall *et al.* 1997). Additionally, several native *Lepidium* species from the western United States are either listed as endangered or are being considered for listing (Young *et al.* 1995). Fifteen species of *Lepidium* are native to California, including four that are considered rare and endangered by the California Native Plant Society (Skinner and Pavlik 1994). Acknowledging these difficulties, Birdsall *et al.* (1997) point out the limitations of herbicidal control and suggest that *L. latifolium*-specific biocontrol agents, either insects or fungi, be sought in the many European countries with other native *Lepidium* species.

Chemical Control

Attempts have been made to control perennial pepperweed with chemical herbicides in California, Oregon, Wyoming, Idaho, and Utah. The most effective herbicides appear to be chlorsulfuron (as Telar®), metsulfuron methyl (as Escort®), and imazapyr (as Arsenal®) based on field trials of one to four years (Cox 1997). Neither Escort® nor Arsenal® is registered for use in California at this time.

Trumbo (1994) reports that tests of chlorsulfuron, triclopyr, and glyphosate at Grizzly Island Wildlife Area in Suisun Marsh, California, showed that each of these compounds can provide significant control of perennial pepperweed. Chlorsulfuron (as Telar®) was most effective, with one application resulting in a reduction in cover of more than 95 percent after two years. Telar® was applied at the recommended rate of 0.75-1 oz/acre, mixed in 30 gallons water with 0.5 percent non-ionic surfactant. It is selective against broadleaved plants. This was advantageous at Grizzly Island Wildlife Area because desirable grasses were not affected. After the initial test, large-scale use of Telar® at Grizzly Island Wildlife Area has confirmed its effectiveness; however, retreatment may be necessary because of the regenerative ability of perennial pepperweed. Telar exhibits some residual soil activity, and its use is not permitted near water.

Triclopyr as Garlon3A® and Garlon4® provided moderate to good control after one year in tests at Grizzly Island Wildlife Area. Garlon3A® was applied as a 2 percent solution with 0.5 percent non-ionic surfactant added. Garlon4® was applied as a 1.5 percent solution with 0.5 percent non-ionic surfactant added. Currently, neither formulation of Garlon® is registered for use over water in California. Triclopyr is broadleaf-specific, so it generally does not affect grasses. Garlon4® does not show residual soil activity. As with chlorsufuron, retreatment may be needed to maximize control.

Glyphosate as Rodeo® and Roundup® provided fair to moderate control after one year in tests at Grizzly Island Wildlife Area. Roundup® was effective as a 2 percent solution. Rodeo® was also used as a 2 percent solution with the addition of 0.5 percent non-ionic surfactant. Rodeo® can be used over water, but Roundup cannot. Roundup® and Rodeo® are broad-spectrum herbicides that control most plants, including grasses. At Grizzly Island Wildlife Area resprouting of pepperweed the year following treatment indicated that several follow-up treatments likely are needed for full control.

In Lassen County, California, Young et al. (in press) tested the effectiveness of 2,4-D, glyphosate, and chlorsulfuron against perennial pepperweed. They found that, while 2,4-D and glyphosate greatly reduced top growth and eliminated seed production in the year of application, they provided no permanent control, since cover returned to 100 percent by the second year after application of these compounds. One application of chlorsulfuron provided up to three years of nearly complete control of perennial pepperweed.

In Nevada, Young and others (1997) found that chlorsulfuron is effective in controlling perennial pepperweed. The highest level of control was obtained from applications during the bud stage. However, in the native hay meadows where the studies were conducted, excellent control was possible with early spring or late fall applications.

At Malheur National Wildlife Refuge in Oregon, chlorsulfuron and metsulfuron methyl were tested alone and in combination with either fire or disking. The herbicides were more effective when used alone, with chlorsulfuron reducing *L. latifolium* densities by 100 percent in all three sites tested, and metsulfuron methyl resulting in density reductions of 90 to 100 percent.

In Idaho herbicides used to control perennial pepperweed include metsulfuron methyl, 2,4-D, dicamba (as Vanquish®), imazapyr, chlorsurfuron, and picloram (Cox 1997). Metsulfuron methyl is the most commonly used and is described as "quite effective." Other compounds noted as "successful" in controlling this species include imazapyr and chlorsulfuron (Cox 1997).

In Wyoming metsulfuron methyl and chlorsulfuron proved most effective in controlling perennial pepperweed. Either compound, used at the recommended rate of 0.75-1.0 oz/acre, resulted in stand reductions of 90 percent or more that persisted for four to five years.

Leucanthemum vulgare Lam.

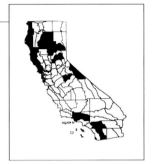

Common names: ox-eye daisy, marguerite, moon daisy, dog daisy

Synonymous scientific name: *Chrysanthemum leucanthemum*

Closely related California natives: 1

Closely related California non-native: 7

Listed: CalEPPC red alert; CDFA nl

by Maria Alvarez

HOW DO I RECOGNIZE IT?

Distinctive Features

Ox-eye daisy *(Leucanthemum vulgare)* is a prostrate herb with stems that sprout laterally from a creeping rootstock. When in flower, the plant's height ranges from one to three feet. The white-petaled flower-like inflorescences have yellow centers. Leaves are dark green on both sides, one to two inches long, smooth, and pinnately lobed or toothed. The number of flower stalks ranges from one to forty per plant.

Description

Asteraceae. Perennial herb. Stems: simple or branched and prostrate. Flower stems 1-3 ft (30-90 cm) tall. Leaves: entire to pinnately lobed or toothed along stems. Lower basal and middle leaves longer, <5 in (<12 cm), obovate to spoon-shaped, upper leaves borne along a stem, more oblong, sessile, and shorter. Petioles wingless. Inflorescence: each solitary flowerhead composed of numerous yellow disc flowers +/- 0.1 in (2.5 mm) and about 20 white ray flowers 0.7-0.8 in (18-20 mm) long. Fruit: flat seed 0.08 in (2 mm) long, 10-ribbed, dark gray at maturity with no pappus. Up to 200 seeds per flowerhead. Number of flower stalks ranges from 1-40 per plant (description from Hickman 1993, Anderson 1987).

WHERE WOULD I FIND IT?

Ox-eye daisy is found in both the North Coast Range and northern Sierra Nevada from sea level bluffs and canyons to "alpine" mountain meadows to 7,000 feet (2200 m) and from central

California into Oregon. It is also common from the northeastern seaboard through the Midwest. Ox-eye daisy is also a problem at Rocky Mountain National Park (USGS 1999). It is a common weed of disturbed areas such as roadsides, fields, and pastures and former homesteads (Cowell 1973, Peck 1993). It readily spreads into wildlands. Ox-eye daisy is found in a variety of plant communities including, prairie, scrub, wet meadows, riparian forests, and open-canopy forests. It thrives in a wide range of conditions

and in full sun to semi-shade. Plants are shallowly rooted to three inches (7.5 cm) deep and tolerate a wide range of soil moisture conditions, but do particularly well in soils that are heavy and damp (Parsons 1992).

WHERE DID IT COME FROM AND HOW DOES IT SPREAD?

Ox-eye daisy is native to Europe (Polunin 1969). It was probably introduced to North America as an ornamental early in the twentieth century. It is currently used as an ornamental, and is it often sold commercially in seed packets labeled as wildflower seed. Ox-eye daisy spreads through abundant seed production and vegetatively by rooting underground stems (rhizomes) (Griswold 1985). Seeds have no special adaptations to aid dispersal, but are small and fall to the ground up to two meters from the parent plant. When the flowerheads are dry, the seeds drop or are shattered easily by touch or movement. Water, human and animal foot traffic, and cultivating and earth-moving machinery can carry seeds into new areas.

WHAT PROBLEMS DOES IT CAUSE?

Ox-eye daisy displaces native plant species, growing so densely it excludes other vegetation. It is not known to be used as forage by animals in California. While not considered poisonous to cows, it does impart a disagreeable taste to their milk. Ox-eye daisy is a host for several viral diseases affecting crops, including the yellow dwarf virus of potatoes (Parsons 1992). It is difficult to control or eradicate because of its large seedbank, long viability of seed, and ability to resprout if not completely removed.

HOW DOES IT GROW AND REPRODUCE?

Ox-eye daisy is capable of reproduction the first summer after it becomes established, regardless of plant size. Plants one inch in diameter have been observed bearing a single flower. Stem growth is prostrate and creeping until development of erect flowering stalks one to three feet (30-90 cm) tall. Flowering commences in late spring (May) and continues until late summer (August). Seed production is prolific when water is adequate. Most ox-eye daisy seeds remain viable for twenty years in the soil, and can remain viable after passing through digestive tracts of animals (Parsons 1992). Seeds germinate continuously as long as there is adequate moisture, fall through late spring in coastal regions. Plant growth slows during periods of flowering and low water availability.

Seed germination is inhibited by continuous darkness but otherwise not affected by variation in light (Thompson 1989). Studies have indicated that ox-eye daisy seedling germination and frequency are greater under increased moisture in hollows versus ridges, but dense groundcover can prevent ox-eye daisy establishment (Reader 1991). Ox-eye daisy can grow year-round, and its lifespan is indeterminate. Maximum growth has been observed in coastal regions at onset of cool fall weather, through winter and spring, just before flowering stalks shoot up. Aerial growth dies back after seed release. Vegetative growth slows in summer during and after flowering (Cowell 1973).

HOW CAN I GET RID OF IT?

Little information has been published on mechanical, cultural, or biological control of ox-eye daisy. An important consideration is that seeds remain viable in soil for at least two years.

The first step in the control of ox-eye daisy is to develop a containment strategy. Removal methods will depend on environmental variables and the type of plant community infested. Primary methods of dispersal besides seed drop should be identified. If plants are growing along trails, shoes and hooves picking up seeds in wet soil may disperse them. People also pick the bright flowerheads, from which ripe seeds may fall as they hike through the region. Information about the daisies should be posted to alert wildland users. Small outlying populations should be treated first.

Complete eradication of a large, well established, and geographically widespread population of ox-eye daisies can be difficult because of their small size and abundant seed production.

Prolific seed set and the ability of rhizomes to resprout make successful removal dependent on appropriately timed treatment and persistent follow-up. Removal sites should be inspected before plants have set a new crop of seed in June. If the infestation is small it may be difficult to locate the previous year's removal site, so the site should be mapped and marked with colored flagging or pin flags, especially if follow-up will be done by someone else. If plants are mulched, the mulch will serve as an effective indication of the location of the infestation. It is much easier to locate daisies after flowering begins, which is typically by mid-June in coastal California populations.

Physical Control

Manual methods: A combination of hand removal and mulching is used to control ox-eye daisy in the Golden Gate National Recreation Area (GGNRA). If the infestation is small (less than 0.25 acre) or widely scattered, hand removal may be efficient. Using a small hand pick, chip around the base of the plant several inches deep to loosen the plant. Then lift the entire plant out intact without leaving any stem pieces (rhizomes) behind. Check for rhizome fragments, since an entire plant can regenerate from them. A round-point shovel is effective for scooping out whole plants. If the soil is flat and compacted, a sharp garden spade can also be used to scrape the plant out of the soil. A hula hoe is also handy for scraping away abundant masses of seedlings or small plants.

Mechanical methods: In Australia shallow cultivation of less than six inches (15 cm) was found to have little effect and was likely to spread roots. Cultivation greater than six inches in summer exposes roots to desiccation. Subsequent shallower cultivations kill seedlings. This technique opens the soil to infestation by other weeds and must be combined with dense revegetation with desirable seed (Parsons 1992).

Mulching: The most successful non-chemical method found for removing large infestations in GGNRA is to mulch heavily. Habitat Restoration volunteers at the GGNRA have successfully eliminated masses of mature and immature plants through the application of rice straw. One application 3-4 in (7.5-10 cm) thick when compacted was successful in two plant communities in the Marin Headlands: coastal scrub and wetland. Straw should be applied in fall at the onset of the growing season. One bale will cover approximately 100 square feet. The site should be monitored in early spring. If any live plants are found under the straw, or any light can reach the soil, then another thick layer should be applied before flowering begins in May. Native perennial plants at the Marin Headlands site came up through the straw, while the ox-eye daisy did not. This is because ox-eye daisy is a prostrate plant except for its flower stalks. Ox-eye daisy was observed to rot under the dense mulch maintained throughout the winter. If the infested area has a lot of woody plants they can make it difficult to mulch thickly and lopping or brush cutting may need to be undertaken first. Certified rice straw was used to avoid introduction of terrestrial weed seeds. Other mulches have not been tried. Wood chips might also be effective if they are applied thickly enough.

Winter monitoring is critical for mulched treatments in order to assess the condition of the mulch before growth surges in spring. If ox-eye daisies are seen growing through the mulch, it may have to be applied again. Check to see if the plants are seedlings or adults. When mulch is adequately applied the first time seedlings should not grow through it. Humans or animals may have passed through the area and disturbed the mulch. Where mulch is thinning, it should be re-applied, especially if there is a month or more of wet weather to come, or if the site is a wet habitat. Once a mature population is removed from the area, a crop of seedlings will take their place. Therefore, the length of time the area should remain mulched depends on the size of the seedbank and longevity of ox-eye daisy seeds in it. Along edges that are difficult to mulch, spot removal can be done by hand.

Prescribed burning: This approach has not been assessed for ox-eye daisy.

Biological Control

Insects and fungi: Biological control has not been investigated for this species.

Grazing: Intensive cattle grazing is an effective control for ox-eye daisy. Although cattle tend to avoid it because of its high acidity, under high stock density in an intensive grazing system, cattle eat this species (Wallander *et al.* 1991).

Chemical Control

Picloram, imazapyr, sulfometuron methyl, and dicamba are effective at label concentration when applied in the early flowering stages, but these herbicides persist in the soil (Parsons 1992). Ox-eye daisy is moderately resistant to MCPA, 2,4-D, and dicamba (Stubbendieck *et al.* 1992), and these herbicides may damage non-target species.

Lupinus arboreus Sims

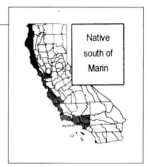

Native south of Marin

Common names: yellow bush lupine, coastal bush lupine

Synonymous scientific names: *Lupinus eximius, L. macrocarpus, L. propinquus, L. rivular*

Closely related California natives: 70

Closely related California non-natives: 0

Listed: CalEPPC A-2; CDFA nl

by Andrea J. Pickart

HOW DO I RECOGNIZE IT?

Distinctive Features

Yellow bush lupine (*Lupinus arboreus*) is a bushy shrub to six feet tall, usually with bright yellow (sometimes blue) sweet-smelling flowers and green, sparsely pubescent (appearing glabrous), palmately compound leaves. It occurs as an invasive species in northern California coastal dunes. It appears to be native to California from Sonoma County south, where plants often have a hairy upper leaf surface. Yellow bush lupine is predominantly a dune species, but it can be found along roadsides and in disturbed areas. Flowers appear in May to July in northern California (earlier in the southern part of its range). Yellow bush lupine hybridizes with the native *L. littoralis*, which is smaller (less than two feet) and more prostrate-decumbent, has purple and white flowers, and always has a hairy upper leaf surface. Intermediates usually are smaller and more pros-

trate than *L. arboreus*, with blended yellow, purple, and white flowers (Wear 1998). In the southern portion of its range yellow bush lupine can be easily distinguished from blue bush lupine (*L. chamissonis*), which can achieve the same height but has silver, densely hairy leaves that appear gray-blue and light-violet to blue flowers.

Description

Fabaceae. Bushy shrub to 6 ft (1.8 m) tall. Leaves: cauline with stipules 0.2-0.5 in (8-12 mm), petiole 0.8-1.2 in (2-3 cm). Each leaf palmately compound with 5-12 leaflets, 1-3 in (25-75 mm) long. Foliage sparsely pubescent, appearing green-glabrous in northern California and glabrous to hairy in southern portion of range. Inflorescence: panicle 4-12-in (10-30-cm) long. Flowers: bright yellow (rarely lilac to purple), pea-shaped flowers 0.6-0.7 in (14-18 mm). Calyx upper lip two-toothed, 0.2-0.4 in (5-9 mm), lower lip entire, 0.2-0.3 in (5-7 mm). Upper keel margin ciliate from claw to tip, and lower keel margin glabrous, as is banner back. Fruit: brown-black pods, hairy and 1.6-2.8 in (4-7 cm) long. There are 8-12 seeds/pod, each 0.16-0.2 in (4-5 mm), black to tan or striped lighter (Hickman 1993).

WHERE WOULD I FIND IT?

Yellow bush lupine occurs from the mouth of the Ventura River in California northward to at least Vancouver Island, British Columbia. It is primarily a coastal plant, but has been found inland east of Berkeley, California (Sholars, pers. comm.). Yellow bush lupine is native in the southern part of its current range, although the demarcation between native and naturalized populations is still disputed. Davy (1902) reported its range as Point Reyes south. Hickman (1993) suggests that populations as far north as Bodega Bay are native.

Naturalization of yellow bush lupine and subsequent gene flow within the genus has resulted in taxonomic and range confusion. Under the current classification (Hickman 1993), there are both purple and yellow forms of *Lupinus arboreus*. For the purposes of identifying yellow bush lupine as an invasive weed in Humboldt County and north, the problem is simplified by the fact that the invasive form has yellow flowers. However, there are intermediate hybrids between the yellow-flowered form and the purple-flowered dune native *L. littoralis* (found from Sonoma County to British Columbia). Where *L. arboreus* occurs off dunes (e.g., along roadsides), it also hybridizes with the purple-flowered *L. rivularis* (found from Mendocino County to British Columbia). Yellow bush lupine is not yet present on the Ten Mile Dunes at MacKerricher State Park in Mendocino County. However, a blue-flowered, woody species identified by by Teresa Sholars as *L. arboreus* is common at Manchester Dunes in Mendocino County. These dunes have already been extensively invaded by *Ammophila arenaria*, and it is not clear whether yellow bush lupine is increasing at this site.

WHERE DID IT COME FROM AND HOW DOES IT SPREAD?

Yellow bush lupine is native to southern and central California. It was introduced repeatedly to many dune systems as a sand stabilizer during the early to mid-1900s. The introduction of yellow bush lupine to the Humboldt Bay dune system was traced by Miller (1988). In 1908 the operator of a fog signal station on the north spit of Humboldt Bay gathered seeds of yellow bush lupine from the Presidio (where it had previously been introduced) and planted them around the station. In 1917 seeds from the new signal station population were collected and scattered beside railroad tracks along the spit. From these and later plantings, the extent of yellow bush lupine has increased from 244 acres (98 ha) in 1939 to over 1,000 acres (400 ha) (Pickart and Sawyer 1998). Yellow bush lupine now dominates 28 percent of the total vegetation cover on Humboldt Bay dunes (Pickart and Sawyer 1998).

As do other members of the genus, *Lupinus arboreus* has relatively large seeds with corresponding high seedling survival. Once a population becomes established, it spreads short distances by rodents or by seeds rolling from parent plants down dune slopes.

WHAT PROBLEMS DOES IT CAUSE?

Yellow bush lupine invades coastal dunes in northern California, where no other large, shrubby, native lupines are found (although *Lupinus littoralis* is sometimes classified as a subshrub). The seeds of yellow bush lupine are long-lived and form a persistent seedbank, creating the need for repeated removal. However, more serious problems are caused by yellow bush lupine's ability to cause ecosystem-level changes. As a nitrogen-fixer, bush lupine readily colonizes the open, mat-like vegetation of northern California dunes (known as dune mat, or the Sand-Verbena/Beach Bursage series). Once the lupine has been present for more than a few years, it causes elevated nitrogen levels that facilitate invasion by non-native weedy grasses (Pickart *et al.* 1998).

Maron and Connors (1996) also observed this phenomenon on coastal prairies at Bodega Bay, where yellow bush lupine is believed by some botanists to be native (there is a relict native dune scrub community containing *Lupinus chamissonis* at Bodega Bay). Eventually, desirable native species in invaded areas are almost entirely displaced by a combination of lupine shrubs, weedy grasses, and/or adventive natives such as *Scrophularia californica*. This assemblage of species has been labeled the Yellow Bush Lupine series by Sawyer and Keeler-Wolf (1995). Eventually, the Yellow Bush Lupine series may give way to the Coyote Brush series, which otherwise would not occur on northern California dunes. In both of these situations, plant cover is much higher than that of the native dune community, so the dune system becomes overstabilized. As a result, sand is not able to move from the foredunes to the backdunes, and physical processes are disrupted. This can result in elongation of the deflation plain behind the foredunes and/or stabilization of the backdunes. Although dunes are naturally subject to cyclic stabilization and rejuvenation in response to major tectonic events, exotic species such as yellow bush lupine and European beachgrass (*Ammophila arenaria*) can greatly accelerate stabilization and could conceivably replace dune mat altogether in the post-disturbance stage of the cycle.

HOW DOES IT GROW AND REPRODUCE?

Yellow bush lupine reproduces solely by seed. It is self-compatible and is pollinated by bumble bees (Wear, pers. comm.). Flowering takes place from May to July in northern California, and seed dispersal occurs in late summer and fall. Pods open explosively, propelling seeds for short distances. Seeds are characterized by a hard seed coat, typical of the genus. However,

seed coat dormancy is overcome in the dune environment, probably by the abrasion caused by sand and wind, and seedlings emerge from November to March (Pickart and Sawyer 1998).

Plants grow rapidly, reaching reproductive maturity in one to two years. Yellow bush lupine plants are short-lived for their size, generally living seven years or less (Davidson and Barbour 1977).

HOW CAN I GET RID OF IT?

Control of yellow bush lupine has so far been limited to manual and mechanical means. The cyclic die-offs observed in both Humboldt Bay and Bodega Bay populations suggest that biological control may be effective, but more research is needed. Chemical control has not been attempted. The observed effects of wildfires in the dunes indicate that yellow bush lupine can resprout after burning, although a very hot fire might kill these plants. Regardless of the method of removal, additional steps may be needed to restore soils to a condition suitable for native vegetation.

Physical Control

Manual methods: If yellow bush lupine has been present only a short time, and native vegetation is still intact beneath the shrubs, restoration of dune mat can be accomplished by cutting mature lupines at the base of the trunk and splitting the trunk to discourage resprouting. Alternatively, a weed wrench can be used to remove the root intact, but this may prove awkward on steep, sandy slopes. Small plants can be pulled by hand. Plants should be removed prior to seed set. It may be easier to detect plants during flowering, but in a large population this could have detrimental impacts on pollinators. Plants are usually piled on a nearby bare area and burned after a few weeks of drying. Treatment will need to be repeated for as many years as plants emerge from the seedbank. To ensure that all plants have been removed, recheck the area during the flowering period, when smaller plants are readily seen.

If yellow bush lupine has been present long enough to alter soils, non-native grasses or other plants not normally found in dune mat will be present, and a duff layer will have accumulated. In this case lupine can be removed using the method described above, but this action will not be sufficient to restore dunes. In addition to removing lupine, all associated non-native or adventive native plants must be removed, and the duff layer should be scraped off to reveal the mineral soil (Pickart *et al.* 1998a). This treatment will need to be repeated for up to four years. The disturbance caused by this treatment stimulates germination from the seedbank, so bush lupine and other weeds in the seedbank will be depleted sooner than if bush lupine alone were being removed. Depending on the amount of remnant native vegetation, native species may need to be reintroduced.

Mechanical methods: Dune areas invaded by yellow bush lupine have been restored using heavy equipment in small-scale experiments at Humboldt Bay dunes (Pickart *et al.* 1998b). Areas in which few native species remain, and that are relatively flat and accessible, are suited to this treatment. In the experiments, all vegetation was removed using a brush rake, followed by scraping off the duff layer with a plough blade. Weed mat was then placed over the soil surface, fastened with staples, and left for two years. One year after weed mat removal, recolonization by lupine and other non-natives was low. The weed mat apparently killed seeds in the seedbank. Revegetation with native species is essential for this treatment.

Biological Control

No USDA approved biological agents are known for control of yellow bush lupine. However, at Bodega Bay, where the species is believed by some botanists to be native, some coastal prairie populations of *Lupinus arboreus* have been found to be cyclic in nature, with cover increasing from zero to as much as 60 percent, then plunging back to zero within three to ten years (Strong *et al.* 1995a). Population fluctuations have been linked to herbivory by subterranean ghost moths (*Hepialus californicus*), whose populations are in turn controlled by entomopathogenic nematodes (Strong *et al.* 1995b). In naturalized populations at Humboldt Bay, similar cyclic die-offs have been observed. No ghost moths have been detected, but many other herbivores, as well as fungal pathogens, are present. In 1996-97, during a major dieback, the "VGC-2" strain of the fungus *Colletotrichum gloeosporoides*, was collected from affected plants. This fungus has caused significant mortality in cultivated yellow bush lupine in New Zealand (Dick 1994).

Chemical Control

No chemical control techniques have been investigated for this species.

Lythrum salicaria L.

Common name: purple loosestrife

Synonymous scientific names: none known

Closely related California natives: 1

Closely related California non-natives: 0

Listed: CalEPPC B; CDFA; noxious by 16 states

by Carri Benefield

HOW DO I RECOGNIZE IT?

Distinctive Features

Purple loosestrife (*Lythrum salicaria*) is a perennial wetland herb, typically less than five feet tall with showy spikes of reddish purple flowers. Each flower has five to seven pink-purple petals surrounding a small yellow center. Plants enlarge at their bases each year with more stems, sometimes becoming a rounded bush-like clump of thirty to fifty stems arising from a single root stock. Leaves are long and narrow, opposite to whorled, and closely attached to the four-sided stem.

Purple loosestrife may be distinguished from similar looking species by its generally larger size, opposite to whorled leaves, and less widely spaced flowers. It may be confused with other native species such as fireweeds, vervain, or the closely related *Lythrum californicum*, so accurate identification should be made before control measures are attempted. Fireweed (*Epilobium angustifolium*) has spikes of magenta flowers, but these spikes are narrower (four to five inches wide at the base) and more conical. Fireweed stems are round, and the leaves are alternate. Blue vervain (*Verbena hastata*) has smaller purple flower spikes, and the edges of the leaves are toothed. *L. californicum* has smaller, narrower leaves and half (six versus twelve) the number of stamens.

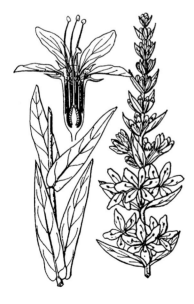

Description

Lythraceae. Perennial herb. Stems: 1-5 ft (0.3-1.5 m), occasionally to 10 ft (3 m) tall, erect stems; four-sided, smooth to lightly hairy, woody with age, persisting through winter. Leaves: opposite or whorled, 2-6 in (5-14 cm) long, generally two-ranked, lanceolate to narrowly oblong, edges smooth, sessile (Hickman 1993). Inflorescence: terminal, spike-like, composed of numerous flowers closely arranged in leaf axils of ascending flower-bearing branches (Rawinski *et al.* 1982). Flower: trimorphic in regard to the relative lengths of stamens and style; hypanthium cylindric; sepals <0.5 in (4-5 mm), deltate; corolla bright red-purple with 5-7 petals <0.5 in (8-14 mm); stamens generally 12, included or exerted. Fruit: dark brown seed capsules containing numerous small seeds <0.05 in (1 mm) (Hickman 1993).

WHERE WOULD I FIND IT?

Purple loosestrife can be found in scattered freshwater wetlands in northern and central California. Counties with infestations include Humboldt, Mendocino, Modoc, Shasta, Siskiyou, and counties in the Sacramento Valley and San Francisco Bay Area. Other infestations occur along rivers in Fresno and Kern counties, such as along the Kern River in the southern Sierra. It is also widespread in the eastern United States. It is common in disturbed wetland habitats, such as stream and river banks, edges of ponds, lakes, and reservoirs, flooded areas, ditches and road-sides, but it can colonize fairly pristine wetland areas, including marshes, wet prairies, meadows, pastures, and bogs. While loosestrife generally grows best in full sun in moist, organic soils, it can tolerate a wide range of soil moisture regimes (intermittent or continuous flooding up to 1.25 feet or 0.4 m), soil types (clay, sand, silt, and muck; pH 4-9), nutrient levels, temperature, and light (up to 50 percent shade) (Skinner *et al.* 1994). If water levels drop, loosestrife can easily become established, and populations can expand on exposed soils.

WHERE DID IT COME FROM AND HOW DOES IT SPREAD?

Purple loosestrife is native to Eurasia and was introduced into the northeastern United States in the early 1800s in ships' ballast, as an herbal, as an ornamental plant, and by bee-keepers. It spreads primarily by seed, but can grow from broken-off stem and root segments and can resprout from the root crown (Bender *et al.* 1987). Seed dispersal occurs primarily by wind and water (most rapidly by flowing water, but also by permanent standing water), and in mud attached to wildlife, boats, vehicle tires, and footwear (Heidorn 1990, Skinner *et al.* 1994).

WHAT PROBLEMS DOES IT CAUSE?

Purple loosestrife is a hardy perennial that can rapidly degrade wetlands, diminishing their value for wildlife habitat. Once established, it forms extensive monotypic stands that displace native vegetation relied on by wetland species for food and habitat (Nelson and Getsinger 1994, Bender 1987). Infestations jeopardize the federally listed endangered bog turtle (*Clemmys muhlenbergi*) among other species, though not in California. Purple loosestrife also clogs water-ways and wetlands used for boating and other recreational activities. Impacts on agriculture include changes in hydrology and soil conditions of wetland pastures and meadows and clogged irrigation systems. Purple loosestrife has the potential to infest rice fields. An estimated 190,000 hectares of wetlands, marshes, pastures, and riparian meadows are affected in North America each year, with an economic impact of millions of dollars.

HOW DOES IT GROW AND REPRODUCE?

Purple loosestrife flowers from late June through September. Seed set begins in mid-July and continues into late summer. New shoots arise the following spring from buds at the top of the rootstocks. Long-lived seeds are copiously produced (more than two million seeds per ma-ture plant) and shed gradually from capsules through the winter, contributing to an immense seedbank. It has not been determined how long purple loosestrife seeds remain viable, but it is thought to be at least several years. Most seedlings establish in late spring to early summer and produce a floral shoot up to one foot (30 cm) high the first year. Moisture is the most critical

habitat requirement for seed germination. Critical temperatures for germination range from 60 to 72 degrees F (15-20 degrees C) (Skinner *et al.* 1994).

HOW CAN I GET RID OF IT?

The best method for removal of purple loosestrife depends on the site, the age and size of the infestation, the importance of impacts to non-targeted species, and the type and quantity of resources available to gain control. Ideally, an integrated pest management (IPM) approach should be developed to take advantage of the best available control methods. All methods require appropriate timing and follow-up monitoring to be successful.

In natural communities with no known invasion, prevention and early detection are the best approaches. Potential habitat should be searched annually for newly established populations in late July through August, when blooming plants are easily spotted from a distance. In addition, public education efforts and avoidance of unnecessary disturbances of wetland habitat are desirable (Heidorn 1990, Mountain 1994).

Physical Control

Manual methods: In areas with individual plants and small localized stands (up to 100 plants), younger plants (one to two years old) can be pulled by hand. Plants should not be pulled after flowering, because this will scatter seed. Older plants, particularly those in loose soils, can be dug out by teasing roots loose with a cultivator, but great care should be taken to bag and re-move all plant parts from the site because broken-off pieces can re-root (Heidorn 1990, Skinner *et al.* 1994). Dispose of plants by burning (preferable) or by drying.

Several methods have been shown to be ineffective and should be avoided: Cutting at water level and below water level has resulted in resprouting and reestablishment. Mowing, if timed correctly, has been shown to prevent flowering and seed set and reduce carbohydrate reserves, but it can result in dissemination of seeds and re-rooting of stem fragments (Heidorn 1990). Flooding is ineffective unless plants can be inundated throughout the summer, but this is detri-mental to most other emergent species and promotes spread of loosestrife to shallow areas (Skinner *et al.* 1994). Disking and flooding has shown some success, but most areas cannot be drained and dried enough to support heavy equipment (Skinner *et al.* 1994). Burning does not kill buried rootstocks, which resprout later. Herbicide plus burning was attempted without success (Skin-ner *et al.* 1994).

Biological Control

Insects and fungi: Since the 1980s the U.S. Fish and Wildlife Service and the International Institute of Biocontrol have been screening, testing, and importing natural purple loosestrife enemies. In the early 1990s the USDA and Washington and Oregon state departments of agri-culture gained approval for the release of five agents, three weevils and two leaf eating beetles (*Galerucella* spp.). California required extra testing (for potential impacts on crape myrtle, a widely used ornamental) before obtaining a permit in 1998 to conduct test releases of leaf-eating beetles in California. Considerable damage to purple loosestrife from the leaf-eating beetles has been reported in other states and in California test release sites. However, further research is necessary before approval of large-scale release of *Galerucella* spp. in California. Other states have reported that it can take four to six years for *Galerucella* spp. to become fully established.

Plant competition: Research on reseeding with competitive species is still in its early stages. It has been difficult to find species that compete successfully with purple loosestrife in a wide variety of conditions and that are not themselves invasive. Japanese millet plantings in Minne-sota have been successful in the short term, but this non-native species regenerates poorly and must be replanted every year (Bender *et al.* 1987).

Chemical Control

In areas up to four acres with clusters in excess of 100 plants, the most effective herbicides available are glyphosate (as Roundup® and Rodeo®) applied in spot applications. Glyphosate is non-selective, so care should be taken to avoid contact with non-targeted species that are critical in recolonizing the site and inhibiting loosestrife reestablishment (Bender 1987, Skinner *et al.* 1994). Herbicides should be applied to the foliage during the blooming season (late June to early September). Effective applications have been reported at both flower initiation (usually late June) (Heidorn 1990, Nelson and Getsinger 1994) and after peak bloom (usually late August)

(Balogh 1986, Rawinski 1982). Timing of application is essential for maximizing chemical effectiveness and preventing seed production. Glyphosate should be applied by hand sprayer as a 1.5 percent solution (2 oz/gal water). Rodeo®, registered for aquatic use, should also be applied as a 1.5 percent solution, with the addition of a surfactant approved for use over water, as specified on the label (Heidorn 1990).

2,4-D approved for aquatic use is the second most commonly used herbicide, despite its inconsistent ability to control purple loosestrife. This herbicide's selectivity for broadleaf plants and low cost make it appealing. Applied before flowering (late May to early June), it is most effective in controlling first-year seedlings and preventing seed production in mature plants, but it does not kill the plants and many will likely resprout (Bender 1987; Skinner *et al.* 1994).

Triclopyr (as Garlon 3A®or Renovate®) is not not yet approved for aquatic use, but it has great potential because it is selective for broadleaf plants and does not harm grasses and most other monocots, which are important in wetland habitats. While some reports point to the inconsistency of Garlon 3A® in the field (Katovich *et al.* 1996), others indicate that it is effective and promotes regrowth by grasses and other wetland monocots (Nelson and Getsinger 1994). Garlon 3A® continues to undergo evaluation for its use in aquatic environments.

Mentha pulegium L.

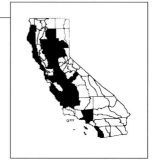

Common name: pennyroyal

Synonymous scientific name: *Pulegium vulgare*

Closely related California natives: 1

Closely related California non-natives: 3

Listed: CalEPPC List B; CDFA nl

by Peter J. Warner

HOW DO I RECOGNIZE IT?

Distinctive Features

Pennyroyal (*Mentha pulegium*) is a perennial mint with a variable habit, ranging from low-growing, spreading plants to lanky, upright subshrubs. The pale or deeper pink, blue, or violet flowers are clustered in dense whorls at the upper nodes. The plant has a powerful and pungent minty odor. The stems are square in cross-section, ascending from rhizomes. Branches and simple leaves are opposite on stems. Unlike many mints, the flowers are not strongly bilateral, but only slightly two-lipped.

Description

Lamiaceae. Perennial herb growing from rhizomes. Stems: subglabrous to short-hairy, decumbent to ascending, branched, 4-angled, 4-35 in (10-90 cm) high. Leaves: generally 0.25-1 in (1-2.5 cm) long, reduced upward, lower leaves petioled, those higher on stems subsessile; blade narrowly ovate to elliptic, base tapered to obtuse, leaf tip rounded, margin entire to finely ser-

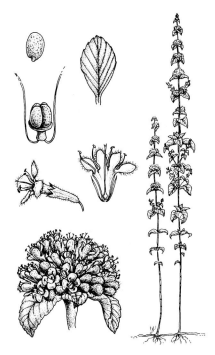

rate, lower surface short-hairy. Inflorescence: collectively spike- or panicle-like, individually axillary and head-like, subtended by reflexed leaves or smaller, leaf-like bracts. Flowers: calyx 0.1-0.25 in (2-4.5 mm), subradial; 5 lobes (teeth) ciliate, lower 3 subulate, upper 2 shorter, wider; corollas 0.25-0.38 in (4-8 mm), violet to lavender, pink or blue, often paler with age, upper lip notched, lower three-lobed; 4 stamens, subequal, exceeding corolla. Carpels: ovary superior, 4-lobed, 2 chamber, each with 2 ovules; 1 style arising from junction of ovary lobes; 2 stigmas, unequal. Fruits: 4 nutlets, seeds 0.07 in (0.75 mm), pale brown, ovoid to oblong (Hickman 1993, Tutin *et al.* 1976).

WHERE WOULD I FIND IT?

Although considered uncommon in much of California (Hickman 1993), pennyroyal occurs in the Sierra foothills, Central Valley, and most coastal counties from the Mexican border to Oregon. It is common as an obligate wetland indicator species in seasonally inundated soils of valley bottomlands, usually below 1,640 feet (500 m) elevation (Reed 1988). It is locally abundant in the San Francisco Bay region, including moist meadows and vernal pools of the Laguna de Santa Rosa floodplain (Sonoma County), along roadsides, and on the perimeters of freshwater and brackish marshes (e.g., Estero Americano, Petaluma River).

Pennyroyal grows in vernally flooded or seasonally wet areas: seeps, streamsides, vernal pools and swales, marshes, and ditches (Hickman 1993; Warner pers. observation). It appears to grow predominantly in heavy clay or silty soils. The plant flourishes on frequently disturbed sites, such as heavily grazed pastures, or sites with milder disruptions, such as seasonal depositions of silt and organic debris. While tolerant of some alkalinity, seasonal drought, and warm temperatures, optimal development of pennyroyal appears to occur where other vegetation shades stems and rhizomes (underground stems) and contributes to sustained moderation of soil moisture and temperature.

WHERE DID IT COME FROM AND HOW DOES IT SPREAD?

Pennyroyal is native to Ireland, across southern and central Europe, to the Ukraine (Tutin *et al.* 1976). A European folk name for the plant is grows-in-the-ditch (Polonin 1969). Repeatedly introduced into North America since European settlement, pennyroyal is now found naturalized in wildlands throughout the world (Grieve 1959). It can spread via seed dispersal or

fragmentation of stolons. Dispersal of seeds or stems by water and/or animals has contributed to its successful establishment beyond domestic gardens. It has hairy calyces containing nutlets that are well adapted to dispersal by sticking to animal fur or wool. Nutlets are also spread via mud on hooves or may be ingested by animals and deposited elsewhere. Humans have contributed to the dispersal and subsequent establishment of pennyroyal in non-cultivated areas by propagating and growing it for its purported medicinal or herbal values (Duke 1985). It is also spread by vehicles traveling from infested to non-infested areas. Daughter plants develop along stolons, spreading it vegetatively locally and forming new patches if the stolons are fragmented, as by cultivation (Parson 1992).

WHAT PROBLEMS DOES IT CAUSE?

Although pennyroyal is considered moderately invasive in wetlands (CalEPPC 1996), its ecological impacts are not well documented. It clearly prospers in habitats that were once dominated by native plants, suggesting that it may have displaced some species. In particular, the flora of vernal pools may have suffered loss of habitat through the introduction of pennyroyal. However, its capacity to displace native plant species is uncertain, especially given the frequent disturbance of habitats that it invades. Pennyroyal can be a nuisance for ranchers, since livestock can be poisoned by this unpalatable rangeland weed (Fuller and McClintock 1986).

HOW DOES IT GROW AND REPRODUCE?

Inconspicuous in spring during vegetative growth, pennyroyal's flowering peak occurs in summer. Timing of flowering appears correlated with soil moisture levels or diurnal maximum temperatures, with populations on drier and warmer sites flowering earlier. Earliest blooming starts in June, with plants in milder or wetter (e.g., coastal, estuarine) sites often flowering through November. Seeds are produced abundantly from the numerous and dense inflorescences from late spring through fall. By late autumn the flowering stems have dried, but seed heads may remain intact through the winter. Water and stock animals and ungulates are principal vectors (Parsons 1992); little is known about the role of granivores in seed dispersal.

Seedlings appear in heavy, silty clays once pools of water formed in winter have almost dried (Warner, pers. observation). Seeds germinate after exposure to alternating temperatures, and they have a light requirement for germination. A seedbank usually develops. Seeds can germinate in water, and seedlings survive readily even in areas with prolonged shallow inundation (Parsons 1992). Seedlings and rhizomes (underground stems) both develop fibrous, net-like root systems (Parsons 1992).

Pennyroyal is a ground-hugging perennial, with herbaceous above-ground tissues regenerating in spring and dying in late fall in established plants. Established plants overwinter as rhizomes just below the soil surface, with roots and flowering stems growing from conspicuous nodes in mid-spring. These rhizomatous ramets may persist for several years. Shoots often assume a decumbent, nodal rooting habit in drying pools, suggesting that this may be a mechanism for optimizing water uptake during growth. Shoot and leaf production begin as water recedes in spring, and the most vigorous growth occurs in late spring and early summer. Once soils have dried, inflorescences are formed on most shoot tips (Bailey 1964).

HOW CAN I GET RID OF IT?

Considering the shortage of scientific literature about *Mentha pulegium*, any effort to document control efforts would probably yield important management information. Monitoring should include data collection on plants and evaluation of various control alternatives, as well as compilation of environmental data. Subsequent reporting of results, however informal, would significantly increase knowledge about the management of pennyroyal. Experiments designed to assess the effectiveness of various control alternatives would benefit management strategies for habitat preservation or restoration.

Physical Control

Mechanical methods: Pennyroyal's brittle stems and propensity for resprouting probably rule out soil tilling or hand pulling as effective control methods. Late spring or early summer mowing, repeated over several years, may weaken plants by depleting photosynthetic reserves. Definitive research data is needed.

Prescribed burning: This may be a viable control method, although high soil moisture levels in most pennyroyal habitats may limit the effectiveness of burning. In habitats where the timing of fire would not harm native species, well designed experiments to investigate the effects of various burn prescriptions are needed.

Biological Control

Insects and fungi: No research has been conducted on biological control agents for pennyroyal.

Grazing: Control by livestock grazing appears unlikely. Pennyroyal is unpalatable as forage for cows or sheep and sometimes causes gastrointestinal irritation when ingested (Parsons 1992; Warner, pers. observation).

Chemical Control

Seedlings are susceptible to ester-formulated 2,4-D. Mature plants are resistant to 2,4-D, but can be killed with label-recommended concentrations of glyphosate or triclopyr (Parsons 1992). However, these herbicides pose hazards to non-target species in wetlands, including desirable plants, animals, and microorganisms. Triclopyr is not currently registered for use in wetlands. Broadcast applications would not be acceptable in many situations. Cut-stem applications would be extremely labor-intensive, even if effective in reducing environmental contamination to an acceptable level.

Mesembryanthemum crystallinum L.

Common names: crystalline iceplant, common iceplant

Synonymous scientific name: *Cryophytum crystallinum*

Closely related California natives: 0

Closely related California non-natives: 8

Listed: CalEPPC B; CDFA nl

by Jonathan J. Randall

HOW DO I RECOGNIZE IT?

Distinctive Features

Crystalline iceplant (*Mesembryanthemum crystallinum*) is a succulent, low-growing herb spreading over ground with flat, fleshy leaves. Leaves and stems are covered with distinctive tiny, clear, blister-like outgrowths. Flowers are small (one-fifth of an inch across) and radial, with many narrow petals that range from white to pinkish, depending on the age of the flower. The stems can range from green to red and usually trail along the soil surface.

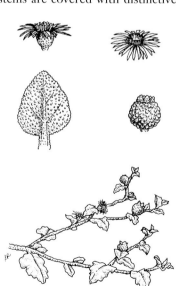

Description

Aizoaceae. Prostrate annual (biennial) herb. Stems: trailing, forked, and <3.3 ft (<1 m) long. Leaves: 0.8-8 in (2-20 cm) long, petioled, blades flat, fleshy, ovate to spoon-shaped with wavy margins. Stems and leaves papillate (covered with tiny, clear, blister-like protuberances). Inflorescence: flowers usually sessile and can either be axillary or in cymes. Flowers: radial, 0.3-0.4 in (7-10 mm) diameter, with a round, fleshy hypanthium that becomes red

with age; 5 fleshy, equal sepals and numerous linear petals, white, aging to pink; stamens many; ovary inferior, surmounted by 5-20 styles. Fruit: coarsely papillate capsule with 5 valves, opening when moist. Stems, leaves, and fruits edible (Hickman 1993).

WHERE WOULD I FIND IT?

Crystalline iceplant occurs along the immediate coast from the San Francisco Bay region south into Baja California, Mexico. It can also be found on all the California

Channel Islands (Junak *et al.* 1995, Munz 1974). A few plants have been observed up to eight miles (13 km) inland (Vivrette and Muller 1977). It is found primarily in saline soils on coastal strand, coastal sage scrub, coastal bluffs and cliffs, and other disturbed ground. It tolerates saline soils, but not frost.

WHERE DID IT COME FROM AND HOW DOES IT SPREAD?

Crystalline iceplant is native to South Africa and is thought to have been introduced to California, perhaps as early as the 1500s, in sand used as ships' ballast. About fifty years ago, the California Department of Transportation began using iceplant for roadside landscaping and erosion control. This practice was discontinued in 1969 when it was discovered that the plant was killed by hard freezes and not effective for erosion control on steeper slopes. Many homeowners still use iceplant as a groundcover or for erosion control (Moss 1994). The constant erosion of coastal bluffs creates an open habitat that encourages colonization and spread of iceplant, as does grazing disturbance in coastal grasslands.

WHAT PROBLEMS DOES IT CAUSE?

Crystalline iceplant can invade coastal bluff areas and compete with native species. Because iceplant has an exceptional ability to absorb moisture from the soil, it can outcompete most other species for water. A high level of nitrate build-up has been found to occur underneath crystalline iceplants, and this can be detrimental to survival of grassland seedlings growing there. Accumulation and release of these salts by crystalline iceplant also prevents or retards reestablishment of native species (Vivrette and Muller 1977).

HOW DOES IT GROW AND REPRODUCE?

Crystalline iceplant usually flowers from March to June, although, because germination is staggered, individual plants can be found flowering most of the year. Fruiting generally occurs from June to August, after which the plant gradually dries from the base upward, fruits being the last to dry (Vivrette and Muller 1977). Rabbits and mice aid in dispersal of seeds.

Seeds germinate with the first rains in autumn and with each succeeding rain or heavy fog. Seedlings continue rapid vegetative growth until spring. Vegetative stages are characterized by broad, flat leaves. Vegetative growth slows and then stops as the hot, dry summer season progresses.

HOW CAN I GET RID OF IT?

Information specific to the control of crystalline iceplant is not available. Control techniques for *Carpobrotus edulis*, another invasive species in the same family (Aizoaceae) with similar growth form, would likely be effective.

Myoporum laetum Forster f.

Common names: myoporum, ngaio tree (New Zealand)

Synonymous scientific names: none known

Closely related California natives: 0

Closely related California non-natives: 0

Listed: CalEPPC A-2; CDFA nl

by Jo Kitz

HOW DO I RECOGNIZE IT?

Distinctive Features

Myoporum is a small, bright green, evergreen tree or shrub with a broadly spreading crown.

The rather narrow, sub-fleshy leaves are bright green and shiny with translucent dots. Young twigs and leaves are bronze-green and sticky. Branches are stout and spreading, and trunks have thick, furrowed bark. This plant's rapid growth, branching and spreading habit, and dense foliage make it useful as a thick, high screen or hedge. Growing alone, it is a handsome multi-trunked plant.

Description

Myoporaceae. Tree or shrub from 9 to 30 ft (3-10 m) high and 20 ft (6 m) wide with trunks 8.7-20 in (23-50 cm) in diameter. Stems: much branched, broadly spreading; twig tips and young leaves bronze-green, sticky. Leaves: 3-4 in (7-10 cm) long and 0.7- 1.3 in (2-3.5 cm) wide, alternate, generally lanceolate to oblong-lanceolate, acute to acuminate; margins crenulate-serrulate in upper half, sinuate in lower half or sinulate throughout (different forms found on same plant), bright green, +/-fleshy with conspicuous translucent pores, and petioles. Inflorescence: axillary cymes on peduncles up to 0.6 in (15 mm) long with 2-6 flowers. Flowers: bisexual, radial to bilateral; calyx 5-lobed, persistent; corolla 5-lobed, bell-shaped, about 0.4 in (1 cm) in diameter, white with purple spots; generally 4 stamens, epipetalous; ovary superior with 4 locules, 1 style, 2 stigmas. Fruit: seed in clusters of 2-6; each 0.5 in (1 cm), in ovoid drupe, fleshy, pale to dark reddish purple (Hickman 1993).

WHERE WOULD I FIND IT?

Myoporum flourishes in coastal areas in the San Francisco Bay region and in Los Angeles, Marin, and Orange counties, and may be found along the coast from Sonoma County to San Diego County. It has naturalized in Hawaii. It is most common in urban, disturbed areas, below 900 feet (300 m) elevation, where it forms dense monocultures if not controlled. It grows well in heavy, alkaline, brackish, and sandy soils. Its invasive tendencies have not been observed in interior regions.

WHERE DID IT COME FROM AND HOW DOES IT SPREAD?

This myoporum (*Myoporum laetum*) is native to New Zealand; it is one of thirty-two species in the genus (Allen 1982). Myoporum species are found widely in Australia and the South Pacific. *M. laetum* was introduced to California as a horticultural species (Griffiths and McClintock, 1971). It spreads via prodigious seed production. The drupes are attractive to birds, which disperse them over long distances. *M. laetum* is not known to spread by vegetative means, but it can resprout from stumps.

WHAT PROBLEMS DOES IT CAUSE?

Myoporum's heavy seed production results in dense monocultures that outcompete other species. If not controlled, it will take over large areas. Extending outward, the area it shades enlarges each year. Seed dispersal by birds results in rapid expansion of infested areas. Slower-growing native species near myoporum become stunted or fail to grow until myoporum is removed. Leaves and fruits are toxic and may be fatal to livestock. The toxin (ngaione) is a furanoid sesquiterpene ketone that constitutes 70 to 80 percent of oil of ngaio, an essential oil (Fuller and McClintock 1986). The fruit is less toxic than the leaves, and toxin is released with leaf fall (Salmon 1986).

It has been reported that, because the interior of large plants contains an accumulation of dead branches, myoporum burns with an intensity that seems in contradiction to its

lush, dark green foliage. This is of concern in areas prone to wildfires.

HOW DOES IT GROW AND REPRODUCE?

Little information is available on the growth and reproduction of myoporum. It reproduces only by seed. Plants bloom in early spring. In areas with water available, it grows rapidly into large, dense stands. It is surprisingly drought tolerant. Myoporum is sensitive to frost, and temperatures more than five degrees below freezing cause severe damage and dieback (Poole 1990).

HOW CAN I GET RID OF IT?

Annual monitoring is necessary when seed sources are plentiful, and annual seedling removal is essential. Post-fire monitoring is essential in areas prone to myoporum incursion.

Physical Control

Mechanical removal: Seedlings can be pulled, but they have long, strong taproots, and pulling must be done when the soil is moist and the plant is small. If the root remains in the ground the plant will resprout with vigor. Seedling removal with cut-stump treatment produces the best results.

Prescribed burning: No information is available about the efficacy of prescribed burning in infested areas. However, specimens have resprouted after wildland fires, and post-fire seedling recruitment competes with native plants for reestablishment.

Biological Control

Insects and fungi: This has not been investigated for myoporum.

Grazing: Grazing is not an option because of the toxicity of this species.

Chemical Control

Trunks should be cut at ground level and saturated with concentrated glyphosate. Leaving any amount of stump may allow resprouting. Cut surfaces must be monitored and retreated as needed. Contact a certified herbicide applicator for additional information on any newly registered herbicides that may be appropriate and approved for this species.

Myriophyllum aquaticum (Vell. conc.) Verde

Common names: parrot's feather, parrot feather watermilfoil, Brazilian water milfoil

Synonymous scientific names: *Myriophyllum brasiliense, M. proserpinacoides, Enydria aquatica*

Closely related California natives: 3

Closely related California non-natives: 1

Listed: CalEPPC B; CDFA nl

by Kris E. Godfrey

HOW DO I RECOGNIZE IT?

Distinctive Features

Parrot's feather (*Myriophyllum aquaticum*) is a stout aquatic perennial that forms dense mats of intertwined brownish stems (rhizomes) in water. These stems grow to six and a half feet in length and resemble bright green bottlebrushes emerging from the water. The bottlebrush appearance results from the fact that the leaves appear in whorls of four to six at each node and each leaf is feather-like, the blade divided into twenty-four to thirty-six thread-like segments. Upon close inspection the leaves look gray-green. Parrot's feather also has leaves below the water surface, appearing reddish, feathery, and limp. Unlike other milfoils (*Myriophyllum* spp.), parrot's feather stems may grow as much as eight inches above the water surface (Orchard 1981, Wester-dahl and Getsinger 1988).

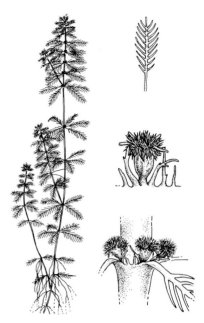

Description

Haloragaceae. Parrot's feather is a perennial, submersed/immersed aquatic plant. Stems: long and unbranched, rooting freely from lower nodes. Leaves: in whorls and slightly dimorphic. Submerged leaves in whorls of 4-6, oblong in appearance, 1.4-1.6 in (3.5-4.0 cm) long, 0.3-0.5 in (0.8-1.2 cm) wide; pectinate with 25-30 linear pinnae to 0.27 in (0.7 cm) long. Lower leaves usually deteriorate rapidly. Emergent leaves glaucous, in whorls of 4-6, oblong, 1-1.4 in (2.5-3.5 cm) long, 0.27-0.31 in (0.7-0.8 cm) wide; pectinate with 18-36 linear pinnae in the upper 80 percent of leaf. Lower 2-2.75 in (5-7 cm) of leaf rachis naked. Inflorescence: indeterminate spike with flowers singly in axils of upper emergent leaves. Flowers: only female flowers known from parrot's feather plants in the United States. Female flowers on pedicel 0.008-0.02 in (0.2-0.4 mm) long; 4 white sepals, 0.02 in (0.5 mm) long, 0.012 in. (0.3 mm) wide, denticulate with one to several small teeth on each margin; no petals or stamens, 4 clavate styles 0.004-0.008 in (0.1-0.2 mm) long, stigmas white and densely fimbriate; ovary pyriform, 0.02-0.03 in (0.6-0.7 mm) long, 0.02 in. (0.6 mm) wide; 4-ribbed longitudinally between sepals. Fruit: none in United States (Orchard 1981).

WHERE WOULD I FIND IT?

Both parrot's feather and spike watermilfoil can be found in freshwater lakes, ponds, and canals with slow-moving waters in northern and central California (Anderson 1990). Parrot's feather can be found throughout much of the United States from New England to Florida and westward to California and Washington. Typically, it is found rooted at depths to 6.5 feet (2 m), but emergent stems may elongate and spread over deeper waters or to pond edges.

WHERE DID IT COME FROM AND HOW DOES IT SPREAD?

Parrot's feather is native to South America and was introduced into the United States in the late 1800s for use in aquaria and water gardens (Kane *et al.* 1991). It was first collected in the United States near Washington D.C. in 1890. It was reported from South Africa in 1918 or 1919, Japan in 1920, New Zealand in 1929, Australia in the 1960s, and England in the 1970s. A population was reported in western Washington in 1944 (Washington Water Quality Program 1998).

Parrot's feather is capable of sexual reproduction in its native range, but the spread of parrot's feather in the United States results solely from vegetative reproduction. The stems of parrot's feather are brittle and fragment easily. These fragments settle in sediments and produce new plants (Orchard 1981, Kane *et al.* 1991). Fragments can be spread by boats, trailers, and by dumping aquarium plants in waterways. They can also be spread by waterfowl and other wildlife, as well as by moving water.

WHAT PROBLEMS DOES IT CAUSE?

Parrot's feather may compete with native aquatic plants, eliminating them or reducing their numbers in infested sites. It forms dense mats that can entirely cover the surface of the water in shallow lakes and other waterways. These mats clog waterways, making them unusable for navigation or recreation and causing flooding out of the channel. It can block irrigation pumps and water intakes, and it provides optimal habitat for mosquitoes (Orr and Resh 1989, Systma and Anderson 1990; Parsons 1992). In California this species is becoming an increasing problem in irrigation and drainage canals. A 1985 survey of irrigation, mosquito abatement, flood control, and reclamation agencies in California indicated that parrot's feather infested nearly 600 miles of waterways and over 500 surface acres (Washington Water Quality Program 1998).

While parrot's feather may provide cover for some aquatic organisms, it can significantly alter the physical and chemical characteristics of lakes and streams. Infestations can alter aquatic ecosystems by shading out algae in the water column that serve as the basis of the aquatic food web. It also alters habitats for aquatic organisms, waterfowl, and other wildlife.

HOW DOES IT GROW AND REPRODUCE?

Reproduction of parrot's feather in the United States is believed to be entirely by vegetative means, resulting from stem fragmentation and/or regrowth from sections of rhizomes (underground stems) (Jacot Guillarmod 1979, Kane *et al.* 1991). Even in South America, virtually all parrot's feather plants are female. Male plants are unknown outside South America, so no seeds are produced in North American populations.

With its tough rhizomes, parrot's feather can be transported long distances on boat trailers. Any rhizome or stem sections with at least one node, even as small as 0.2 inch (5mm) long, can root and establish new plants. Rhizomes stored under moist conditions in a refrigerator survived for one year. Once rooted, these new plants produce rhizomes that spread through sediments and stems that grow until they reach the water surface (Orchard 1981). The result is a dense,

tangled mass of parrot's feather in the water column.

Growth is most rapid from March until September. In spring shoots begin to grow rapidly from overwintering rhizomes as water temperature increases. Rhizomes function as a support structure for adventitious roots and provide buoyancy for emergent growth in summer. Emergent stems and leaves extend from a few inches to over one foot above the water surface. Underwater leaves tend to senesce as the season advances. Plants usu-

ally flower in spring, but some plants may also flower in fall. The inconspicuous flowers form where emergent leaves attach to the stem.

In fall plants typically die back to the rhizomes. In some areas, parrot's feather may maintain considerable winter biomass. Because the plant lacks tubers, turions, and winterbuds, rhizomes serve all those functions. Parrot's feather does not store phosphorus or carbon in its rhizomes, and this may explain its failure to invade areas with severe winters.

HOW CAN I GET RID OF IT?

Parrot's feather is difficult to remove from an aquatic system, so it is best to prevent it from establishing in the first place. The public must be made aware of the problems caused by parrot's feather and how it can be spread by dumping unwanted plants from water gardens or aquaria or by boats, trailers, and fishing equipment that are not cleaned before being moved to a new waterway. If parrot's feather becomes established, only chemical and mechanical control methods are available.

Physical Control

Mechanical methods: Parrot's feather can be removed by mechanical harvesters. In Washington, workers use a dragline to remove parrot's feather plants. A truck-mounted crane with a special attachment plucks weeds out of the ditch. The dragline operation is conducted annually from August to December, with control generally lasting for one growing season (Washington Water Quality Program 1998). Care must be taken to ensure removal of all plant parts during harvest, since even tiny stem or rhizome fragments can root and establish new plants. Because of this, mechanical harvesting often results in the spread of parrot's feather rather than its elimination or suppression.

Biological Control

Parrot's feather has a high tannin content, so most grazers, including grass carp (*Ctenopharyngodon idella*), find it unpalatable. Grass carp also prefer soft plants, such as *Elodea canadensis*, and the tough, woody parrot's feather stems are avoided. USDA approved biological control agents are not currently available. Potential agents do exist, but they have yet be tested for host specificity. A complex of insects feed on parrot's feather in its native habitat. *Lysathia flavipes*, a flea beetle found on parrot's feather in Argentina, causes moderate damage under field conditions. Also found in Argentina is a weevil, *Listronotus marginicollis*, that apparently feeds only on parrot's feather in its native range. Other insects have been found on parrot's feather in Florida. *Lysathia ludoviciana*, a flea beetle native to the southern United States and the Caribbean, uses parrot's feather as a host plant for larvae under laboratory conditions. However, the flea beetle is not often found on parrot's feather in the field. Two members of the Tortricidae family, *Argyrotaenia ivana* and *Choristoneura parallela*, have also been found on parrot's feather in Florida, but their effect on the plant is unknown. In addition, larvae of the caterpillar, *Parapoynx allionealis*, mine parrot's feather leaves, but the impact of these larvae is unknown.

Fungal control options exist as well. An isolate of *Pythium carolinianum* collected in California has shown some promise as a potential biocontrol agent. Parrot's feather stems experimentally inoculated with this fungus produced significantly less growth than control plants (Washington Water Quality Program 1998).

Chemical Control

The underwater and above-water foliage of parrot's feather make herbicides difficult to deliver effectively. Emergent stems and leaves have a thick, waxy cuticle that inhibits herbicide uptake, and a wetting agent is required to penetrate it. Often the weight of the spray will cause emergent vegetation to collapse into the water, where the herbicide is washed off before it can be translocated throughout the plant. The most recent version of an herbicide label will give recommended rates and information about whether the compound is registered for use in specific situations. Herbicide use is more highly regulated in aquatic systems than in terrestrial systems.

Westerdahl and Getsinger (1988) report excellent control of parrot's feather with 2,4-D, diquat, diquat and complexed copper, endothall dipotassium salt, fluridone, and endothall and complexed copper. Diquat is used on emergent parrot's feather, as well as in the water to kill rhizomes. Copper complexes are used only on submersed plants. Diquat is not legal for use in aquatic systems in California. Fair control was obtained with acrolein and glyphosate. Acrolein is used only in non-fisheries water, and glyphosate, formulated as Rodeo, is used only on emergent parrot's feather. The Monsanto Company suggested that applying a 1.75 percent solution of Rodeo® with surfactant to the plants in summer or fall when water levels are low would give about 95 percent control. Control of parrot's feather may be achieved with low-volatility ester of 2,4-D at 4.4-8.9 kg/ha, sprayed onto emergent foliage. The granular formulation of 2,4-D was needed to control parrot's feather for periods greater than twelve months. It is more effective when applied to young, actively growing plants (Washington Water Quality Program 1998).

In practice, weed control efforts report little success with herbicides to control parrot's feather. Glyphosate causes emergent vegetation to turn black, but within two weeks the plants have recovered. An experimental fall application of triclopyr also proved ineffective (Washington Water Quality Program 1998).

Myriophyllum spicatum L.

Common names: Eurasian watermilfoil, spike watermilfoil

Synonymous scientific name: *Myriophyllum exalbescens*

Closely related California natives: 3

Closely related California non-natives 1

Listed: CalEPPC A-1; CDFA nl

by Carla Bossard

HOW DO I RECOGNIZE IT?

Distinctive Features

Eurasian watermilfoil (*Myriophyllum spicatum*) is an aquatic plant with branching stems near the water surface, creating dense floating mats. The feather-like leaves grow in whorls of four around the stem, with each leaf divided into nine to twenty-one paired leaflets. This watermilfoil is distinguished from the dioecious parrot's feather (*M. aquaticum*) by the fact that it is monoecious, does not have flattened midribs, has only submersed foliage, and has longer stems at maturity (Hoffman and Kearns 1997)

Description

Haloragaceae. Perennial, submersed aquatic plant. Stems: branching and leafy, 19.5-28 in (50-70 cm) long, generally >6.5 ft (>2 m) long, reddish or olive-green when dry. Stems thicken below inflorescence and double in width further down, often curving to lie parallel with the water surface. Stem fragments root freely. Leaves: submersed, feather-like leaves in whorls of 4; each leaf 0.6-1.6 in (1.5-4 cm) long, most often with 14-24 pairs of filiform divisions. Inflorescence: terminal spike, 4-20 cm long and emergent, often pink. Spike erect at anthesis, parallel to water surface at fruit set. Flowers: lower flowers pistillate, upper flowers staminate; flowers verticillate in 4s, whorls 2-ranked, adjacent whorls rotated 45 degrees, occasional hermaphrodite flowers in transition zone. Lower 2-4 whorls of floral bracts usually pictinate and often longer than flowers; upper bracts entire, broader than long and shorter than flowers.

Female flowers lack perianth; gynoecium 4-lobed with pink, tufted, recurved stigmas. Male flowers with pink, cauduceous petals; 8 stamens. Fruit: subglobose schizocarp, 0.08-0.13 in (2-3mm) long, 4-sulcate with two rather wrinkled ridges adjacent to lines of dehiscence (from Aiken *et al.* 1979, Hickman 1993).

WHERE WOULD I FIND IT?

Eurasian watermilfoil can be found in freshwater lakes, ponds, and canals with slow-moving waters in northern and central California, particularly in the San Francisco Bay and San Joaquin Valley regions (Anderson 1990) and Lake Tahoe (Goldman, pers. comm.). It is found throughout much of the United States from New England to Florida and westward to California and Washington. Typically, it is rooted at water depths of three to ten feet (1-5 m), but can spread to waters to thirty-three feet (10 m) deep.

This watermilfoil can grow on sandy, silty, or rocky substrates but grows best in fertile, fine-textured, inorganic sediments. It is an opportunistic species that prefers disturbed substrates with much nutrient runoff. High temperatures promote multiple periods of flowering and fragmentation (Hoffman and Kearns 1997).

WHERE DID IT COME FROM AND HOW DOES IT SPREAD?

Eurasian watermilfoil is native to Greenland, North Africa, Europe, and Asia. It was introduced from Europe to the eastern United States in ships' ballast before 1940 and had spread to California by the 1960s (Couch and Nelson 1985, Aiken *et al.* 1979).

Eurasian watermilfoil produces some long-viable, often dormant seed, but its spread results primarily from vegetative reproduction. The stems are brittle and fragment easily. These fragments can settle in sediments, producing new plants. Fragments are spread by boats or trailers, in bait buckets, or by floating downstream. They can also be spread by waterfowl and other wildlife (Aiken *et al.* 1979). Dumping surplus plants from aquatic gardens or aquariums into waterways is another means of spread for this species (Hoffman and Kearns 1997).

WHAT PROBLEMS DOES IT CAUSE?

Eurasian watermilfoil grows and spreads rapidly, creating dense mats on the water surface. These monotypic mats outcompete native aquatic plants, reducing species diversity. Loss of nutrient-rich native plants reduces food sources for waterfowl, impacts fish spawning grounds, and disrupts predator-prey relationships by fencing out larger fish (Hoffman and Kearns 1997). Infestations can alter aquatic ecosystems by shading out algae in the water column that serve as the basis of the aquatic food web. The mats inhibit recreational use of waterways for boating, fishing, and swimming and cause flooding out of the waterway. Mats also can block irrigation pumps and water intakes and provide optimal habitat for mosquitoes (Aiken 1979, Parsons 1992). Millions of dollars annually are spent in the United States and Canada on efforts to control Eurasian watermilfoil.

HOW DOES IT GROW AND REPRODUCE?

Colonization of new sites is usually by fragments. Once established in an aquatic habitat, Eurasian watermilfoil grows rapidly in spring (March-April). Stolons, lower stems, root crowns, and roots persist over the winter in California. In waters where temperatures do not drop below

50 degrees F (10 degrees C) there is little seasonal die-back (Aiken *et al.* 1979). Root crowns store starch that fuels early takeover of the water column (Grace and Wetzal 1978). During the growing season this plant undergoes auto-fragmentation, with fragments often developing roots before separation from the parent plant. Sloughing of plant parts is common after flowering.

Flowering usually occurs in spring, but some plants flower in fall as well (Pullman 1993). Flowers generally are wind-polli-

nated. Spikes lie parallel to the water as fruits mature. Fruits float for some hours, allowing for some water dispersal. Fruits do not release the seeds. Seeds exhibit prolonged dormancy, and seedlings, although viable, are rare in nature (Cobble and Vance 1987; Aiken *et al.* 1979). Vegetative reproduction is far more important in this species, with fragments being released throughout the growing season (Madsen *et al.* 1988). Small axillary buds detach from the root crown in late winter and may establish new plants during the growing season (Aiken *et al.* 1979)

HOW CAN I GET RID OF IT?

Like other milfoils, Eurasian watermilfoil is difficult to remove from an aquatic system once established, so monitoring and prevention are the most important methods of control. Efforts to make the public aware of the need to remove weed fragments at boat landings and a watershed management program that keeps nutrients from unnaturally enriching aquatic systems, which stimulates this species, are critical to its control (Hoffman and Kearns 1997).

Physical Control

Mechanical methods: Eurasian watermilfoil can be removed by mechanical harvesters. However, native vegetation tends to be removed by this method simultaneously, eliminating beneficial competitors. Also, fragments are usually created, which contributes to dispersal. Harvesters should be used only where colonies are widespread (Hoffman and Kearns 1997, Pullman 1993). Hand pulling, while time-consuming, can be used to control colonies smaller than one acre (0.6 hectare). All fragments, including roots, must be carefully removed. Where possible, anchored bottom screens can be used to prevent new sproutings, but these must be cleaned of sediments once a year (Hoffman and Kearns 1997).

Raising or lowering water levels can inhibit this species. Raising the water level can deprive plants of access to light. This technique can be augmented with light-limiting dyes or shade barriers. Lowering water level can dehydrate plants or, in winter, freeze them to death (Bowen 1995).

Biological Control

There are no USDA approved biocontrol agents for Eurasian watermilfoil. A North American weevil, *Eurhychiopsis lecontei*, feeds on Eurasian watermilfoil in the midwestern United States. It has been shown to cause extensive damage in some populations, with a 50 percent reduction in biomass and no significant effect on biomass of ten native plant species (Sheldon and Creed 1995). The weevil's effectiveness for controlling watermilfoil on a large scale is under investigation in twelve Wisconsin lakes (Hoffman and Kearns 1997).

The fungus *Mycoleptodiscus terrestris* is also being researched as a potential biocontrol agent in Michigan and Massachusetts (Hoffman and Kearns 1997). Grass carp (*Ctenopharyngodon idella*) used as a biocontrol on other invasive aquatic species find watermilfoil unappealing.

Chemical Control

Herbicide use is more highly regulated in aquatic systems than in terrestrial systems and, because of dilution effects, more difficult to use. The most recent version of herbicide labels will give recommended rates and information about whether the compound is registered for use in a specific situation. Two herbicides effective on Eurasian watermilfoil are not legal for use in aquatic systems in California (diquat and 2,4,D butoxyethenol ester, 20 percent attaclay). Since these herbicides are non-selective, the potential disruption to and contamination of aquatic ecosystems is deemed more problematic than the damage caused by watermilfoil infestation.

Under some conditions, the herbicide fluridone has been approved for lake or pond treatments in California, in low concentrations (10-13 ppb), applied early in the season when water temperatures are low and most native vegetation is not actively growing. Fluridone inhibits photosynthesis and synthesis of important protein compounds in plants susceptible to it, as is watermilfoil. The degree of impact is related to the concentration and the amount of time the plant is in contact with the herbicide (Anderson 1981). Fluridone is taken up by roots and shoots, but appears to be translocated only from root to shoot (Marquis *et al.* 1981).

Glyphosate and triclopyr were found to be ineffective in control of Eurasian watermilfoil in studies in Washington (Thurston County 1995).

Pennisetum setaceum Forsskal

Common name: fountain grass, crimson fountaingrass

Synonymous scientific names: *Pennisetum ruppelii, Phalaris setaceum*

Closely related California natives: 0

Closely related California non-natives: 2

Listed: CalEPPC A-1; CDFA nl

by Jeffrey E. Lovich

HOW DO I RECOGNIZE IT?

Distinctive Features

Fountain grass (*Pennisetum setaceum*) is a coarse perennial grass with a densely clumped growth form and erect stems usually one and a half to five feet tall. The flowerheads are prominent, nodding, and feathery. They resemble bottlebrushes six to fifteen inches long, with many, small, light pink to purple flowers.

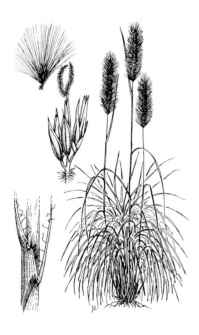

Description

Poaceae. Perennial, caespitose grass 16-59 in (40-150 cm) high. Leaves: blades elongate, 0.8-2.6 in (20-65 cm) long, 0.08-0.14 in (2-3.5 mm) wide. Leaf sheath smooth. Several named cultivars, including 'Rubrum,' with rose-colored foliage, 'Cupreum,' with reddish foliage, and 'Atrosanguineum,' with purple foliage (Hammer 1996). Green cultivars reported to be triploid, purple cultivars hexaploid. Purple cultivars may have resulted from chromosome doubling in green cultivar (Simpson and Bashaw 1969). Inflorescence: dense, panicle-like, 3.1-11.8 in (8-30 cm) long. Color varies from light pink to purple. Spikelets on main inflorescence usually 0.18-0.26 in (4.5-6.5 mm) long, about 0.04 in (1 mm) wide; lower glume absent or minute, upper glume generally less than 0.6 times length of spikelet; 2 florets, lower generally sterile, upper fertile; lemmas 3-veined with acuminate tip; paleas absent.

WHERE WOULD I FIND IT?

In California, where fountain grass is still spreading (Rejmánek and Randall 1994), it is found along the coast from the San Francisco Bay Area to the South Coast and Baja California. It also occurs inland in the Sacramento-San Joaquin Delta, southern San Joaquin Valley, and interior areas of the South Coast and Imperial County. It occurs in grasslands, deserts, canyons, and disturbed areas along roadsides, especially adjacent to urban centers. In Hawaii fountain grass invades many types of natural areas, from bare lava flows to rangelands (Tunison 1992). It has a wide elevational range, but is limited to areas with a median annual rainfall of less than fifty inches (<130 cm). In California it is commonly found below elevations of 325 feet (100 m), but it grows up to elevations of at least 2,000 feet (610 m) near Palm Desert. In Hawaii this plant is found from sea level to subalpine sites at 9,184 feet (2,800 m) (Tunison 1992, Williams *et al.* 1995). Fountain grass has a narrower altitudinal distribution in its native range (Williams *et al.* 1995).

WHERE DID IT COME FROM AND HOW DOES IT SPREAD?

Originally native to Africa and the Middle East (Williams *et al.* 1995), fountain grass has been introduced to many areas, including Arizona, California, Florida, Hawaii, Fiji, South Africa, and Australia (Williams *et al.* 1995). It has spread in large part because of its popularity as an ornamental plant (Neal and Senesac 1991, Hammer 1996). It is easily dispersed by vehicles, humans, livestock, and, over short distances, by wind (Cuddihy *et al.* 1988), by water, and possibly by birds (Tunison 1992).

The remarkable spread of this species into a broad range of habitats in Hawaii is attributed to its ability to adapt, physiologically and morphologically, to different environments. Growth and reproduction traits change most prominently with altitude, whereas physiological traits are more strongly affected by physical environment. For example, photosynthetic rates are higher at higher altitudes (Williams *et al.* 1995).

WHAT PROBLEMS DOES IT CAUSE?

Fountain grass is well adapted to fire (D'Antonio and Vitousek 1992), and plants can recover to pre-burn density, even increase in density, following a burn. Fire can actually contribute to the spread of fountain grass (Smith and Tunison 1992). Fountain grass raises fuel loads, which increases the intensity and spread of fire and results in severe damage to native, dry forest species adapted to less extreme fire regimes (Benton 1998). By enhancing the fuel load, fountain grass endangers native plant communities that are not as fire-tolerant (Tunison 1992). In Hawaii, where it alters the natural fire regime, fountain grass is a major threat to some critically imperiled plant species and natural communities (Benton 1998). Fires fueled by fountain grass impact ground-nesting birds and terrestrial animals as well.

Thick infestations of fountain grass interfere with regeneration of native plant species. For example, fountain grass grows faster than the native Hawaiian grass *Heteropogon contortus* and can outcompete it in the arid lowlands of Hawaii. Fountain grass produced 51 percent more total biomass, allocated 49 percent more biomass to leaves, and had higher net photosynthetic rates on a leaf area basis than the native Hawaiian grass. Both are C_4 species, and both produced less biomass and allocated more mass to roots in response to drought. Fountain grass showed no greater flexibility in response to drought than the native grass. Higher net photosynthetic rates

and greater biomass allocation to leaves, however, gives fountain grass a greater growth rate relative to *H. contortus*, which gives it a competitive advantage as an invader in the arid lowlands of Hawaii (Bruegmann 1996).

HOW DOES IT GROW AND REPRODUCE?

Fountain grass can reproduce by either fertilized or unfertilized seeds (Simpson and Bashaw 1969, Dujardin and Hanna 1989). Plants flower from July through October. Fountain grass is apomictic, meaning that it can reproduce asexually by producing seeds from the cells of female plants other than egg cells (Simpson and Bashaw 1969, Dujardin and Hanna 1989). It may also reproduce by seeds produced following pollination and subsequent fertilization of a female egg cell. A green cultivar set approximately 10 percent seed following self-pollination, while a purple cultivar set 0.05 percent. In one study the purple cultivar set as high as 18 percent seed following application of pollen from *Pennisetum ciliare* (Simpson and Bashaw 1969). Seeds stored in a laboratory, under dry conditions, for eighteen months decreased in viability from 80 percent to 44 percent (Tunison 1992). Seeds may remain viable in the soil for at least seven years (Tunison *et al.* 1995). No information is available on conditions favoring germination.

Growth rates of fifty plants with an average basal diameter of 1.1 inch (2.7 cm) were moni-tored for forty-nine months in Hawaii. Average leaf length increased 4.6 inches (11.6 cm) a year, and basal diameter increased three inches (7.74 cm) a year. In five years individual plants can grow leaves up to four feet long and increase their basal diameter to twelve inches (Tunison, 1992).

Fountain grass growing on dunes in Hawaii had vesicular-arbuscular mycorrhizae fungi associated with their roots (Koske 1988).

HOW CAN I GET RID OF IT?

Fountain grass is difficult to eliminate. The long-lived seeds make control extremely diffi-cult, and continued monitoring is essential. Control efforts are most effective when concen-trated first on peripheral or satellite populations to control spread, and then on the core of the infestation (Tunison *et al.* 1994). Use of preemergent herbicides may be useful following re-moval of an infestation. Surveys on horseback are more effective than surveys on foot. Helicop-ter surveys are not efficient, as only large plants can be spotted, and large infestations of small plants could be overlooked (Tunison 1992).

Physical Control

Mechanical methods: Small infestations of fountain grass can be removed by uprooting or cutting with weed eaters (Bruegmann 1996). Heavy tools such as a pick or mattock may be needed to uproot large plants with a basal diameter over six inches (>15 cm). If inflorescences are present, they should be cut and placed in plastic bags, then destroyed to prevent spread of seeds (Tunison *et al.* 1994). Removal by hand may need to be repeated several times a year.

Prescribed burning: Fountain grass seems to be stimulated by fire, so burning is not recom-mended (D'Antonio and Vitousek 1992).

Biological Control

No pathogens of fountain grass have been reported in recent literature (Tunison 1992). Biocontrol efforts would likely also affect related forage and crop species. The congeneric kikuyu grass (*Pennisetum clandestinum*) is a valuable forage grass in some areas (Tunison 1992), and pearl millet (*P. glaucum*) is a cultivated species that can hybridize with fountain grass (Dujardin and Hanna 1989). Fortunately, male hybrids are sterile and females have poor fertility because of obligate apomixis.

Grazing: In Hawaii cattle eat fountain grass only when no other grasses are available (Bruegmann 1996).

Chemical Control

Extensive infestations of fountain grass are probably best controlled with the help of herbi-cides, especially those with some systemic activity. Ten percent liquid hexazinone (as Velpar) at <5.14 kg ai/ha can be used as a post-emergent or preemergent herbicide once a year in areas with high densities. Preemergent herbicides are necessary to control plants in areas with high seedling recruitment. Hexazinone at 10.09 kg ai/ha is effective as a preemergent for nine to twelve months (Tunison *et al.* 1994). Tunison *et al.* (1994) recommend using hexazinone in areas with high densities of fountain grass, medium to shallow soils, away from watercourses, and away from trees (because of their sensitivity to the herbicide).

Other herbicides are less effective in controlling fountain grass. Foliar applications of glyphosate (as Roundup®) are not consistently effective (Tunison *et al.* 1994). Under greenhouse conditions (Neal and Senesac 1991), preemergent herbicide toxicity was greatest using metolachlor granular application (6.7 kg ai/ha) or spray (6.7 kg ai/ha), napropamide 50 percent wetable powder (4.5 kg ai/ha), and oryzalin (0.48 kg/l aqueous suspension at 4.5 kg ai/ha). All of these combinations injured and reduced growth of fountain grass, but did not kill it at the rates used. Not all of these are registered for use in wildlands in California. Mechanical methods combined with chemical techniques may be more effective than either method alone.

Phalaris aquatica L.

Common name: Harding grass

Synonymous scientific names: *Phalaris stenoptera, P. tuberosa* var. *stenoptera*

Closely related California natives: 4

Closely related California non-natives: 5

Listed: CalEPPC B; CDFA nl

by Kerry C. Harrington and W. Thomas Lanini

HOW DO I RECOGNIZE IT?

Distinctive Features

Harding grass (*Phalaris aquatica*) is an erect, waist-high, stout perennial grass with grayish to bluish green leaves. Flowering heads are dense, spike-like, and usually two to five inches long.

It is slow to develop from seed, but can form large bunches after several years. Harding grass is similar to three other species: littleseed canary grass (*Phalaris minor*), canary grass (*P. canariensis*), and reed canary grass (*P. arundinacea*). Unlike these other species, the base of the Harding grass stem often produces a reddish sap when cut. The other three species have winged glumes, although the wing is widest in the upper third of the glumes of littleseed canary grass and canary grass (Wheeler *et al.* 1982). The wing margins of littleseed canary grass usually have tiny teeth, whereas the margins are entire in the other two species. Both littleseed canary grass and canary grass differ in that they are annuals, they commonly have shorter inflorescences, and they lack the tuberous swelling often found at the base of the stem of well established Harding grass plants. Reed canary grass (*P. arundinacea*) differs from Harding grass in having more distinct rhizomes and an inflorescence that is compact at first but later becomes more open as the branches spread. Hybrids of Harding grass and reed canary grass have been produced.

Description

Poaceae. Perennial, tufted, erect, deep-rooted, rather stout grass, 3-4 ft (1-1.3 m) tall. Leaves: blades grayish to bluish green, rolled at emergence, 4-15 in (10-40 cm) long, 0.25 to 0.75 in (6-18 mm) wide, hairless, with membranous ligules 0.10 to 0.40 in (3-10 mm) long and no true auricles. Inflorescence: dense, cylindrical, spike-like, 2-5 in (5-12 cm) long and 0.6 in (1.5 cm) broad, becoming narrower toward the tip. Primary and secondary panicle branches very contracted, with numerous short-pediceled spikelets. Spikelets 0.20-0.25 in (5.5-6.5 mm) long, laterally flattened, all similar, and at maturity not falling entire but disarticulating above persistent glumes. Three florets per spikelet, only the uppermost fertile, lower two reduced to sterile lemmas. Glumes almost equal, membranous, the length of the spikelet and enclosing it, 3-nerved, strongly keeled along the mid-nerve with a row of short spines on the back in the upper half; lateral nerves also prominent, green, margins papery and white. Two sterile lemmas reduced, narrow, pointed, curved, and hairy, about 0.06 in (1.5 mm) long, not supporting paleas. Upper lemma 5-nerved, rounded, and fairly densely hairy on the back, becoming shiny at maturity, about 3/4 the length of glumes. Palea faintly 2-nerved, nerves close together along the mid-line, hairy in upper half on back between nerves. Three anthers, 0.10 in (2.5 mm) long (Tothill and Hacker 1983).

WHERE WOULD I FIND IT?

Harding grass is widespread in California because it has been used as a forage species and for revegetating after fires. It is most common in coastal valley and foothill grasslands from Oregon to the Mexican border. It is also found in the Sacramento and San Joaquin valleys at elevations below 4,000 feet (1,200 m). Harding grass is typically found along roadsides that are seldom defoliated, allowing this tall, erect, leafy plant to dominate neighboring vegetation. It is also frequently found beside ditches and streams because it tolerates wet soil conditions. However, it also tolerates dry conditions because of its deep root system. It can be found on a wide range of soil types, growing best in high-fertility conditions but tolerating low-fertility soils (Lambrechtsen 1992).

WHERE DID IT COME FROM AND HOW DOES IT SPREAD?

Native to the Mediterranean region, it has been dispersed throughout the world by agronomists and farmers for its value as forage in pastures. Its main agronomic value is its ability to tolerate conditions of low moisture, heavy grazing, and winter pugging by livestock (Langer 1990). Once planted widely for forage, it continues to colonize new areas through spread by seed. Seeds are disseminated short distances primarily by wind and by animals, while long-distance spread is through human activity.

WHAT PROBLEMS DOES IT CAUSE?

In wildland habitats Harding grass outcompetes and displaces native plant species. Tall stands of its dry foliage can present a fire hazard in summer. Although valued as a forage plant, it can cause a condition known as "staggers" in sheep. Staggers is characterized by respiratory distress, poor muscle coordination, and even death in some animals. Harding grass is less palatable to animals when plants are mature (Langer 1990).

HOW DOES IT GROW AND REPRODUCE?

Harding grass reproduces by seed. Flowering occurs in May and June, as soils dry after winter rains, with viable seed formed between May and September. A significant amount of seed is produced each year by established plants, up to 40,000 seeds per square meter under some conditions (Reddy *et al.* 1996). However, seed production varies considerably with plant density, soil type, and weather conditions.

It is not known how long seed remains viable in the soil. Seeds will germinate whenever moisture is available, although germination rates decrease as temperatures drop below 50 degrees F (10 degrees C) or rise above 85 degrees F (30 degrees C) (Charlton *et al.* 1986). Seedlings will establish successfully only if there is minimal competition. Although Harding grass is an aggressive competitor once established, it has weak growth as a seedling.

Harding grass can tolerate some shade but prefers open ground. Although it can tolerate dry conditions, it typically goes dormant over summer if moisture is limited, as in most areas of California, then recommences active growth with fall rains. Plants grow actively through autumn, winter, and spring,

producing much new seed in spring before the onset of dry conditions. Individual plants can spread laterally by production of short rhizomes (underground stems), allowing clumps to form, but this spread is minimal compared with reed canary grass, which has a much more extensive rhizome system.

HOW CAN I GET RID OF IT?

Physical Control

Mechanical methods: Cultivation is generally not an effective method of control because Harding grass produces an abundant seedbank and can also regenerate from short pieces of rhizome left in the ground. Repeated cultivation when plants are actively growing would be necessary. Active growth corresponds to the time of frequent rainfall, which limits the ability to cultivate. However, cultivation may be used to remove a flush of seedlings and reduce the seedbank.

Mowing: Kay (1969) observed that close mowing or clipping late in the growing season can greatly reduce the vigor of Harding grass. Mowing should be done when plants are still green but seasonal soil moisture is almost exhausted.

Prescribed burning: Although most studies on burning Harding grass were seeking to increase its productivity, burns made after mid-January were injurious to this species. Injury may have resulted from damage to young shoots. Recovery from fire was slow.

Mowing and irrigation can be used to stimulate new growth of Harding grass. New growth can then be treated with glyphosate or fluazifop, resulting in high mortality. Grazing can be used in place of mowing, but in either case, at least ten to twelve inches (25-30 cm) of regrowth is needed before an herbicide application.

Biological Control

Insects and fungi: No insects or fungi are known to be effective in controlling Harding grass.

Grazing: Livestock or geese can help to reduce the abundance and/or vigor of Harding grass. Grazing alone would not be expected to provide sufficient control. However, mob stocking (high numbers of livestock in a small area) or confined grazing with geese (grasses are the preferred food of geese) could be used in combination with other treatments to reduce Harding grass infestations.

Competition: Once Harding grass plants have been killed, new seedlings will soon establish if the soil is left bare. However, young seedlings are susceptible to competition, which is why Harding grass pastures can be difficult to establish (Langer 1990). To prevent reestablishment in habitats where Harding grass is not wanted, desirable species should be planted that will smother seedlings as they germinate. If broad-leaved species or fine fescues are planted, fluazifop can be used to selectively remove any young Harding grass seedlings that manage to emerge.

Chemical Control

Post-emergence control: Spot treatment with a 2 percent solution of glyphosate applied as a foliar spray to actively growing plants will kill Harding grass (Parsons 1992). A broadcast rate of 1.5 to 2.0 lb ai/acre is effective for large infestations. Ideal timing for this treatment is either at the early heading stage of development (mid- to late spring) or in early fall. Fluazifop at 0.25

to 0.375 lb ai/acre will kill actively growing Harding grass. Harding grass that is suffering from water stress will not be controlled. Activity is slow with this herbicide, often taking two to four weeks. With both fluazifop and glyphosate, repeat applications should be made if regrowth occurs or to control plants not killed by the first treatment.

Preemergence control: Sulfometuron applied at 2.25 to 3.75 oz ai/acre provides good control of Harding grass. Applications to the soil must be made prior to the start of seasonal growth. This highly active herbicide must be moved directly into the soil by rainfall or irrigation to prevent offsite movement. Bromacil, hexazinone, and simazine are other preemergence herbicides that provide good control of Harding grass. Both bromacil at 5.5 to 8.5 lb ai/acre and hexazinone at 3.0 to 6.0 lb ai/acre can provide control of seedling and established Harding grass (O'Conner 1996), while simazine is effective only for germinating seedlings. Use of these preemergents is limited primarily to non-crop sites, and they must be applied prior to the initiation of Harding grass seasonal growth.

Retama monosperma Bailey

Common name: bridal veil broom

Synonymous scientific names: *Genista monosperma, Spartium monosperma, Lygos monosperma*

Closely related California natives: 0

Closely related California non-natives: 3

Listed: CalEPPC red alert; CDFA nl

by Edie Jacobsen

HOW DO I RECOGNIZE IT?

Distinctive Features

Bridal veil broom (*Retama monosperma*) is a large shrub, ten feet tall or more, and when full grown may be twenty feet across. Plants are gray-green, almost entirely leafless, with slender branches drooping in weeping willow fashion. Young plants are usually wispy, with a single stem and strong taproot. Young plants are most visible in late summer and early fall when annual grasses are brown. Flowers are small, white, and pea-like, appearing in short clusters from the stems. There is usually one seed per pod, as indicated by its specific epithet. Spanish and Scotch broom can be easily distinguished from this plant by their larger golden yellow flowers. Spanish and Scotch broom also have more seeds per pod, four to six and five to twelve, respectively.

Description

Fabaceae. Tall ornamental woody shrub, to 10 ft (3 m) or more. Stems: unarmed, deciduous, slender grayish branches, nearly leafless; multi-stemmed, with branch diameters to 6 in (15 cm). Leaves: small, simple or rarely trifoliate, blades generally linear or linear-spatulate, silky. Inflorescence: flowers in short lateral racemes. Flowers: papilionaceous, white, fragrant; corolla silky, 0.4-0.6 in (10-17 mm); calyx 2-lipped 0.12-0.2 in (3-5 mm), with upper lip deeply 2-parted; style incurved, purple; keel cuspidate. Fruit: a legume; pod broadly oval, 1-2 seeded, dehiscent (Bailey 1933, Tutin *et al.* 1980).

WHERE WOULD I FIND IT?

Bridal veil broom is known to occur in the wild in California only on and adjacent to the Naval Ordnance Center, Pacific Division, Fallbrook Annex in San Diego County. It occurs in granitic sandy loam soils that are well drained. Associated plant communities are Diegan coastal sage scrub, southern perennial grassland, and disturbed annual grassland. It occurs mostly in disturbed annual grassland and disturbed Diegan coastal sage scrub. The plant is suited to mediterranean climates on rocky slopes and sandy banks. Planted as an ornamental species in south and central coastal California, it may escape and become established elsewhere. It is unknown at this time if other coastal habitats are threatened by this plant.

WHERE DID IT COME FROM AND HOW DOES IT SPREAD?

Bridal veil broom is native to the Mediterranean region, chiefly Spain (Iberian Peninsula) and North Africa, where it occurs on sandy soils near the coast and at border marshes. In its native habitat it is a minor component of the coastal scrub community (Valdesi *et al.* 1987). It was cultivated as an ornamental in England as early as 1690 and has been known to escape from cultivation there. There are a few records indicating some cultivation of this plant in California since 1916, with one record from San Diego in 1917 (McClintock, pers. comm.). A new invasive exotic in San Diego County, it was first noted six years ago on the Naval Ordnance Center near the city of Fallbrook. At that time it was growing in a few scattered patches covering less than ten acres. When eradication measures were implemented in 1996, it occupied over 2,000 acres of the Annex. It is believed that this plant escaped from a nearby nursery. Bridal veil broom spreads primarily by seed. Each plant produces hundreds, and on larger plants thousands, of seeds. It appears that ants and birds are the primary seed dispersers. At present, little is known about its physiology and reproductive mechanisms.

WHAT PROBLEMS DOES IT CAUSE?

Bridal veil broom crowds out other vegetation and can dominate grasslands and disturbed habitats. It is a threat to the habitat of the federally endangered Stephen's kangaroo rat and the threatened California gnatcatcher. As it becomes dominant in the landscape it reduces the availability of vegetation to support sensitive species. No nesting or foraging in and among these plants has been noted. There is no direct evidence of allelopathy at this time. It is unknown whether this plant increases fire risk. It has been in the natural landscape for such a short time that there has been no large die-off of bridal veil broom or large build up of dead litter on the ground in infested areas. However, its growth form of large, finely branched limbs suggests that plants would burn readily.

HOW DOES IT GROW AND REPRODUCE?

Specific data regarding reproduction is unavailable. However, peak flowering is February-April, and plants have been observed flowering year round. Each plant produces large numbers of seed. Seeds are dispersed from the pods ballistically and then probably further dispersed by ants since they have elaiosomes (Bossard, pers. comm.) and rain runoff. Seeds remain viable in the soil for several years.

Bridal veil broom is a shrub, but detailed data regarding growth patterns is unavailable. Rapid growth is apparent during peak growth periods in spring and early summer. Annual growth rates are not known, but up to three feet (1 m) of growth has been noted within one year following cut-stump treatment. Young plants appear to grow one to two feet in the first year. The plant develops deep taproots. Bridal veil broom spreads readily and can resprout from the crown. Large stands of over 200 plants per acre have been noted. Where solid, homogeneous stands had not developed prior to eradication efforts, it appeared likely that they would develop in short order.

The plant appears to readily decompose upon cutting, and there is no direct evidence that it spreads by shoot fragments.

HOW CAN I GET RID OF IT?

A preferred method of control has not yet been developed. This species is known in California from only one population, and current eradication methods appear to be eliminating it there. Continued monitoring and control efforts will be necessary until the seedbank is depleted. Annual monitoring in summer following chemical treatment is recommended. Young plants are most easily seen in summer, and sensitive species such as the California gnatcatcher are not nesting at that time.

Physical Control

Mechanical removal: Cutting stems off near the ground with saws (brush saw and chain saw) or loppers will stress the plant. Cut plants resprout vigorously, so cutting alone will not kill the plants (Jacobsen, unpubl. data).

Prescribed burning: Burning has not been investigated for this species, but may be useful for clearing the seedbank as for other brooms (Bossard 1993).

Biological Control

No USDA approved biological controls are available. However, some plants were killed by gophers eating the roots. At this time nothing is known about insects or fungi that could be used to control bridal veil broom. No browsing by cattle has been noted.

Chemical Control

Foliar application of glyphosate (as Roundup®, 2.5 percent) will kill young plants less than about two feet tall. Medium (up to six feet) and large (twenty feet in diameter) plants will be stressed but not killed. Foliar application of triclopyr (as Garlon 4®, 10 percent) on young and medium-sized plants will achieve approximately 50 percent kill. Basal bark application of triclopyr (as Garlon 4®, 25 percent) yielded the same results but is less likely to impact non-target species.

A combination of cutting with chain saws and applying 10 percent triclopyr (as Garlon 4®) solution to stump surfaces severely stressed the plants but killed only 50 percent of them. A subsequent summer application achieved approximately 90 percent mortality.

Ricinus communis L.

Common name: castor bean

Synonymous scientific names: none known

Closely related California natives: 0

Closely related California non-natives: 0

Listed: CalEPPC B; CDFA nl

by Cindy Burrascano

HOW DO I RECOGNIZE IT?

Distinctive Features

Castor bean (*Ricinus communis*) is a perennial shrub, sometimes tree-like, three to fifteen feet tall, with large, palmately lobed leaves and sharply toothed leaf margins. The leaves are usually deep green, but in some strains they have a reddish cast. They have an odor when crushed. The stems are smooth, round, and frequently red, with clear sap. The flowers are small and greenish, with both male and female flowers on the same plant. The fruit is a quarter-sized, round, spiny capsule, often reddish, containing up to three shiny, smooth, mottled seeds that resemble ticks.

Description

Euphorbiaceae. Shrub to 15 ft (5 m). Leaves: simple, 4-16 in (10-40 cm) broad, palmately 5-11 lobed with serrate margins. Leaves are alternate on the stem and peltate (petiole attached inside the edge of the blade rather than along the edge as in most species). Petioles with conspicuous glands and stipules fused and sheaf-like. Inflorescence: terminal panicle, 4-12 in (10-30 cm) long. Flowers: plants are monoecious (separate male and female flowers on the same plant). Male flower has many stamens on much-branched filaments and a 3-5 parted calyx; female flowers (located above male flowers) have 3 red styles united at the base and a calyx that falls early. Styles are bifid, plumose, and red. There are no petals, nectaries, or disks on flowers of either sex. Fruit: capsule with three 2-valved carpels, 0.4-0.8 in (1-2 cm) in diameter with soft spines. One seed in each carpel. Seeds glabrous, shiny, 0.12-0.16 in (3-4 mm) wide, 0.4-0.9 in (9-22 mm) long, and flattish, oblong-ellipsoid, variously marked and colored more or less mottled lustrous-silvery and brown with a fleshy appendage at one end (Hickman 1993, Parsons 1992).

WHERE WOULD I FIND IT?

Castor bean is widespread in the southern United States, where it has been introduced and naturalized. It is grown as a cultivated crop in California, Illinois, Missouri, Kansas, Oklahoma, and Oregon. In California, outside cultivation, castor bean has naturalized below 1,000 feet (300 m) elevation in the southern San Joaquin Valley, along the central and south coast, in the San Francisco Bay Area, and in Trinity County. It grows as a shrub in mild climates such as coastal southern California, but can grow as an annual in colder climates (Munz 1974).

Castor bean is frequently found in riparian areas, especially along the south and central coast, where it invades and displaces native vegetation. This plant is also common as an escape in abandoned fields, drainages, ditches, and along roadsides and railroad tracks. It is killed by low temperatures (Robbins *et al.* 1941), and as little as twenty-four hours at 2 degrees F is sufficient to produce visible impacts on cellular membranes of seedlings at any stage of germination (Breidenbach *et al.* 1974). Distribution is limited by castor bean's intolerance of cold temperatures and the inhibitory effect of low humidity or water stress on photosynthesis (Dai *et al.*

1992). It is tolerant of a wide range of soil types and conditions. Plants tend to germinate more profusely in full sun (Kitz, pers. comm.).

WHERE DID IT COME FROM AND HOW DOES IT SPREAD?

Castor bean is native to warmer parts of Asia and Africa (Robbins *et al.* 1941). This plant has been cultivated as an oil crop (Whitson 1992) and as an ornamental (Hogan 1992). It has been used as a purgative and an industrial oil, and seeds are currently sold as a gopher deterrent. Castor bean spreads by seed and is capable of resprouting from the root crown if cut. The seed pods dehisce when ripe, spreading seeds near the parent plant (Parsons 1992). Seeds can be carried to new locations by moving water or by transport of soil. Seeds also may be spread by road maintenance machinery. The plant is not known to spread by root fragments.

WHAT PROBLEMS DOES IT CAUSE?

Castor bean displaces native plant species in riparian areas and drainages. Its seeds are among the first to germinate following fire. Plants colonize disturbed areas, and they grow rapidly, shading out native seeds and seedlings and producing monospecific stands in areas with previously healthy native vegetation.

Castor bean seeds are highly toxic to humans, cattle, horses, rabbits, sheep, pigs, goats, gophers, cats, dogs, and poultry (Robbins *et al.* 1941, Cooper and Johnson 1984, Reynolds 1996). Ingestion of two beans can be lethal to humans, and the toxin and its mode of action have been well characterized and utilized in cancer treatment (Saelinger 1990). Ricin, the toxic water-soluble protein that can make castor bean deadly, is at its highest concentration in the seeds, but is also found in the leaves. The seed coat must be damaged to allow water to penetrate the seed interior for ricin to be absorbed in the intestines (Cooper and Johnson 1984). Most reports of animal deaths are associated with livestock, but several thousand ducks in the Texas Panhandle reportedly died from ingesting castor beans in fall and winter 1969-71 (Jensen and Allen 1981). Aphids are susceptible to poisoning by ingesting the phloem, and European corn borer and southern corn rootworm larvae were killed when exposed to feed painted with 2 percent ricin

(Olaifa *et al.* 1991). Parasitic soil nematodes have been shown to decrease in number in soil associated with castor bean plants (Fuller and McClintock 1986)

Castor oil is associated with allergic reactions (Lodi *et al.* 1992), and farm workers exposed to castor beans in Brazil and India have developed allergic asthma and undergone anaphylaxis from castor bean dust (Mendes 1980, Challoner and McCarron 1990).

HOW DOES IT GROW AND REPRODUCE?

Castor bean reproduces by seed. Plants become reproductive in the first season (within six months) and are capable of flowering year round in a frost-free environment. A single large plant 10.2 feet (8 m) diameter was found to produce 150,000 seeds, while a smaller plant thirty-nine inches (1 m) diameter produced only 1,500 seeds. Seeds generally germinate in December-April, but depending on weather and soil moisture, plants may germinate at other times of year.

The plant can grow 6.5 feet (2-5 m) in a single season in full sun with plenty of heat and moisture (Hogan 1992). Castor bean is a tropical plant that has a high photosynthetic capacity with high humidity (Dai *et al.* 1992). Low humidity strongly inhibits photosynthesis because of stomatal closure. Castor bean is not found in arid places. Freezing temperatures kill seedlings, but plants in coastal areas tend to grow and bloom year round as perennials. In areas subject to frost, plants grow as annuals. A long frost-free period is needed for seeds to develop. Shade tends to inhibit germination and produce smaller, slower-growing plants. Areas of native vegetation subjected to fire have produced solid stands of castor bean, although castor bean plants had been absent from the area for more than ten years, suggesting that seeds of castor bean are long-lived (Kitz, pers. comm.). Plants resprout from root crowns when cut.

HOW CAN I GET RID OF IT?

The best method to remove castor bean depends on the size of the plant, soil type and moisture, and the importance of avoiding impacts to non-target species. Follow-up monitoring is always appropriate. If mature plants containing seeds are not removed from the site, efforts should be made to remove resulting seedlings.

Physical Control

Mechanical methods: Pulling plants by hand when small or in wet sandy soils is a feasible technique in most riparian areas. The bulk of the root should be removed. Plants broken at the root crown will regenerate with multiple shoots. Weed wrenches can be used to remove small to medium-sized plants. If soil is so dry that plants break off from the main root, chemical treatment is needed. Gloves should be worn for hand pulling.

Prescribed burning: Burning is not recommended for castor bean removal in coastal areas as it creates ideal conditions for habitat conversion. Seeds in or on the soil readily germinate after fire, and seedlings grow so rapidly that they outcompete other species, dominating the area and driving out desirable natives. Studies of the impacts of burns and multiple burns are not known. In areas subject to frost prescribed burns might be useful if they are timed appropriately.

Biological Control

Since castor bean is a crop in some places in the United States, there is no biocontrol program. A large number of diseases and pests are known to impact castor bean crops (Kranz *et al.* 1977). Mung moth, pink bollworm, scab, wilt, leaf spot, seedling blight, inflorescence rot, pod rot, rust spot, graymold, crown rot, stem canker, leaf blight, bacterial wilt, and angular leaf spots are known to impact castor bean crops, but are rarely seen in riparian wildland plants. Leaves rarely show evidence of herbivory, although occasional leaf browse recently has been seen in the Santa Monica Mountains (Kitz, pers. comm.). Grazing is not recommended because of the plant's toxicity to livestock and other animals.

Chemical Control

Foliar-sprayed 2 percent glyphosate (as Roundup®) can be used to kill mature shrubs. Foliar spray can impact non-target species. Cut-stump treatment with loppers or saws and 25 percent glyphosate can also be used to kill mature shrubs, eliminating collateral damage to non-target species and reducing herbicide introduced to the system, especially if the plant is large. Small saws (hand or chain) will be required for larger plants.

Robinia pseudoacacia L.

Common name: black locust

Synonymous scientific names: none known

Closely related California natives: 1

Closely related California non-natives: none

Listed: CalEPPC B; CDFA nl

by John Hunter

HOW DO I RECOGNIZE IT?

Distinctive Features

Black locust (*Robinia pseudoacacia*) is a tree that grows up to 100 feet tall. Its dark brown bark is very rough and has interlaced ridges. The dark green leaflets are smooth, and the margins lack teeth or serrations (entire), forming pinnate leaves eight to fourteen inches long. Leaflets have short, slender spines at the base. The fragrant, cream-colored, pea-like flowers are in elongate drooping clusters. The fruit is a flattened brownish pod two to four inches long, each containing four to eight seeds. Black locust differs from the native New Mexican locust (*Robinia neo-mexicana*) in the absence of hairs on leaves, flowers, and fruits.

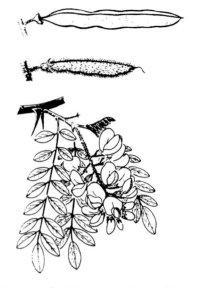

Description

Fabaceae. Deciduous tree 30-100 ft (10-30 m). Bark: brownish, thick, and furrowed, with forked ridges. Branches lack a terminal bud; generally have stipular spines alongside leaf scars. Leaves: alternate, odd-pinnate, once-compound, with 7-19 leaflets. Leaflets entire, elliptic to oval, and at the tip either rounded or truncate with a short spiny point. Inflorescences: pendant racemes borne in leaf axils of current year's growth. Flowers: generally whitish, <1 in (1.5-2 cm), papilionaceous (pea-like); calyx bell-shaped with 5 lobes; corolla consists of 5 petals, with larger upper petal (banner) reflexed; 9 of 10 stamens have fused filaments. Fruit: flat dehiscent pod with narrow wing and 4-10 seeds. Seeds dark brown to black, 0.15 to 0.25 in (0.4-0.6 cm) long (description based on Bailey 1949, Harlow *et al.* 1979, Huntley 1990, and Hickman 1993).

WHERE WOULD I FIND IT?

Black locust is widespread, particularly in northern California, below about 6,300 feet (1,910 m) elevation (Hickman 1993). It is also common in the Great Basin area. It can grow on a wide range of sites, but grows best on rich, moist, limestone-derived soils. It does not do well on heavy or poorly drained soils, although it appears to be tolerant of some flooding (Huston and Smith 1987, Huntley 1990). In the northeast United States it is found on floodplain sites with a 40 to 100 percent probability of flooding in any given year. Acceptable soil pH ranges from 4.6 to 8.2 (Huston and Smith 1987). Vogel (1981) reported the lower pH limit for black locust growth as 4.0. It does not tolerate shade, but otherwise occurs in many habitats, particularly on heavily disturbed lands such as roadsides and in stream bottoms and ravines. It is found in disturbed woodlands and ravines in Tahoe National Forest and John Muir National Monument.

WHERE DID IT COME FROM AND HOW DOES IT SPREAD?

Black locust is native to eastern North America from Pennsylvania and southern Indiana south to Georgia and Louisiana and west to Iowa, Missouri, and Oklahoma (Huntley 1990). The species has several desirable traits that led to its cultivation: showy fragrant flowers, strong rot-resistant wood, and nitrogen-fixing root nodules. Because of these traits, it was planted widely beyond its original range and has become naturalized throughout the United States and southern Canada (Huntley 1990). It was probably brought to California by the gold-rush era settlers, although it is not included in lists of early introductions (Jepson 1911, Frenkel 1970). It spreads by seed dispersal (by gravity and wind) and by root sprouts.

WHAT PROBLEMS DOES IT CAUSE?

Through root sprouts and seedling establishment, black locust creates large stands that displace native vegetation. Its seeds, leaves, and bark are toxic to humans and livestock (Hickman 1993).

HOW DOES IT GROW AND REPRODUCE?

Black locust reproduces both by seed and by root sprouts. It flowers in May-June. Fruits ripen in fall and open on the tree, dispersing seeds throughout fall and winter (Olson 1974). Seeds remain viable for ten years or more and require scarification for germination (Olson 1974, Strode 1977). Seedlings are intolerant of shade and herbaceous competition, but, once established, they are capable of growing over 3.3 feet (1 m) per year on better sites (Huntley 1990). Saplings begin producing seed at about six years. The best seed production occurs between fifteen and forty years of age. Seed production continues until about age sixty. Heavier seed crops are produced at one- to two-year intervals (Olson 1974, Huntley 1990).

Black locust leafs out in late spring, and shoots elongate rapidly. Trees reach mature heights in twenty to forty years (Vines 1960), but are short-lived. Decadence may begin at forty years of age, and trees rarely live more than 100 years (Strode 1977, Collingwood 1937).

Black locust produces root and stump sprouts. Sprout production is stimulated by top damage. Root suckers usually are more important to reproduction than are seedlings. Root suckers first appear when stems are four to five years old. Sprout production is greatest in full sun (Huntley 1990). Sprouting is an important mechanism for colonizing areas that have herbaceous plant cover but no woody canopy. Grasses form a sod that prevents establishment of black locust seedlings, but root sprouts are able to colonize these areas (Hardt and Forman 1989).

HOW CAN I GET RID OF IT?

If mechanical or chemical control is attempted, sites should be monitored at least twice each growing season. New root sprouts should be removed, and monitoring should be continued for one year after the last sprout is removed. Because there is also a dormant seedbank, sites should be checked for new saplings at least every other year for ten years.

Physical Control

Mechanical methods: Cutting or girdling a black locust stem will result in prolific root suckering. Mechanical removal therefore will be ineffective in controlling black locust unless all stems are cut several times per year. Repeated cutting of sprouts can kill the tree. Cutting probably will need to be repeated for several years. Mowing may not be effective in controlling seedlings and sprouts. More effective control can be obtained by immediately brushing the freshly cut surface of the stem with herbicide.

Prescribed burning: Burning has not been effective in controlling black locust. Fire may kill main stems, but this will result in prolific sprouting. Fire also may stimulate seed germination and create favorable conditions for seedling establishment.

Biological Control

Black locust suffers considerable damage from insects, particularly the black locust borer,

Megacylline robinine. However, no USDA biological control program for black locust has been attempted, and no USDA approved biocontrol agents exist for this species. Black locust suffers some browse herbivory, particularly the young growth of sprouts, which may aid eradication efforts (Huntley 1990, Luken 1992).

Chemical Control

Black locust has been effectively controlled with herbicides (Gouin 1979, Liegel *et al.* 1984, Scheerer and Jackson 1989, Smith 1993). Herbicide applications should be most effective in spring, just after leaves are fully expanded. Smaller sprouts may be controlled by spraying all foliage with 4 percent glyphosate (Chemical & Pharmaceutical Press 1997). Young stems may be killed by generously applying 15-20 percent triclopyr (as Garlon®) to the bark from the stem base to twenty inches above the ground (Gouin 1979, Chemical & Pharmaceutical Press 1997). The thicker bark of larger stems interferes with uptake of herbicide, and therefore, to kill larger plants, the stem needs to be frilled (have an encircling ring of bark removed) and the herbicide applied to the freshly exposed surface.

Applying herbicide to freshly cut stumps is probably the most effective means of controlling black locust. Wiping the stump with 100 percent glyphosate (as Roundup Ultra®) within fifteen minutes of cutting should reduce or even eliminate subsequent root suckering (Chemical & Pharmaceutical Press 1997).

Rubus discolor Weihe & Nees

Common names: Himalayan blackberry, Himalayaberry

Synonymous scientific name: *Rubus procerus, R. armeniacus*

Closely related California natives: 11

Closely related California non-natives: 5

Listed: CalEPPC A-1; CDFA nl

by Marc C. Hoshovsky

HOW DO I RECOGNIZE IT?

Distinctive Features

Himalayan blackberry (*Rubus discolor*) grows as a dense thicket of long, bending branches (canes), appearing as tall, ten-foot mounds or banks, particularly along watercourses. Canes have hooked prickles. Flowers are white, yielding black berries that usually ripen later than native blackberries.

Description

Rosaceae. Sprawling, essentially evergreen, glandless, robust shrub. The shrubs appear as "great mounds or banks" (Bailey 1945). Stems: some canes to 10 ft (3 m) tall, others decumbent, trailing, or scandent to 20-40 ft (6-12 m) long (Bailey 1923), frequently taking root at tips. Primocanes pilose-pubescent, becoming nearly glabrous with age, very strongly angled and furrowed, bearing well spaced, heavy, broad-based, straight or somewhat curved prickles 0.24-0.4 in (6-10 mm) long. Leaves: 5 foliolate, glabrous above when mature and cano-pubescent to cano-tomentose beneath. Hooked prickles on petioles and petiolules. Leaflets large and broad; terminal leaflet roundish to broad-oblong. Leaflets abruptly narrowed at the apex, unequally and coarsely serrate-dentate. Floricane leaflets 3-5 foliolate and smaller than on primocanes. Inflorescence: a large terminal cluster with branches in lower axils. Peduncles and pedicels cano-tomentose and prickly. Flowers: white or rose, 0.8-1 in (2-2.5 cm) across, with broad petals. Sepals broad, cano-tomentose, conspicuously pointed and soon reflexed, +/- 0.28-0.32 in (7-8 mm) long. Fruit: roundish, black, and shiny, up to 0.8 in (2 cm) long, with large succulent drupelets. Fruit

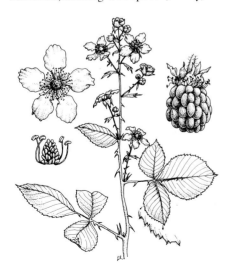

ripens late compared with native blackberries and over a considerable interval (Bailey 1945), from mid-summer to fall (Bailey 1923).

WHERE WOULD I FIND IT?

Himalayan blackberry occurs in California along the coast in the Coast Ranges, Central Valley, and the Sierra Nevada (Dudley and Collins 1995). It forms impenetrable thickets in wastelands, pastures, and forest plantations. It grows along roadsides, creek gullies, river flats, fence lines (Parsons and Amor 1968), and right-of-way corridors. It is common in riparian areas, where it establishes and persists despite periodic inundation by fresh or brackish water. Periodic flooding can produce long-lived early seral communities conducive to the growth and spread

of blackberries. Himalayan blackberry is one of few woody plants that pioneer certain intertidal zones of the lower Sacramento River (Katibah, *et al.* 1984).

Blackberries grow well on a variety of barren, infertile soil types (Brinkman 1974). These shrubs tolerate a wide range of soil pH and texture, but do require adequate soil moisture. Himalayan blackberry prefers disturbed and wet sites even in relatively wet climates. It prefers areas with an average annual rainfall greater than 76 cm on both acidic and alkaline soils (Amor 1972). It appears to be tolerant of periodic flooding by brackish or fresh water (Willoughby and

Davilla 1984). It grows at elevations of over 6,000 feet in Arizona and to 5,000 feet in Utah (Kearney *et al.* 1960, Welsh *et al.* 1987)

WHERE DID IT COME FROM AND HOW DOES IT SPREAD?

Himalayan blackberry is native to western Europe (Hickman 1993). There is no botanical evidence to show that it is native to the Himalayan region. It may have found its way there as a cultivar. Himalayan blackberry probably was introduced to North America in 1885 as a cultivated crop (Bailey 1945). By 1945 it had become naturalized along the West Coast. By this time it also occurred in nursery and experimental grounds along the East Coast and in Ohio (Bailey 1945). It seeds heavily, and seeds are readily dispersed by mammals and birds. Seeds can be spread considerable distances by streams and rivers (Parsons 1992). It also spreads vegetatively by rooting of cane tips.

HOW DOES IT GROW AND REPRODUCE?

Reproductive versatility is well represented in the genus *Rubus*, with sexual reproduction, parthenogenesis (development of the egg without fertilization), pseudogamy (a form of apomixis in which pollination is required), and parthenocarpy (production of fruit without fertilization) occurring widely. These modes of asexual reproduction contribute to the aggressive spread of blackberries.

Flowering begins in May and continues through July. Fruit is produced from July to September. Most blackberries produce good seed crops nearly every year. Immature fruit of Himalayan blackberry is red and hard, but at maturity fruit becomes shiny black, soft, and succulent.

Himalayan blackberry thickets can produce 7,000 to 13,000 seeds per square meter (Amor 1974). When grown in dense shade, however, most species of blackberry do not form seeds (Brinkman 1974). Seeds of blackberries are readily dispersed by gravity and by many species of birds and mammals. The large, succulent fruits are highly favored and, after they mature, rarely remain on the plant for long (Brinkman 1974). A hard seed coat protects the embryo even when seeds are ingested. Passing through animal digestive tracts appears to scarify seeds and may enhance germination. Prompt invasion of cut-over lands by Himalayan blackberry suggests that dispersed seeds can remain viable in the soil for several years (Brinkman 1974). Seeds germinate mainly in spring.

Blackberry seeds have a hard, impermeable coat and a dormant embryo (Brinkman 1974). Consequently, germination is often slow. Most blackberries require, at a minimum, warm stratification at 68 to 86 degrees F (20 to 30 degrees C) for ninety days, followed by cold stratification at 36 to 41 degrees F (2 to 5 degrees C) for an additional ninety days (Brinkman 1974). These conditions are frequently encountered naturally as seeds mature in summer and remain in the soil throughout the cold winter months.

In Australia Himalayan blackberry seedlings receiving less than 44 percent of full sunlight did not survive (Amor 1974). The slow growth of seedlings and their intolerance of shading suggest that few seedlings would be expected to survive in dense pastures or forest plantations. Blackberry thickets are also poor sites for seedling development. Amor (1972) counted less than 0.4 seedlings per square meter near thickets. Establishment of Himalayan blackberry seedlings depends on the availability of open habitats such as land neglected after cultivation, degraded pastures, and eroded soils along streams (Amor 1974). Although seedlings show the potential for rapid growth under laboratory conditions, they grow much more slowly in the field and are easily surpassed by the more rapid growth of daughter plants.

Himalayan blackberry can form roots at cane apices. Amor (1974a) observed canes growing to a height of 40 cm before they arched over and trailed on the ground. Daughter plants developed where these canes rooted, forming only on first-year canes. All canes produced berries in the second year and then died, senescence commencing near the middle and at the apices of canes without daughter plants. Reentry of canes into the center of the thicket resulted in an impenetrable mass of prickly canes within two and a half years. Individual canes may live only two to three years yet reach a density of 525 canes per square meter. A large quantity of litter and standing dead canes develops in old thickets.

Canes of Himalayan blackberry can grow to twenty-three feet (7 m) long in a single season. At one site observed by Amor (1974a), the mean horizontal projection of fifty first-year canes was eleven feet (3.3 m). Ninety-six percent of these canes had daughter plants at their apices. Lateral branches on some canes had also formed daughter plants.

The root crown on Himalayan blackberry, from which many lateral roots grow at various angles, can be up to eight inches (20 cm) in diameter. One root had a maximum depth of almost 3 feet (90 cm) and was more than thirty-three feet (10 m) long (Northcroft 1927). Adventitious shoots (suckers) are occasionally formed on the roots and may emerge from a depth of 45 cm. Blackberries also readily propagate from root pieces and cane cuttings (Amor 1974a). In less than two years a cane cutting can produce a thicket sixteen feet (5 m) in diameter (Amor 1973).

WHAT PROBLEMS DOES IT CAUSE?

Himalayan blackberry colonizes areas initially disturbed and then neglected by humans and can dominate range and pasture lands if not controlled. Himalayan blackberry is a strong competitor, and it rapidly displaces native plant species. Blackberries are highly competitive plants. Thickets produce such a dense canopy that the lack of light severely limits the growth of other plants. Because plants are prickly, livestock, particularly sheep and cattle, avoid grazing near them, effectively decreasing the usable pasture area. Young sheep and goats that get tangled up in the canes have been known to die of thirst and hunger. In wet areas blackberries may hinder medium-sized to large mammals from gaining access to water. The impenetrable nature of blackberry thickets reduces access for maintenance of fence lines and for forestry practices, as well as recreational pursuits. Dense thickets around farm buildings and fence lines are a considerable fire hazard.

HOW CAN I GET RID OF IT?

Mechanical removal or burning may be the most effective ways of removing mature plants. Subsequent treatment with herbicides should be conducted cautiously for two reasons. Himalayan blackberry often grows in riparian areas, where the herbicide may be distributed to unforeseen locations by running water, and some herbicides promote vegetative growth from lateral roots.

Reestablishment of Himalayan blackberry may be prevented by planting fast-growing shrubs or trees, since the species is usually intolerant of shade. Regrowth has also been controlled by grazing sheep and goats in areas where mature plants have been removed.

Physical Control

Mechanical methods: Most mechanical control techniques, such as cutting or using a weed wrench, are suitable for Himalayan blackberry. Care should be taken to prevent vegetative reproduction from cuttings. Burning slash piles is an effective method of disposal.

An advantage of cane removal over use of foliar herbicides is that cane removal does not stimulate sucker formation on lateral roots. Amor (1974b) provides evidence that herbicides such as picloram are not much more effective than cane removal. However, removal of canes alone is insufficient to control Himalayan blackberry, as root crowns will resprout and produce more canes.

Manual methods: Removing rootstocks by hand digging is a slow but effective way of destroying Himalayan blackberry, which resprouts from roots. The work must be thorough to be effective because every piece of root that breaks off and remains in the soil may produce a new plant. This technique is suitable only for small infestations and around trees and shrubs where other methods are not practical.

Himalayan blackberry plants may be trimmed back by tractor-mounted mowers on even ground or by scythes on rough or stony ground. Perennial weeds such as Himalayan blackberry usually require several cuttings before underground plant parts exhaust their reserve food supply. If only a single cutting can be made, the best time is when plants begin to flower. At this stage the reserve food supply in the roots has been nearly exhausted, and new seeds have not yet been produced. After cutting or chopping with mechanical equipment, Himalayan blackberry may resprout from root crowns in greater density if not treated with herbicides.

Prescribed burning: Burning is suitable for removing large thickets, but requires follow-up to control resprouts.

Biological Control

Insects and fungi: The USDA will not support introduction of herbivorous insects to control Himalayan blackberry because of the risk posed to commercially important *Rubus* species.

Grazing: Sheep, cattle, and horses can be effective in reducing the spread of Himalayan blackberry (Amor 1974). In New Zealand infestations have been controlled by the grazing of large numbers of goats. This method has been effective in preventing canes from covering large areas (Featherstone 1957). Crouchley (1980) mentions that blackberry is readily eaten by goats throughout the year, even when there is an abundant supply of other plants. In many areas of California the use of angora and Spanish goats is showing promise in controlling Himalayan blackberry (Daar 1983).

Chemical Control

Picloram suppresses cane regrowth of Himalayan blackberry but stimulates the development of adventitious shoots. Picloram is currently not registered for use in California wildlands. Foliage spraying is more effective in summer than in winter.

Many other herbicides have been used in efforts to control Himalayan blackberry with varying degrees of effectiveness. Fosamine can be effective (Shaw and Bruzzese 1979), and blackberry control has also been accomplished with amitrole-thiocyanate (Amor 1972), and triclopyr ester (as Garlon®) (McCavish 1980). Not all of these are currently registered for use in California.

Schinus terebinthifolius Raddi

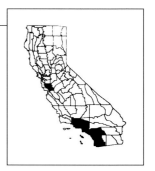

Common names: Brazilian pepper tree, Christmas-berry tree, Christmasberry, Florida holly

Synonymous scientific names: *Schinus mucronulata, S. antiarthriticus*

Closely related California natives: 0

Closely related California non-natives: *Schinus molle*

Listed: CalEPPC B; CDFA nl

by Jonathan J. Randall

HOW DO I RECOGNIZE IT?

Distinctive Features

Brazilian pepper tree (*Schinus terebinthifolius*) is a many-stemmed shrub or small tree. Its branches form a nearly impenetrable tangle down to ground level. The flowers are small, less than one-tenth of an inch in diameter, and have five green-tinged, white petals. The tree is covered with round, bright red fruit from December to February. The odd-pinnate compound leaves have a narrowly winged rachis and rounded, often toothed leaflets that give off a strong smell of turpentine when crushed (Tomlinson 1980).

Description

Anacardiaceae: Shrub or small tree to 40 ft (13 meters) tall with a multi-stemmed short trunk that is curved with grayish and often scaly bark. Leaves: 0.32-0.6 in (8-15 cm), generally opposite, with a somewhat resinous coating and emitting a distinctive odor. Usually 7-9 leaflets,

odd-pinnate, sessile or short-stalked, 1-2.8 in (2.5-7 cm) long, 0.4-1 in (10-25 mm) wide, elliptic to oblong, with edges entire to toothed. Inflorescence: panicle, either axillary or terminal. Flowers: 0.04-0.09 in (1.2-2.5 mm) long, with greenish white, oblong to egg-shaped petals and pedicels 0.08-0.16 in (2-4 mm). Fruit: drupes, 0.16-0.28 in (4-7 mm) in diameter, generally pink to red, arranged in dense bunches (Ewel *et al.* 1982). Each drupe contains a single seed (Ewel *et al.* 1982).

WHERE WOULD I FIND IT?

While it can be found in Santa Clara County in northern California, Brazilian pepper tree is far more common and problematic in southern California from Riverside to the coast, including Ventura and San Diego counties. Usually it is found below 200 meters elevation, especially in canyons and washes (Hickman 1993). Brazilian pepper is capable of surviving a broad range of hydrologic conditions, but does best in well drained sites (Ewel 1979). This species is a pioneer of disturbed sites, such as highway rights-of-way, fallow fields, and drained bald cypress stands. It can also establish in undisturbed wildlands (Woodall 1982). It causes severe problems in southern Florida, where it invades pinelands, mangrove forests, and marshes, and it is a major invader in croplands under restoration in Everglades National Park.

WHERE DID IT COME FROM AND HOW DOES IT SPREAD?

Brazilian pepper tree is native to Argentina, Paraguay, and Brazil, where it occupies dry savannah (Nilsen *et al.* 1980). It is a sparsely distributed species in its native range, where it rarely acquires the dominance that it achieves in California and other areas in which it has become naturalized. It was introduced into Florida in 1891 as an ornamental, although there is evidence that it arrived in the United States fifty years earlier (Gogue *et al.* 1974). It is still used as an ornamental in California, Texas, and Louisiana. Other than escaping from human cultivation in gardens, Brazilian pepper tree spreads via small mammals and birds, especially robins, which eat the drupes and then distribute the seeds during their travels. Mammals such as raccoons and possums consume the fruits and deposit the seeds with fecal materials, giving the seeds a nutrient-rich microsite in which to establish (Ewal *et al.* 1982). This species has infiltrated the islands of Hawaii, Bermuda, and the Bahamas, as well as Florida and, more recently, California. In addition to spreading by seed dispersal, it can resprout, especially after fire (Elfers 1995).

WHAT PROBLEMS DOES IT CAUSE?

Outside its native home, Brazilian pepper tree spreads aggressively. Typically, dense monospecific stands form within a few years after trees invade an area. This often creates a dense canopy and can shade out most competing vegetation, posing a serious threat to natural vegetation and organisms that depend on them (Bennett *et al.* 1988, Doren and Whiteaker 1990, Ewel 1979). The tenacity of Brazilian pepper seedlings impairs competition by native vegetation. Brazilian pepper also seems to produce allelopathic chemicals in the soil that inhibit growth of other plants (Bennett *et al.* 1988). The plant is moderately salt tolerant, withstands flooding, fire, and drought, and resprouts quickly after being cut. Its ability to resprout and grow rapidly allows it to quickly dominate burned areas.

HOW DOES IT GROW AND REPRODUCE?

Brazilian pepper tree reproduces by seed. Flowering begins in September, and by mid-October almost every tree is in flower. Most flowering ceases in early November. A small fraction of the population flowers in March-May. Fruit ripening follows close behind flowering, with most occurring between December and February (Ewel *et al.* 1982). Mature female trees are prodigious seed producers, which, combined with a viability rate of 30 to 60 percent, results in a vast number of seedlings (Elfers 1995). It has been suggested that female flowers may mimic male flowers, attracting foragers in search of pollen (Ewel *et al.* 1982). Fruit production occurs in winter (November to February), at which time the branches of female trees are heavily laden with red fruits.

Ripe fruits are retained on the tree for up to eight months, and all are dispersed before the

next flowering season (Ewel *et al.* 1982). The attractive fruits are readily eaten and transported by birds and mammals, with water and gravity serving as less important dispersal agents. Seed dispersal by native and exotic birds, such as catbirds, mockingbirds, American robins, and red-whiskered bulbuls, accounts for the presence of Brazilian pepper tree in almost every terrestrial plant habitat in southern Florida (Austin 1978, Ewel *et al.* 1982, Ewel 1986). Robins, when present, are believed to consume and transport more Brazilian pepper tree seed than all other dispersal agents combined. Raccoons and possibly possums are known to ingest the fruits (Ewel *et al.* 1982). The fact that little else is fruiting during the winter months when Brazilian pepper tree seeds are dispersed has been suggested as a possible explanation for the success of this plant in southern Florida (Ewel 1986).

Seeds are generally not viable beyond five months after dispersal. However, Ewel (1979) reported seed germination in late fall. Under certain conditions; seeds apparently retain their viability during the wet season and germinate when water levels drop later in the year.

Most seed germination takes place between January and February, but the range is from November to April. Seeds germinate within about twenty days. Seedlings have been found to grow quickly in most young successional communities and slowly in most older communities. Well watered conditions favor germination. Survival of Brazilian pepper tree seedlings, even of mature forest trees, is 66 to 100 percent (Ewel 1986). Ewel *et al.* (1982) concluded that the growth plasticity of Brazilian pepper tree seedlings makes this species especially difficult to manage. Seedlings can grow slowly under the dense shade of mature stands and then exhibit vigorous growth when the canopy is opened after a disturbance. In exposed, open areas, such as young successional communities, their growth rate is as high as twelve to twenty inches (30-50 cm) per year.

Under favorable growing conditions, Brazilian pepper tree can reproduce within three years after germination. This species can occur individually or as extensive stands; in areas with warm, tropical conditions, it occurs primarily in dense stands.

Like many hardwood species, Brazilian pepper tree can resprout from above-ground stems and root crowns following stump cutting, bark girdling, fire that girdles a stem, and herbicide application (Woodall 1979). Resprouting is often profuse, and growth rates of sprouts, which originate from dormant and adventitious buds, are high. Brazilian pepper tree's generally shallow root system also favors the production of underground root suckers. Root suckers form without evidence of damage to a tree or its root system and can develop into another plant. The clumping of trees often seen during the early stages of invasion can be explained by this suckering mechanism (Woodall, 1979).

HOW CAN I GET RID OF IT?

The severity of the problem is an important consideration when designing a control strategy for Brazilian pepper tree. In many cases, it is unrealistic to attempt eradication where this species is widespread; instead the goal should be to prevent colonies from establishing new infestations. Because this species is an invasive, habitat-altering weed in other states, especially Florida, research on control techniques has been underway for some time (Elfers 1995). With all control and removal techniques, a lack of sprouting for one or even two years may not guarantee that the sprouting potential of the roots is exhausted. Yearly monitoring for at least three years following control efforts is recommended. Monitoring should occur in late spring to determine if any shoots have survived or any resprouting has occurred in treated areas.

Physical Control

Manual/mechanical methods: Entire saplings, including root systems, can be pulled up by hand, but by the time the plant is several feet tall, hand pulling may be impossible. If as much as one-quarter of the root system is left in the ground, the plant may resprout. Using heavy equipment such as bulldozers, the entire plant, including the root system, can be removed.

Soil removal: A study in south Florida found that removing soil with a bulldozer down to the limestone bedrock effectively prevented the return of Brazilian pepper tree. It also stimulated the growth of a diversity of plant species. However, this site had little or no soil before it was rock-plowed and farmed, so removing the soil was part of a logical restoration program. Removing soil can also stimulate the growth of non-natives that thrive in disturbed areas.

Prescribed burning: Results of burning were mixed. Because they cannot tolerate heat, Brazilian pepper tree seeds will not germinate following fire, although basal trunk and root sprouting is aggressive. Once saplings attain a height of one meter, most are able to survive fire by regrowth from the roots. Since it can grow more rapidly than competing native hardwoods, Brazilian pepper tree can establish or reestablish dominance (Loope and Dunevitz 1981). Repeated burning does not prevent this plant from reinvading an area in which it was previously dominant (Doren and Whiteaker 1990).

Flooding: Prolonged submergence may result in increased seedling mortality (Ewel *et al.* 1982).

Biological Control

Insects and fungi: No USDA biological control agents have yet been approved by the USDA, but research is underway, especially in the heavily infested states of Florida and Hawaii. In 1988 a seed-eating wasp, *Megastigmus transvaalensis*, was reared on Brazilian pepper tree fruit collected from Palm Beach, Florida. Part of the controversy surrounding the development of plant-eating fauna for this species is the concern that insects that attack *Schinus terebinthifolius* will also attack the closely related *S. molle*, a common ornamental. A study in Hawaii reported that a caterpillar, *Epsismus utilis* and a beetle, *Bruchus atronotatus*, had the potential to limit seed production, but field tests found that these insects did not significantly control the plant (Clausen 1978).

Chemical Control

Triclopyr (as Garlon 3-A®), applied at 100 percent and using the frill-cut method, has been shown to kill mature trees and prevent regrowth. Depending on the setting, it may be advantageous to cut down the trees down and then apply herbicide to the stumps to prevent regrowth (Bily, pers. comm.).

Successful treatments for full-sized plants also include basal spot applications of bromacil and hexazinone, which kill by blocking photosynthesis. For widely scattered plants, where access to the main stem is difficult, basal spot treatments are easily applied and effective. Bromacil and hexazinone are selective, so nearby vegetation is not harmed. Since Brazilian pepper tree's transpiration rate per unit leaf area is unusually high, and since it generally occupies an emergent canopy position, it acts as a strong sink for soil-applied herbicides, thus minimizing leaching losses and off-target damage. When it is growing in the shade of other plants, this generalization does not hold (Woodall, 1982).

Foliar herbicides are the fastest acting with the least residual activity, although their probability of success is relatively low. Foliar herbicides that have been used with some success include 15 percent triclopyr (as Garlon 3A®) diluted with water or diesel, 2 percent triclopyr (as Garlon 4) diluted with water, and 2 percent dicamba (as Banvel 720®) diluted with water. Ammonium-sulfate and glyphosate are also foliage absorbed. All of these herbicides are most effective when applied to seedlings. The only herbicide that resulted in consistently killed roots as well as shoots was picloram (Woodall 1982), but this is not currently registered for use in California.

Schismus arabicus Nees   *Schismus barbatus* (L.) Thell.

Common names: Mediterranean grass, Arabian schismus, schismus, split grass

Synonymous scientific name: *Festuca barbata* (for *S. barbatus*)

Closely related California natives: 0

Closely related California non-natives: 0

Listed: CalEPPC B; CDFA nl

by Matthew L. Brooks

Schismus arabicus

HOW DO I RECOGNIZE IT?

Schismus barbatus and *S. arabicus* are so genetically and morphologically similar (Faruqui and Quarish 1979, Faruqui 1981, Bor 1968), with similar geographic ranges and habitats in California (Hickman 1993), that they are treated together here.

Distinctive Features

Mediterranean grass (*Schismus barbatus* or *S. arabicus*) is a small, tufted annual grass with erect or spreading, green, smooth culms to eight inches, often with brown nodes. Individual plants can remain rooted and upright for up to two years following death,

S. barbatus

but eventually detach at ground level and blow across the ground like a tumbleweed, falling apart in the process. Mediterranean grass is most common in the spaces between shrubs, often producing a carpet of green that turns purplish at maturity and fades to a light straw color soon after death.

Schismus arabicus

Description

Poaceae. Annual grass with green culms that ascend or spread to 20 cm. Leaves: usually inrolled; smooth except near the orifice where there is a ring of rigid hairs to 0.1 in (3 mm); blade <0.1 in (2mm) wide, thread-like. Inflorescence: a dense, narrow panicle; rachis disarticulating above glumes and between florets; green when young, aging to purplish. Spikelets laterally compressed; 5- to 10-flowered; glumes subequal, persistent, scarious margins, lanceolate, larger than lowest lemmas, 3-5 veined; lemma 9-veined, membranous, 2-

S. barbatus

Schismus sp.

lobed apex; palea shorter or as long as lemma; awnless. Seed: roundish, translucent, loosely bounded by palea and lemma; embryo to half the length of the caryopsis. Description adapted from Californian (Hickman 1993), Eurasian (Bor 1968, Tutin *et al.* 1980), and worldwide (Conert and Turpe 1974) treatments of the species.

Spikelet morphology is used to distinguish between *Schismus arabicus* and *S. barbatus*. In *S. arabicus* the apical lobes of the lemma gradually taper to sharp, narrow

Schismus sp.

points and are 30 to 50 percent of the total length of the lemma, whereas in *S. barbatus* the lobes are wider and have broadly pointed to rounded points that are 15 to 25 percent of the total length of the lemmas. The lowest floret in the spikelet should be examined, because the morphology of the lobes on the lemma becomes highly variable in higher florets. In *S. arabicus* the palea does not generally extend beyond the base of the fissure on the lemma, whereas in *S. barbatus* the palea extends beyond the base of the fissure and may be as long as the lemma lobes. The length of the glumes in the terminal spikelet of the inflorescence is generally <0.2 in (5mm) in *S. arabicus* and >0.2 in (5mm) in *S. barbatus*. These characteristics were adapted from Bor (1968).

WHERE WOULD I FIND IT?

Mediterranean grass is found below 4,250 feet (1300 m) elevation in disturbed and undisturbed areas of the central and southern coastal regions, the Central Valley, and the deserts of California. *Schismus arabicus* is generally more common in arid regions, whereas *S. barbatus* is more common in semi-arid shrublands, extending into the northern coast to Mendocino County. Mediterranean grass is most common in spaces between shrubs where it is not shaded by taller plants. It is widespread and common in the desert (Brooks 1998); its presence in coastal shrubland may not be readily apparent except on bare soil or following fire, where it can appear in great numbers.

WHERE DID IT COME FROM AND HOW DOES IT SPREAD?

Mediterranean grass is native to southern Europe, northern Africa, and the Near East (Jackson 1985) and has spread to areas of North America, South America, Australia, and the west coast of Europe where Mediterranean climate regimes occur (Bor 1968). It appears to have spread westward from Arizona into California during the early 1900s (Burgess *et al.* 1991), and was first recorded in California in 1935 (Robbins 1940). Mediterranean grass is particularly abundant where grazing, off-road-vehicle use, or construction of linear corridors has reduced shrub cover and disturbed the soil. Seeds disperse by sheet flooding and by wind and often persist within the inflorescence, detaching after it is blown across the ground for a short distance from the parent plant.

WHAT PROBLEMS DOES IT CAUSE?

Mediterranean grass rose from relative obscurity to become one of the dominant annual grasses in arid and semi-arid regions of California during the 1940s (Clarke, pers. comm.). As Mediterranean grass became more dominant, the similar native annual grass, six-weeks fescue (*Vulpia octoflora*) became less common (Clarke, pers. comm.). Mediterranean grass can compete effectively for limiting nutrients with native annual plants that occupy spaces between shrubs (Brooks 1998).

Fire is readily carried across inter-shrub spaces by the dead stems of Mediterranean grass (Brooks 1998, Brooks in press), which may have contributed to the increasing frequency and extent of fire in recent decades in California deserts.

HOW DOES IT GROW AND REPRODUCE?

Mediterranean grass reproduces by seed only. Seeds are tiny and dust-like (Loria and Noy-Meir 1979, 1980), and disperse into small cracks and depressions in the soil (Gutterman 1994). Only a fraction of the seedbank germinates during a given year, leaving most seeds in reserve for future years when the cohort may die prior to reproduction (Gutterman 1994). This bet-hedging strategy is characteristic of annual plants that have evolved in arid desert regions with locally variable rainfall patterns, and it predisposes Mediterranean grass to successful establishment in California deserts.

Mediterranean grass is a winter annual, germinating in early winter following 0.4 inches (10 mm) or more of rainfall and emerging about two weeks later. It grows little until early spring (typically March), when rainfall and higher temperatures stimulate accelerated growth and flowering. Progression from seedling to flowering stages can occur in as little as two weeks, which makes it one of the fastest-maturing desert annuals. Plants flower from March through May, or until they die of water stress. Mediterranean grass may also germinate in summer when supplied with artificial irrigation, and it can survive with no further irrigation for up to four months (Gutterman and Evanari 1994).

Mediterranean grass is generally intolerant of shading, which may explain its association with inter-shrub spaces. However, it can thrive under perennial shrubs, particularly under south canopies or where tall-statured annual plants are less common (Brooks 1998). The root system can form an extensive mat near the surface where plant litter is present.

HOW CAN I GET RID OF IT?

Physical Control

Mechanical methods: Its small size makes hand thinning of Mediterranean grass impractical. In addition, the extensive mat of roots near the surface of the soil often results in significant disruption of the soil surface when plants are removed, which may promote further weed establishment. Plowing, disking, or scraping may initially reduce surface biomass of Mediterranean grass, but soil disturbance and reduced shading results in improved site conditions for this species.

Prescribed burning: Fire generally promotes the growth of Mediterranean grass. Its small seeds settle near or beneath ground level, where they are protected from high temperatures. Significant seed death occurs only under perennial shrubs where intense burning heats deeper into the soil. After fire has removed plant litter and taller competitors, and in some cases has increased soil nutrients, Mediterranean grass can dominate until plant litter and populations of taller annual plants become reestablished. Prescribed burning generally is counterproductive to the control of Mediterranean grass. However, if the choice is between controlling Mediterranean grass or more problematic annual weeds that may be controlled by certain fire prescriptions, then the latter should usually receive priority.

Biological Control

Insects and fungi: No USDA approved insects or fungi to be used as biocontrol agents exist for these grasses. Ants feed on the seeds of Mediterranean grass in its native range (Gutterman 1993) and in North America (Rissing 1988). Ants generally harvest seeds while they are still concentrated within the inflorescence. After dispersal the seeds spread out across the landscape and settle into small cracks in the soil, thereby hindering predation (Gutterman 1994).

A black smut, *Ustilago aegyptica*, can form on Mediterranean grass, destroying the spikelets (Gilbertson and Blackwell 1988). Natural infestations of smuts do not seem to be widespread or severe enough to significantly affect populations of Mediterranean grass in California. However, inoculation of smuts into areas of the desert could prove to be useful for short-term control of Mediterranean grass and other alien annual grasses, as long as they do not affect native perennial grasses. Especially in years of abundant rainfall, smuts may reduce the density of alien annual grasses prior to revegetation efforts. This would have to be researched and approved by the USDA and the CDFA Biological Control Program before such distribution would be permitted.

Grazing: Livestock grazing can remove biomass of Mediterranean grass. However, the relative biomass of alien annual grasses compared to native annuals tends to increase following moderate to intense grazing. Mediterranean grass is particularly dominant in ephemeral sheep grazing allotments in the California desert. As with mechanical plowing, reduced shrub cover and increased soil disturbance caused by grazing ultimately improve site conditions for Mediterranean grass.

Chemical Control

Various herbicides, including glyphosate, can control Mediterranean grass, but the small surface area of the leaves and culms make application problematic. In addition, broadcast herbicides would negatively affect non-target species.

Senecio jacobaea L.

Common names: tansy ragwort, stinking willie

Synonymous scientific names: none known

Closely related California natives: 36

Closely related California non-natives: 5

Listed: CalEPPC B; CDFA B

by Steve Harris

HOW DO I RECOGNIZE IT?

Distinctive Features

Tansy ragwort (*Senecio jacobaea*) is a perennial herb in the dandelion family that sometimes reaches more than four feet in height. Numerous inch-wide, daisy-like yellow flowerheads with golden or light brown centers form at the tip of each branch from mid-summer to fall. The plant has a basal rosette of leaves, and the upper parts are branched. Leaves are deeply pinnately dissected into irregular segments, giving the plant a ragged appearance. Leaves or segments are wider than long. Ray flowers distinguish this plant from common tansy (*Tanacetum vulgare*).

Description

Compositae. Stems: from taprooted caudex, branched above, 4-50 in (10-120 cm). Leaves: evenly spaced, reduced upward; 2-8+ in (5-20+ cm), ovate, deeply 1-2-pinnately dissected; lower leaves soon deciduous. Inflorescence: heads radiate, 20-60; 13 phyllaries, 0.15-0.2 in (3-5 mm), tips green or black. Flowers: 13 ray flowers; ligules about 0.3-0.5 in (8-12 mm); 40 or more disk flowers. Fruit: glabrous or only edges of fruit hairy.

WHERE WOULD I FIND IT?

In California tansy ragwort is found on the North Coast and into the Klamath and Cascade ranges, also occurring in the Sacramento Valley and San Francisco Bay region. It is commonly found in pastures, on roadsides, and in disturbed places. It grows best in light, well drained soils, but can become established in heavier soils, particularly soils broken up by trampling or frequent cultivation.

WHERE DID IT COME FROM AND HOW DOES IT SPREAD?

Tansy ragwort is endemic to parts of western Europe and western Asia, where it is a weed of minor importance on roadsides and grasslands. It is considered by some authors to be native to the dunes of Holland (van der Meijden 1971). From Europe it has been introduced into Argentina, New Zealand, Australia, Canada, and the western United States (Schmidl 1972a).

Tansy ragwort seeds, like those of other members of the Asteraceae, are assumed to be primarily wind-dispersed (van der Meijden 1971). Plants are commonly found along roadsides in otherwise uninfested areas (Holden 1989).

Animal transport of ragwort achenes seems likely. The pappus and bristle hairs of disk achenes would allow attachment to fur and feathers. Plants have been commonly observed along deer and elk trails through otherwise uninfested regions (Holden 1989).

Tansy ragwort has been transported in infested hay. This mode of dispersal was postulated to explain several isolated infestations observed in eastern Oregon when hunting parties carried hay into previously uninfected areas (Cox and McEvoy 1983).

The plant can also spread via regeneration of root fragments contained in mud or soil adhering to vehicles.

WHAT PROBLEMS DOES IT CAUSE?

Tansy ragwort is highly toxic to livestock. It contains pyrrolizidine alkaloids that cause liver damage in horses and cattle (Cheeke 1979). Cattle and horses are affected seriously; goats may suffer poisoning; sheep are generally not poisoned by this plant. Tansy ragwort easily outcompetes native and naturalized grasses and forbs (Harper 1958). It is estimated to occur on three million acres in western Oregon, where two out of every five acres of pasture are infested. More than 112,000 acres of pasture in western Washington contain tansy ragwort. It has been reported in Idaho and is there considered a serious potential weed problem that has not yet reached a level of economic importance. Most ragwort is on forested and clearcut lands (Bedell *et al.* 1995).

HOW DOES IT GROW AND REPRODUCE?

Tansy ragwort is a biennial species. The first-year plant is a rosette of basal leaves that are raggedly lobed and to nine inches (22.5 cm) long. In the second year the first-year leaves die back and the plant develops tall, leafy flowering stems in summer. It is ordinarily about three and one-third feet (1 m) in height, but plants to ten feet (3 m) are not uncommon, with many branches near the top (Holden 1989).

Tansy ragwort is capable of regeneration from pieces of rootstock. Roots without any portion of the root crown are able to produce new plants, and even severed roots of cotyledons can

grow new shoots (Holden 1989). Poole and Cairns (1940) found that first-year plants, or rosettes, buried five inches (12.5 cm) in the soil sprouted three months later, even in heavy soil.

In addition to vegetative propagation, ragwort also reproduces by seed. Two studies at different sites recorded 4,760 to 174,230 seeds per plant (Camoron 1935) and 6,480 to 137,500 seeds per plant (Poole and Cairns 1940).

Ragwort inflorescences contain two types of flowers, an outer row of radiate (ray) female flowers and many central tubular (disk) perfect flowers. Each flower type produces distinctive single-seeded fruits (achenes) that have different dispersal mechanisms, germination requirements, and germination rates (Harper 1965, Burtt 1977). Seeds are tipped by hair-like plumes that carry seeds in the wind for long distances. Although both fruit types are primarily dispersed by wind, disk achenes are dispersed earlier and farther than ray achenes (Green 1937). Poole and Cairns (1940) found that 60 percent of the total seed shed landed within fourteen feet (4.6 m) of the base of the plant; an additional 39 percent landed between fourteen and twenty feet (4.6 and 9 m) from the plant.

Tansy ragwort seeds remain viable in the soil for several years. In this way the species can wait for favorable conditions to occur. McEvoy (unpubl. data) found 0.75 viable ragwort seeds/cc in the upper 1.2 inches (3 cm) of soil underneath a pasture heavily infested with ragwort. Viable seeds were encountered in samples taken as deep as ten inches (25 cm) below the surface.

On the Oregon coast, tansy ragwort seeds mature in late summer and early fall. Seeds reach maximum germination potential before dispersal; they do not possess an innate dormancy (van der Meijden and van der Waals-kooi 1979). There are two peaks of germination: fall and spring. However, some germination occurs year round (Harper and Wood 1957).

Soil moisture, soil surface humidity, and light are important factors in ragwort germination. Van der Meijden and van der Waals-kooi (1979) found germination of ragwort achenes to be greater at soil moisture levels between 15 and 29 percent and a relative humidity at the soil surface of 100 percent.

Ragwort needs an opening through some form of disturbance to become established (van der Meijden and van der Waals-kooi 1979). Several disturbance agents have been cited, including moles and gophers, ants, rabbits, livestock, and humans (Harper and Wood 1957).

HOW CAN I GET RID OF IT?

Physical Control

Mechanical methods: Deep plowing has been unsuccessful in controlling tansy ragwort (Holden 1989). This technique severs roots and distributes them over a wide area. Plowing also unearths buried seeds and can contribute to a more severe infestation (Holden 1989).

Mowing has been widely used in attempts to prevent the spread of ragwort, but has provided only cosmetic control. Repeated mowing could deplete reserves and eventually exhaust the population. Fields would have to be mowed every six weeks during spring and summer months and be accompanied by moisture stress (Cox and McEvoy 1983). Single mowings during flowering might intensify local infestations. Given that ragwort seeds ripen over a range of time and that severed capitula are able to produce seed, mowing might increase local seedling density while freeing the parent plant from reproductive burdens and engendering basal sprouting (Holden 1989).

Hand pulling has been the most common technique used on small pastures in the early stages of infestation. Soil moisture is critical, as drier soils allow root breakage and pulling in wet soils removes large soil clumps. Partially opened flowers continue to set viable seed if sufficient moisture is available in the cut plant. Therefore, plants must be removed from the treated site and buried or burned (Holden 1989).

Clipping or otherwise deflowering plants eliminates seed set and prevents further spread if flowerheads are collected and burned (Holden 1989). The original plant, however, continues to sprout, grow, and reflower in subsequent seasons.

Prescribed burning: Burning is traditionally used in agricultural croplands as a preemergence weed treatment. In ragwort control, Poole and Cairns (1940) experimented with a flame thrower, but their results were inconclusive and have not been reproduced by others. Mastroguiseppe et al. (1982) conducted several control burns at an infested site in Redwood National Park, California, but the results were inconclusive.

Biological Control

Insects and fungi: Biological control has proven to be effective for long-term control of extensive infestations of tansy ragwort. Several USDA certified biocontrol agents have been released for control of this weed. The cinnabar moth, *Tyria jacobaeae*, is a day-flying moth indigenous to Europe and Asia. Cinnabar larvae feed primarily on the flower buds, but readily consume leaves and stems as well. Unfortunately, the cinnabar moth is subject to disease, predation, and parasitism.

Tansy seedfly (*Pegohylemia seneciella*) is a small muscid fly common in tansy flowers in France and Italy. The seedfly consumes the seeds of tansy, but does not kill or inhibit growth of the host plant.

Tansy flea beetle, *Longitarsus jacobaeae*, is another host-specific pest of tansy that is indigenous to Mediterranean Europe. Flea beetle larvae feed throughout the root crown and sometimes feed externally on lateral roots. Larvae bore into the stem and leaf petioles for two to five inches (5-10 cm), causing wilt and death of the plant. Plants seldom recover from a heavy infestation of the beetle. Adult flea beetles feed on the leaves, causing a characteristic "shot-hole" pattern.

Holden (1989) obtained dramatic results from introduction of the flea beetle; percent absolute cover of ragwort declined from a mean of 22.7 in the fall 1982 to zero in summer 1984. At the beginning of this study (October 1982) ragwort cover in the circular vegetation plots ranged from less than 1 percent to 38 percent. By July 1984, all measured plots showed zero cover.

Vegetative competition: In New Zealand many hectares of land were converted to forests by planting commercial trees over infested sites (Holden 1989).

Grazing: Sheep grazing reduced the density of tansy ragwort, but the effect appears to be temporary. Ragwort reappeared as soon as the sheep were removed (Holden 1989).

Chemical Control

The first chemical used to combat tansy ragwort was the soil sterilant, sodium chlorate, a non-selective inorganic compound. The high cost of sodium chlorate prevented its use as a widespread control agent. Whitson *et al.* (1985) recommends dicamba (as Banvel®) for ragwort control at label concentrations. Ragwort plants treated with herbicides have been found to be more readily eaten by livestock, so grazing must be discontinued. Cattle may reenter the field only after the plants have completely dried and been replaced by other forage species (Holden 1989). Often herbicides have not been effective in killing this plant because the herbicide leached out of the roots without being transported throughout the plant (Poole and Cairns 1940).

Removal of individual ragwort plants or small populations requires immediate attention. To maintain control of tansy ragwort, a combination of control efforts must be diligently applied, and every opportunity must be taken to eliminate outlying individuals and populations.

Spartina alterniflora Loisel.

Common name: smooth cordgrass

Synonymous scientific names: none known

Closely related California natives: *Spartina foliosa, S. gracilis*

Closely related California non-natives: 3

Listed: CalEPPC A-2; CDFA nl

by Curtis C. Daehler

HOW DO I RECOGNIZE IT?

Distinctive Features

Smooth cordgrass (*Spartina alterniflora*) is a perennial, spreading grass from one foot tall in spring to six or eight feet tall in fall. It grows naturally only in intertidal estuarine habitats, and is often found in large, nearly monospecific stands in coastal or bayside marshes. Its large, round stems are hollow in cross section. Leaves are hairless, and leaf tips are sharply pointed. Young, healthy green shoots and leaf sheaths are often streaked with red or purple just below the sediment surface. This species is easily confused with the closely related native California cordgrass (*S. foliosa*), which is usually less than four and a half feet tall in fall and lacks red pigment in green, healthy shoot tissues (it may have red pigment on decaying tissues). Smooth cordgrass is highly variable in size, depending on growing conditions, and may hybridize with native California cordgrass.

Description

Poaceae. Rhizomes: 1/8-1/4 in (4-7 mm) wide, fleshy, whitish. Stems: typically <6 ft (2 m) but occasionally exceeding 8 ft (2.5 m), width at base to 1 in (25 mm), round, hollow, over .3 in (8 mm) diameter. Ligules <0.2 in (4mm) long. Leaves: to 1 in (25 mm) wide, glabrous, generally flat when fresh. Upper leaf surface ribbed, lower leaf surface smooth. Inflorescences: 4-18 in (10-45 cm) long, 1/4-7/8 in (5-22 mm) wide; 3-30 branches (spikes) per inflorescence, with 5-35 spikelets per branch. Flowers (spikelet): 3/8-9/16 in (8-15 mm) long, keels and glume glabrous to hairy. A clonal patch may live >100 years (Stiller and Denton 1995).

WHERE WOULD I FIND IT?

In California smooth cordgrass currently is found in Marin County and south San Francisco Bay, where it is rapidly spreading across open intertidal mud flats. Smooth cordgrass also invades established salt marsh communities, where it is likely to be found in the company of

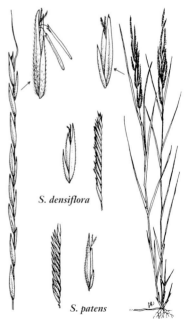

S. densiflora

S. patens

Spartina anglica

S. alterniflora

pickleweed (*Salicornia*) and California cordgrass (*Spartina foliosa*). In Washington smooth cordgrass is rapidly invading Willapa Bay and Puget Sound (Padilla Bay). Smooth cordgrass may be expected in other Pacific Coast estuaries in the future. A small patch was found in Humboldt Bay in the early 1980s, but it has since been eradicated (Daehler and Strong 1996). Smooth cordgrass does not grow on wave-swept Pacific Coast beaches and does not invade freshwater marshes that lack saltwater influence.

WHERE DID IT COME FROM AND HOW DOES IT SPREAD?

Smooth cordgrass is native to the Atlantic and Gulf Coast marshes of North America. It is a dominant component of Atlantic Coast salt marshes, where it forms extensive monospecific stands (Adam 1990). Introductions to San Francisco Bay and Padilla Bay, Washington, were associated with salt marsh restoration and erosion control projects (Spicher and Josselyn 1985, Daehler and Strong 1996 and 1997a). The introduction in Willapa Bay, Washington, appears to have been in association with oyster shipments during the nineteenth century (Sayce 1988).

Vegetative fragments may break off from established plants on eroding banks of tidal sloughs. Dredging an area infested with smooth cordgrass can promote the spread of vegetative fragments. A viable vegetative fragment must contain either root or rhizome material and can be transported with tides. Seeds can float and may also be transported with tides.

WHAT PROBLEMS DOES IT CAUSE?

Open intertidal mudflats are characteristic of Pacific Coast estuaries and provide important feeding grounds for many shorebird species (Daehler and Strong, 1996). Smooth cordgrass transforms open intertidal habitats into monospecific stands of tall grass, reducing shorebird feeding areas. The dense growth of smooth cordgrass also traps and holds sediments and can clog flood control and navigation channels and alter hydrology. The spread of smooth cordgrass threatens productive oystering grounds in Willapa Bay, Washington. Oystering grounds in other estuaries, such as Tomales Bay, California, would likely also be threatened if smooth cordgrass were introduced there. Smooth cordgrass can invade and replace native California cordgrass stands (Callaway and Josselyn 1992, Baron 1994), and the genetic integrity of native California cordgrass is threatened by hybridization with smooth cordgrass (Daehler and Strong, 1997b).

HOW DOES IT GROW AND REPRODUCE?

Smooth cordgrass grows most rapidly from April-September. Winter dieback of large flowering stems begins around October, and large, dead culms are usually washed away by tides in winter. Young, green shoots remain on most plants throughout the winter in California. During the early stages of mudflat invasion, plants grow as isolated, circular patches (clones). Clones spread laterally by vegetative shoots, often three and one-third feet (>1 m) per year (Callaway and Josselyn 1992). Over time, circular patches fuse together, and mudflats are transformed into meadows of smooth cordgrass.

Smooth cordgrass can become reproductive as early as the first year, but more often plants first flower after two to three years. Flowering occurs in late July-September. Copious inflorescences but few seeds are produced in most plants (Daehler and Strong 1994). Most clones require cross-pollination for good seed set, and cross-pollination (by wind) is rare in isolated patches (Daehler 1996a). A few plants with high self-fertility have consistently high seed output

(Daehler 1996a). Ripe seeds fall from inflorescences from October-January. Seeds trapped in mud or wrack germinate in February-May. Most seedlings do not survive the first winter because of winter storms and burial by algae; however, those seedlings that do survive can grow rapidly the following spring (Daehler 1996b). Seeds remain viable for only one year, and they do not tolerate desiccation. Vegetative fragments may be spread year round at sites prone to erosion.

HOW CAN I GET RID OF IT?

Smooth cordgrass can grow on very soft, sometimes hip-deep mud. Some patches may be inaccessible by foot and must be approached by boat or by air.

Physical Control

Manual/mechanical: Hand pulling is the simplest option for small propagules (one to a few plants <0.5 m diameter). On soft substrates, plants can be removed by gently pulling rhizomes and roots from the mud. It is necessary to feel around in the mud for any rhizomes that might have broken off and to remove them as well. On harder substrates, a shovel may be needed, and some rhizomes are likely to be missed. All plant material should be carried to an area well above the high tide mark, where it will dry out and die. The site should be marked with stakes and revisited after a few months to be sure the entire propagule was removed.

Solarization: If the infestation is small (one to a few patches 1-10 m diameter), burial of the plants under geotextile fabric or a black plastic tarp can be a means of eradication. Stems can be mowed with a weed whacker or similar device and then covered with 100 percent shadecloth (geotextile fabric) or heavy-duty black plastic. Covering is best begun in spring. The area covered should exceed the plant diameter by at least one meter, and the covering must be well anchored to the mudflat with sandbags or boards and deep stakes. Completely covered patches may die within four months, but mortality is assured only by waiting one year or more before removing the cover.

Mowing: This technique has sometimes been successful in killing smooth cordgrass outside California (Wijte and Gallagher 1992, Aberle 1993), especially if all shoots are mown to mud level in late fall. In areas where plants grow year round, such as California, mowing eight or more times a year may be required to kill smooth cordgrass. Occasional mowing in spring or summer will not kill the plants and may increase shoot density within patches (Daehler, pers. observation).

Biological Control

Biocontrol of smooth cordgrass using host-specific insects might be possible in Willapa Bay, where plants appear to have reduced resistance to herbivory (Daehler and Strong 1997b). However, no control agents have been USDA approved for release in Pacific Coast estuaries. In California biocontrol is probably not feasible because most suitable agents would be likely to attack endemic California cordgrass as well as smooth cordgrass.

Chemical Control

Large infestations (many patches > 3 ft or >10 m diameter) are best controlled with herbicide. Glyphosate (as Rodeo®) is currently the only herbicide approved for use in estuarine wetlands, and it must be applied by a licensed pesticide applicator. For hand spraying, applications of 2 to 5 percent glyphosate along with a surfactant are recommended. Surfactants that have been used with some success include X-77 and LI-700. To maximize chances of killing smooth cordgrass, Rodeo® must be applied at low tide, when most plants are exposed to the herbicide for eight or more hours. Rodeo® may be applied by spraying or wicking. Chamberlain (1995) found spring applications to be more effective than fall applications. Multiple applications may be required, especially if some leaves are missed during the first application, if exposure time is less than eight hours, if leaves are muddy, or if plants are growing low in the intertidal zone. To date there have been no successful eradications of large infestations of smooth cordgrass, although attempts are in progress in Willapa Bay and San Francisco Bay (Alameda County).

Spartina anglica Lois.

Common name: common cordgrass

Synonymous scientific names: none known

Closely related California natives: 2

Closely related California non-natives: 3

Listed: CalEPPC Red Alert; CDFA nl

by Carla Bossard and Carri Benefield

HOW DO I RECOGNIZE IT?

Distinctive Features

Common cordgrass (*Spartina anglica*) is a perennial, spreading grass from one foot tall in spring to six or even eight feet tall in fall. It grows naturally only in intertidal estuarine habitats; often in large, nearly monospecific stands in coastal or bayside marshes. It forms dense clumps, compared to the more open clumps of smooth cordgrass (*S. alterniflora*). Common cordgrass stems are hollow in cross section, round, and about half the diameter of smooth cordgrass. Leaves are hairless, and leaf tips are sharply pointed. Young, healthy green shoots and leaf sheaths are often streaked with red or purple just below the sediment surface. Common cordgrass also has rows of hairs where the leaf meets the stem (ligules) that are up to a third of an inch long, compared to the less than one-fifth-inch length of smooth cordgrass ligules.

Description

Poaceae. Perennial grass. (See *Spartina alterniflora* description above for differences). Rhizomes: 1/4-1/8 in (4-7 mm) wide, fleshy, whitish. Stems: typically <6 ft (2 m) but occasionally exceeding 8 ft (2.5 m), width at base to 1 in (25 mm), round, hollow, over 5 mm diameter. Ligules to 0.3 in (8 mm) long. Leaves: to 1 in (25 mm) wide, glabrous, generally flat when fresh. Upper leaf surface ribbed, lower leaf surface smooth. Inflorescences: 4-18 in (10-45 cm) long, 1/4-7/8 in (5-22 mm) wide; 3-30 branches (spikes) per inflorescence, with 5-35 spikelets per branch. Flowers (spikelet): 3/8-9/16 in (8-15 mm) long, keels and glume glabrous to hairy. A clonal patch may live >100 years (Stiller and Denton 1995).

WHERE WOULD I FIND IT?

Common cordgrass is found in coastal salt marshes, mud flats, and estuarine habitats similar to that of *Spartina alterniflora* in California coastal counties. In California it is currently found only in South San Francisco Bay.

WHERE DID IT COME FROM AND HOW DOES IT SPREAD?

Common cordgrass is a fertile hybrid between *Spartina maritima* (a British species) and smooth cordgrass (*S. alterniflora*). This hybrid was introduced into Washington in 1967 at Skagit Salt Marsh and dispersed down Puget Sound. It may have dispersed to California with south-flowing ocean currents, but this has not been established.

Common cordgrass seeds are spread by wind and water currents, but can also spread vegetatively by rhizomes and fragments that break off and move downstream.

WHAT PROBLEMS DOES IT CAUSE?

This cordgrass, like other non-native species of *Spartina*, traps sediment, builds marshes from the edge out, and overgrows native vegetation. It alters the course of succession and produces a monospecific stand that has much less value for wildlife than the native marsh flora. It is believed that common cordgrass caused the dieback of the native *S. maritima* in the United Kingdom (Doody 1984).

HOW DOES IT GROW AND REPRODUCE?

Common cordgrass flowers usually after one to two years, in late July-September. It has many inflorescences. Ripe seeds fall from October-January. Seeds germinate in February-May. Common cordgrass has been successful as a result of its rapid rate of growth, high fecundity, and aggressive colonization pattern. Seeds are an important source of invasion, but vegetative fragments may spread year round to enlarge the population locally (Benham 1990).

HOW CAN I GET RID OF IT?

Research has not been conducted on removal of common cordgrass, but recommendations for removal of smooth cordgrass (*Spartina alterniflora*) should be effective.

Spartina densiflora Brongn.

Common name: dense-flowered cordgrass

Synonymous scientific names: none known

Closely related California natives: 2

Closely related California non-natives: 3

Listed: CalEPPC Red Alert; CDFA nl

by Phyllis M. Faber

HOW DO I RECOGNIZE IT?

Distinctive Features

Dense-flowered cordgrass (*Spartina densiflora*) is a perennial cordgrass distinguished from the native cordgrass growing in West Coast coastal salt marshes by its grayish foliage, narrower leaf blades, dense, compact growth form (giving the plant the appearance of growing in distinct clumps), and earlier bloom period (up to a month earlier than native cordgrass).

Description

Poaceae. Perennial cordgrass. If present, rhizomes short and stout, >0.4 in (10 mm) wide. Stems cespitose, 10-60 in (2.7-15 dm) long, slender with firm internodes. Leaves: blades 0.45-17 in (12-43 cm), generally inrolled when fresh, with ridges on upper surfaces. Inflorescence: spike from 2.4-11 in (6-30 cm), 0.15-0.46 in (4-12 mm) wide, compact; 2-20 branches, overlapping, appressed, 0.4-4.1 in (1-11 cm), 2.5-6 mm wide. Flower spikelet 0.3-0.55 in (8-14 mm), glume and lemma keels have short, sharp bristles at least near tip; lower glume 0.15-0.3 in (4-7 mm), upper glume 0.3-0.55 in (8-14 mm), lemma 0.28-0.35 (7-9 mm).

WHERE WOULD I FIND IT?

Dense-flowered cordgrass is found along the central and northern California coast, particularly in Humboldt Bay salt marshes and Corte Madera Creek in Marin County. Eicher (1987) describes its distributional range between 1.3 and 2.42 feet elevation at Humboldt Bay.

WHERE DID IT COME FROM AND HOW DOES IT SPREAD?

This species was introduced in the 1800s from Chile by lumber ships that emptied their ballast water into Humboldt Bay when loading north coast lumber bound for Chile (Spicher and Josselyn 1985). Until 1984, when it was recognized as a distinct and exotic taxon, it had been misidentified as a northern form of the native *Spartina foliosa*. It is the dominant vegetation growing between lower and higher elevational populations of pickleweed (*Salicornia* sp.). It was introduced into a marsh restoration project at Creekside Park in Marin County by a landscape architect. It appears to have traveled across San Francisco Bay to Richmond, where it grows in clumps in parts of East Bay Regional Park.

WHAT PROBLEMS DOES IT CAUSE?

Dense-flowered cordgrass grows at a slightly higher elevation than native cordgrass and thus competes and replaces native plants such as *Frankenia salina*, *Limonium californicum*, and *Jaumea carnosa* in the middle to high tidal zone.

HOW DOES IT GROW AND REPRODUCE?

Spartina densiflora reproduces by seed and also expands vegetatively. It appears to spread by seed. Once established, a caespitose compact clump of cordgrass forms, with many leaves and inflorescences. Over a period of several years a single clump can expand to a meter or more in diameter (Barnhart *et al.* 1992).

HOW CAN I GET RID OF IT?

Little work has been done on techniques for removal of this cordgrass, but methods used to control and eliminate *Spartina alterniflora* should be effective on *S. densiflora*.

Spartina patens (Aiton) Muhlenb.

Common name: salt-meadow cordgrass

Synonymous scientific names: none known

Closely related California natives: 2

Closely related California non-natives: 3

Listed: CalEPPC B; CDFA nl

by Allison M. Brown

HOW DO I RECOGNIZE IT?

Distinctive Features

Salt-meadow cordgrass (*Spartina patens*) is a long-lived grass one to four feet tall, with thin wiry stems and tightly furled leaf blades, both of which are narrow (less than one-fifth of an inch) at the base. The inflorescence is open, composed of two to thirteen spike-like branches, which emerge about sixty degrees from the central axis, the lowest of which rarely overlaps (Hickman 1993). Other California coastal cordgrass species have compact inflorescences with spike-like branches appressed to the main axis. When plants are not blooming, the extremely narrow stem and angled leaf distinguish this plant from other cordgrasses. Salt-meadow cordgrass has a reddish purple base and no basal leaf sheath (Pattens, pers. comm.).

Description

Poaceae. Perennial grass. Rhizomes: if present, very thin, 0.2 in (<2-4 mm) wide. Stems: solitary or in small clumps, with firm internodes <0.2 in (1.5-4 mm) at base. Leaves: blades 4-20 in (10-50 cm) long, inrolled in the middle, with 3 ridges per mm on exterior surface, narrow at base, <0.2 in (1-4 mm). Inflorescence: open, panicle-like, 0.4-4 in (2-9 cm) wide and 2-9 in (5-22 cm) long, with 2-13 spike-like branches spreading approximately 60 degrees from the central axis, each branch 0.4-3 in (1-8 cm) long and 0.08-0.2 in (2-4 mm) wide. Flower: spikelet 0.3-0.5 in (7-12 mm), with upper glume generally larger than lower glume, lemma 0.2-0.3 in (5-8 mm); keels rough-hairy at the tip (Hickman 1993). Seeds: <5 mm (0.2 in) long when obtained from mature spikelets in August (Brown, pers. observation).

WHERE WOULD I FIND IT?

Salt-meadow cordgrass is found in coastal salt marshes, sand dunes, and swales in California coastal counties from Santa Barbara County north to Sonoma County (Calflora reports). In Benecia State Park marsh, east of the Carquinez Strait, it occurs in the high salt marsh zone. It can tolerate a wide range of salinity levels (Silander 1979). In salt marshes it prefers moderately saline to brackish conditions in the middle to upper zone (Chabreck and Condrey 1979). Salt-meadow cordgrass occurs with pickleweed (*Salicornia virginica*), fleshy jaumea (*Jaumea carnosa*), salt grass (*Distichlis spicata*) (Whitlow 1980), and sea lavender (*Limonium californica*), as well as soft bird's beak (*Cordylanthus mollis* ssp. *mollis*) in the mid-upper salt marsh zone (Brown pers. observation). It has invaded two additional

Pacific Coast locations, Suislaw Estuary on Cox Island in Oregon (Frenkel and Boss 1988) and Dosewallips State Park, in Washington on the west side of Puget Sound (Frenkel 1987).

WHERE DID IT COME FROM AND HOW DOES IT SPREAD?

This cordgrass is native to the coastal regions of the Gulf of Mexico and the Atlantic Ocean from Texas to Newfoundland (Frenkel and Boss 1988). It may have been introduced as packing material in association with unofficial plantings of eastern oysters (*Crassotrea virginica*) in Oregon and Washington (Townsend 1896, Frenkel and Boss 1985). The source of the population in California is unknown (Spicher, pers. comm.), but it was first reported thirty-five years ago (Munz 1985). Dense, monotypic circular patches form from a single genetic group, and proliferation and reestablishment of colonies appears to be a function of seed dispersal (Frenkel and Boss 1988). Reduced interspecific competition and the presence of barren or disturbed areas near established colonies also facilitate invasion.

WHAT PROBLEMS DOES IT CAUSE?

Since its introduction to Cox Island, salt-meadow cordgrass has expanded exponentially, with new outlying patches appearing in native vegetation (Frenkel and Boss 1988). As colonies expand, the dense stems impede growth of native species, often collapsing onto adjacent plants and smothering them (Frenkel and Boss 1988). Accretion coupled with expansion of the cordgrass community ultimately may result in conversion of low marsh to monotypic higher-elevation marsh. Dense mats of roots and rhizomes occupying depths of up to eight inches (Clark 1994) may further impede recolonization by native species. Seedlings can easily be overlooked and pose additional problems for eradication.

In August 1997 the Benicia State Park marsh population of salt-meadow cordgrass was localized in a 341 square foot area (Spicher and Brown, pers. observation). Its proximity to the rare plant species *Cordylanthus mollis* ssp. *mollis*, now federally listed, warrants close attention, since current restoration efforts in this marsh may alter conditions and favor expansion.

HOW DOES IT GROW AND REPRODUCE?

Radial expansion of cordgrass colonies is clonal. New plants arise from tillers, which produce abundant rhizomes and roots inside the patch perimeter. Tillers originating from a single plant can extend thirty-three feet (10 m) in one season (Silander 1979). Die-back generally occurs toward the center of the patch, possibly as a consequence of litter accumulation and associated anoxia (Goodwin and Williams 1961). Production of new colonies is thought to be by seed, although vegetative propagules may also facilitate colony formation (Frenkel and Boss 1988).

Salt-meadow cordgrass flowers from July through October. Seed heads form within two to three years after seedling emergence (Pattens, pers. comm.) The number of flowers is proportional to colony size (Frenkel and Boss 1988). Seed production is highly variable and most likely dependent on pollen viability (Frenkel, pers. comm.). In the marsh environment, selection seems to favor genotypes with higher vegetative biomass and lower reproductive output (Silander 1979). In its native area, seeds mature in August or September and germinate in April or May when conditions are warmer (Clark 1994).

HOW CAN I GET RID OF IT?

The best method for removing salt-meadow cordgrass depends on the size of the population and the local conditions in which it grows. Unfortunately, effective means for complete eradication of this species have yet to be discovered. Since the plant often grows among other species in tidally influenced marshes, great care is required in implementing treatments.

Physical Control

Manual/mechanical methods: Combined with excavation of subterranean plant parts, hand pulling may be effective for small patches. Seedlings are easily pulled as the root system is still shallow. Clipping inflorescences and seed heads in small patches may help prevent establishment of new colonies in adjacent areas, although existing clonal patches will be unaffected. Mowing with a weed eater has been used with variable success in Washington, but it must be done before plants set seed (May-June).

Solarization: Geotech fabric (tightly woven plastic mat) has been effective in controlling salt-meadow cordgrass in areas with moderate tidal range in Washington. Cloth is tucked into the ground beyond the tillering edge with a shovel, then staked every ten feet (3 m). Rope or cable is criss-crossed over the surface of the plastic and cinched to the stakes. Sand bands and bricks are then placed onto the cloth to secure it (Clark 1994). The cloth should be left in place for at least four growing seasons. This method is less effective in lower elevations of marsh or in areas where drift wood accumulates (Pattens, pers. comm.).

Biological Control

Potential agents for this species have not been investigated. Salt-meadow cordgrass establishment in barren, disturbed areas may be impeded by planting another vigorously growing native species such as pickleweed or saltgrass.

Chemical Control

Wicking with 20 percent glyphosate (as Rodeo) has been effective for portions of the Washington state population. In California Rodeo® must be applied by a certified pesticide applicator. While this method may be appropriate for small peripheral stands and patches, problems with calibration and excessive dripping have been reported (Moore, pers. comm.). Backpack sprayers using 5 percent Rodeo® may be an effective alternative for larger populations.

Spartium junceum L.

Common names: Spanish broom, gorse, weaver's broom

Synonymous scientific names: none known

Closely related California natives: 0

Closely related California non-natives: 1

Listed: CalEPPC B; CDFA nl

by Erik Tallak Nilsen

HOW DO I RECOGNIZE IT?

Distinctive Features

Spanish broom (*Spartium junceum*) is a perennial shrub more than head high, with cylindrical rush-like branches, green when young, turning brown as branches mature. Leaves are about half an inch long, entire, and elongated to lance-shaped. Leaves are present on the plant from February to early June, so this plant is leafless most of the year. Large (an inch or more), light yellow, pea-like flowers are prominently displayed on terminal racemes of outer canopy branches in late spring. Pods are abundant and dark brown when mature. Pods split when dry, and the two halves of the pod twist into a spiral shape, dropping the seeds.

Description

Fabaceae. Shrub to small tree to 10-15 ft (3-5 m) in height. Stems: cylindrical green stems resemble rush stems (large central spongy pith) when young; mature into woody branches with bark, leading to development of one to several trunks. Leaves: small <0.5 in (1.5 cm) in length, linear to lanceolate; entire leaves placed in a sub-opposite to alternate orientation on new shoots only; ephemeral, lasting 4 months or less; upper surface glabrous; lower surface has appressed

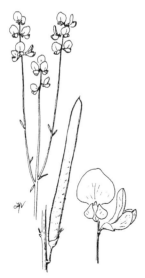

hairs. Inflorescence: raceme with many flowers located on terminal current-year shoots. Flowers: fragrant, yellow, 1.0 in (2.5 -3.0 cm) with 10 free stamens and fused filaments; calyx split almost to base; sepals and petals shaped like a pea flower. Fruit: pod with 10-15 seeds; 2-4 in (5-10 cm) long and 0.2-0.4 in (0.5-1.0 cm) wide; dehiscent longitudinally with halves splaying apart and twisting into a spiral, ejecting seeds (Hickman 1993).

WHERE WOULD I FIND IT?

In California Spanish broom is found in North Coast counties, the San Francisco Bay region, the Sacramento Valley, through South Coast counties to the Mexican border, in the western Transverse Ranges, and the Channel Islands. It is common in disturbed places, particularly eroding slopes, river banks, road cuts, and abandoned or disturbed lands, and can colonize post-burn chaparral and soft chaparral sites.

WHERE DID IT COME FROM AND HOW DOES IT SPREAD?

Spanish broom is native to the southern Mediterranean region of Europe, including Spain, Morocco, the Canary Islands, Madeira, and the Azores (Hickman, 1993). It was introduced into the California ornamental trade in 1848 in San Francisco (Butterfield 1964). Beginning in the late 1930s, Spanish broom was planted along mountain highways in southern California (Hellmers and Ashby 1958). In 1949 *Spartium junceum* had established naturalized populations in Marin County (Howell 1949). This species spreads by abundantly, producing seeds that are transported by any type of erosion or by rain wash.

WHAT PROBLEMS DOES IT CAUSE?

Spanish broom rapidly colonizes disturbed habitats and develops thick shrub communities that prevent colonization by native soft or hard chaparral species. Plants grow to more than head height and form a tangle containing a large amount of dead wood. Mature stands of Spanish broom should be considered a fire hazard during the dry season. It is poor forage for native wildlife.

HOW DOES IT GROW AND REPRODUCE?

Spanish broom reproduces by seed after two to three years of growth. Flowers are pollinated by bees and produced in late March to early April. Pods mature and dehisce in late May through early June. Each inflorescence produces ten to fifteen pods containing approximately fifteen seeds each. One plant can easily produce 7,000 to 10,000 seed in one season. Seeds fall near the plant and are subsequently moved by erosion, rain wash, and possibly ants. Seed viability is at least five years, suggesting that a significant seedbank is present in these stands. No research has been conducted on this plant's seedbank, seed germination, or seedling recruitment.

Shoots of Spanish broom are initiated in late winter and early spring, but most rapid growth occurs in May. Shoots elongate quickly and produce leaves with long internodes. The shoots harden off in late spring and leaves drop. Photosynthesis by stems occurs all year. Although the leaves have twice the photosynthetic rate of stems (Nilsen *et al.* 1993), photosynthesis in stems provides most of the whole plant carbon gain because of their longer life span and larger surface area (Nilsen and Bao, 1990). Spanish broom flowers are popular in the ornamental trade and are used for yellow dye. Stems are used for fibers, which accounts for one of its common names, weaver's broom. This species is also an effective stem sprouter.

HOW CAN I GET RID OF IT?

Information about eradicating Spanish broom is meager because there has been little experimentation with this species compared with that for Scotch broom (*Cytisus scoparius*) or French broom (*Genista monspessulana*). Evaluation of the following mechanisms of renewal is therefore based on the biology of the species rather than on information derived from controlled experiments.

It is likely that the success of any control method will vary with topography, soil chemistry, and climate, age and density of plants in the stand, and the availability of human and technical resources. Since a large and persistent seedbank is predicted from reproductive characteristics, rapid regrowth of the stand is likely following fire or mechanical removal of above-ground plant parts.

A comprehensive monitoring of control effectiveness is critical because there is no scientifi-

cally based knowledge about control of Spanish broom. Experimental manipulations should be monitored at least annually. Each monitoring visit should determine the number of new plants and the size or age distribution of the recovering population. Significant attention should be placed on the proportion of new individuals coming from the seedbank or resprouting from old plants. Monitoring should continue for at least five years following control treatment.

Physical Control

Manual/mechanical methods: Hand pulling of Spanish broom plants will be practical when the stand is one to four years old. When plants are small enough, pulling should be an effective mechanism of control as long as roots are removed and follow-up treatment of seedlings is done. The optimal season for pulling may be July-September when plants are already experiencing water stress (Nilsen and Karpa 1994). When plants have matured to small tree size, they are not amenable to pulling technology such as the weed wrench.

Brush hogs: Use of these machines is not likely to be effective for several reasons. Many of the sites where Spanish broom has invaded are steep slopes that may be difficult to traverse with a brush hog. The trunks of Spanish broom also rapidly increase to a size outside the range of effectiveness for this technology. Because of their twisting motion, brush hogs are more effective in reducing resprouting than is saw cutting, but resprouting should be expected following cutting by brush hog.

Saw cutting: This method would remove individuals in more mature stands when the bases of plants are too large for pulling or for brush hog removal. However, the clean cut from a saw will allow for maximum resprouting. Spanish broom has a great facility for resprouting from a saw cut even when the cut is close to the ground. Among all mechanical methods, saw cutting is the least likely to be effective in preventing resprouting. In general, manual/mechanical removal may be effective only when Spanish broom populations are very young.

Prescribed burning: After low-temperature fires Spanish broom will be able to resprout vigorously from trunk bases and stem meristems. However, a hot fire that kills all above-ground stems and burns hot and close to the ground will completely kill standing individuals and most likely remove some of the seedbank. Seeds of this species are similar in structure to those of Scotch broom (*Cytisus scoparius*). In heterogeneous or low-temperature fires Scotch broom seedbanks were not effectively reduced. Under similar fire conditions it is unlikely that fire will effectively reduce seedbank regeneration of Spanish broom.

Biological Control

There are no USDA approved biological control agents for Spanish broom. In greenhouse situations plants are susceptible to mealy bugs and show significant evidence of viral depression of growth.

Chemical Control

Spanish broom is sensitive to applied pesticides. In greenhouse situations only mild pesticides can be used without detrimentally affecting the plants. It is highly likely that application of chemicals such as glyphosate or triclopyr will drastically reduce population size. The ramifications of applying herbicides to a plant community must be carefully considered, because effects on non-target species are likely, especially when foliage spray methods are used.

A particularly effective control combination may be saw cutting followed by an application of 3 percent glyphosate (as Roundup®) to cut stems. This would remove above-ground plant parts to let light in for other species and also kill the root system, preventing root sprouting. However, if a substantial seedbank of Spanish broom already exists, the described combination will work effectively only if seedlings are also removed for several years, either by hand or by chemical treatments.

Taeniatherum caput-medusae (L.) Nevski

Common name: medusahead

Synonymous scientific name: *Elymus caput-medusae*

Closely related California natives: *Elymus arizonacus, E. canadensis*

Closely related California non-natives: 2

Listed: CalEPPC A-1; CDFA nl

by Tamara Kan and Oren Pollak

HOW DO I RECOGNIZE IT?

Distinctive Features

Medusahead (*Taeniatherum caput-medusae*) is a slender annual grass. The one- to three-inch awns are straight and compressed when green, but, upon drying, the awns twist and spread erratically in a manner reminiscent of the snake-covered head of the mythic Medusa. Medusahead can often be recognized by its color, which stands out against surrounding grasses. It matures from two to four weeks later than most other annual grasses, displaying distinctive patches of green in an otherwise brown grassland. After seed production, the dead medusahead plants bleach to light gray or tan. Medusahead is sometimes confused with foxtail (*Alopecurus* sp.) or with squirreltail (*Elymus elymoides*); however, medusahead's spike head does not break apart as seeds mature.

Description

Poaceae. Stems: 8-24 in (2-6 dm) long, slender, decumbent to ascending. Leaves: sheath glabrous, appendages <0.1 in (3 mm); ligule <0.2 in (0.5 mm), truncate; blade 1-3 mm wide, +/- inrolled, glabrous to puberulent, long-ciliate near collar. Inflorescence: 0.6-2 in (1.5-5 cm) excluding awns. Awns 1-3 in (2.5-7.5 cm), straight and compressed when green, twisting when dry. Spikelet: glumes 0.6-1.6 in (1.5-4 cm), fused at base, awn-like, stiff, generally glabrous; lemma narrowly lanceolate; awn 0.6-2 in (1.5-7.5 cm), flat, straight to curving outward (Hickman 1993).

WHERE WOULD I FIND IT?

As recently as 1950 medusahead was reported from only six counties in northwestern California. It has since spread rapidly throughout California, and currently is known to occur in over twenty counties (Pollak pers. observation). It has been reported from almost every county in northern California and in many areas of central California, extending as far south as Riverside County. It also infests rangeland, grassland, and sagebrush communities in Oregon, Washington, Idaho, Nevada, and Utah.

Medusahead invades grasslands, oak savannah, oak woodland, and chaparral communities. It grows in a wide range of climatic conditions. Clay or clay-loam soils with at least ten inches of rainfall annually are most susceptible to invasion (Dahl and Tisdale 1975). However, medusahead has been found on coarse-textured soils as well (Young 1992).

WHERE DID IT COME FROM AND HOW DOES IT SPREAD?

Medusahead is native to Spain, Portugal, southern France, Morocco, and Algeria. It was introduced to the United States in the late 1800s. The grass reproduces by seed, which is dispersed locally by wind and water. The long-awned seeds cling to the coats of grazing animals, such as sheep or cattle, and in this way are transported to more distant sites. Seeds can also disperse by attaching to machinery, vehicles, and clothing.

WHAT PROBLEMS DOES IT CAUSE?

Medusahead outcompetes native grasses and forbs, and, once established, can reach densities of 1,000 to 2,000 plants per square meter. After seed set, the silica-rich plants persist as a dense litter layer that prevents germination and survival of native species, ties up nutrients, and contributes to fire danger in summer. Because of its high silica content, medusahead is unpalatable to livestock and native wildlife except early in the growing season. The sharp awns can injure the eyes and mouths of livestock.

HOW DOES IT GROW AND REPRODUCE?

Medusahead is predominantly self-pollinating and reproduces by seed. Flowering usually occurs in May. An average of eight to fifteen seeds are produced per spike in late spring or early summer. Seeds usually disperse by mid-summer. Seed dormancy varies from a few weeks to over six months, depending on location. Germination normally occurs with the first rains in fall, and the germination rate is high. The seeds are well adapted to germinating and growing in the dense litter layer. Under these conditions, seeds can germinate even if they are not touching the soil. If they dry out, the primary root dies; but if they are moistened again, a new adventitious root develops (Young *et al.* 1971).

In California medusahead seeds usually germinate in October or November. The shoot system remains small, while the root system develops throughout the cold winter months. Early germination and rapid root growth consumes available water and nutrients, outcompeting slower-growing native species. Medusahead continues to grow, extract soil moisture, and produce seeds after most other annual grasses have turned brown.

HOW CAN I GET RID OF IT?

Physical Control

Mechanical methods: Mowing alone, or in combination with grazing, was found to be effective in reducing infestations. Plowing or discing are also effective means for controlling medusahead (Hilken and Miller 1980).

Prescribed burning: Several studies have shown that burning stands of medusahead prior to seed dispersal is an effective control mea-

sure (Furbush 1953, Hilken and Miller 1980, McKell *et al.* 1962, Murphy and Lusk 1961, Pollak and Kan 1996). Burns should be scheduled for late spring, after seed set but before seed heads have shattered (known as the "soft dough" stage of seed development). Seeds still on the plants are destroyed by the burn, while dispersed seeds lying on or buried below the soil surface are protected from the intense heat of the burn. With few seed reserves in the soil, medusahead abundance can be dramatically reduced if the seed input for even one year is eliminated.

This method takes advantage of the fact that medusahead matures later than most of the surrounding vegetation, so most other species have already dispersed their seeds and are dry enough to carry a burn. At the Jepson Prairie Preserve in Solano County effective control burns were conducted in late May and early June. Proper timing may vary depending on local conditions and weather. Some studies have found medusahead to increase after burning, but most of these studies conducted burns in August, presumably after seed dispersal.

Biological Control

Insect and fungi: No insect or fungal control agents are known. However, some preliminary research has been done on the effect of dry soil conditions on infestations of crown rot (*Fusarium culmoron*), a soil-borne pathogen, on medusahead (Grey *et al.* 1995).

Grazing: Heavy grazing by sheep in early spring (when medusahead is still palatable) can assist in controlling medusahead, but animals should be removed before seed heads form to limit seed dispersal. Early spring grazing is especially effective in areas where dried medusahead litter has been previously burned or grazed. Fertilizing with nitrogen improves the palatability of medusahead (Lusk *et al.* 1961). Properly timed grazing may reduce, but not eliminate, medusahead infestations.

Chemical Control

Small-scale infestations can be controlled by chemical herbicides. Atrazine applied in fall at 2 lbs/acre, was effective in controlling medusahead (Hilken and Miller, 1980). Check with a certified herbicide applicator or the herbicide label for current registration information.

Tamarix ramosissima Ledeb.

Common names: tamarisk, saltcedar

Synonymous scientific name: *Tamarix pentandra*

Closely related California natives: 0

Closely related California non-natives: 5

Listed: CalEPPC A-1; CDFA nl

Tamarix chinensis

Common name: none

Synonymous scientific name: none known

Closely related California natives: 0

Closely related California non-natives: 5

Listed: CalEPPC A-1; CDFA nl

Tamarix gallica

Common name: French tamarisk

Synonymous scientific names: none known

Closely related California natives: 0

Closely related California non-natives: 5

Listed: CalEPPC A-1; CDFA nl

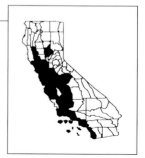

Tamarix parviflora

Common names: smallflower tamarisk, tamarisk, four-stamen tamarisk

Synonymous scientific name: *Tamarix tetrandra*

Closely related California natives: 0

Closely related California non-natives: 5

Listed: CalEPPC A-1; CDFA nl

by Jeffrey Lovich

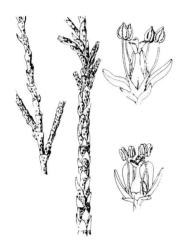

Tamarix ramosissima

HOW DO I RECOGNIZE IT?

Distinctive Features

Four invasive *Tamarix* species have been identified in California: *T. ramosissima*, *T. chinensis*, *T. gallica*, and *T. parviflora*. All four are many-branched shrubs or trees less than twenty-six feet tall with small scale-like leaves, from which comes the name saltcedar. Leaves have salt glands, and salt crystals can often be seen on leaves. Small white to deep pink flowers are densely arranged on racemes. The bark is reddish brown with smooth stems less than an inch in diameter.

Description

Tamaricaceae. *Tamarix ramosissima* is the species described here. Other invasive *Tamarix* species are similar, differing slightly in floral and leaf morphology. Stems: height <26 ft (8 m), usually <20 ft (6 m), reddish brown, glabrous, with jointed stems. Leaves: 0.06-0.14 in (1.5-3.5 mm), ovate, sessile with narrow base, tip acute to acuminate. Inflorescence: spike 0.06-0.28 in (1.5-7 mm) long and 0.12-0.16 in (3-4 mm) wide. Bract longer than pedicel, triangular, acuminate, margins +/- denticulate, mainly in the lower part. Flowers: 5 sepals, 0.02-0.04 in (0.5-1 mm) long, +/- ovate, obtuse to acute, 5 petals, 0.04-0.08 in (1-2 mm) long, elliptic to oblanceolate, nectar globes wider than long, stamens with alternate disk lobes, calyx and corolla pentamerous. Seeds: hairy-tufted, 0.02 in (<0.5 mm) in diameter and <0.01 in (<0.2 mm) long.

Over 50 species of *Tamarix* were recognized by Baum (1978), and 5 are reported from California, including *T. aphylla*, *T. chinensis*, *T. parviflora*, and *T. gallica* (DiTomaso 1996). *T. aphylla* is not an invasive pest under most circumstances. *T. ramosissima* may be synonymous with *T. chinensis* and is sometimes incorrectly referred to as *T. pentandra* (Baum 1978).

WHERE WOULD I FIND IT?

Saltcedar is widely distributed throughout the Mojave and Colorado deserts, Owen's Valley, the Central and South coasts, and the San Joaquin Valley. It occurs in parts of the San Francisco

Bay Area and the Sacramento Valley, particularly Yolo and Solano counties. French tamarisk (*T. gallica*) occurs in the Central Valley, Bay Area, and Central and South coasts. Smallflower tamarisk (*T. parviflora*) has a similar range, but also occurs in Inyo County. Saltcedar is abundant where surface or subsurface water is available for most of the year, including stream banks, lake and pond margins, springs, canals, ditches, and some washes. Disturbed sites, including burned areas, are particularly favorable for saltcedar establishment. It survives, and even thrives, on saline soils where most native, woody, riparian plants cannot.

WHERE DID IT COME FROM AND HOW DOES IT SPREAD?

Tamarix ramosissima is found throughout much of central Asia, from the Near East around the Caspian Sea, through western China to North Korea (Baum 1978). Although saltcedar may have been introduced into North America by the Spaniards, it did not gain recognition in the western United States until the 1800s (Robinson 1965). It was planted widely for erosion control, as a windbreak, for shade, and as an ornamental. It spreads by seed and vegetative growth. Individual plants can produce 500,000 tiny seeds per year (DiTomaso 1996), which are easily dispersed long distances by wind and water. The roots also sprout adventitiously (Kerpez and Smith 1987, Lovich *et al.* 1994).

WHAT PROBLEMS DOES IT CAUSE?

There is debate as to whether saltcedar is a consequence (Anderson 1996) or a cause (Lovich and de Gouvenain 1998) of environmental changes associated with its presence and proliferation. Regardless, the presence of saltcedar is associated with dramatic changes in geomorphology, groundwater availability, soil chemistry, fire frequency, plant community composition, and native wildlife diversity. Geomorphological impacts include trapping and stabilizing alluvial sediments, which results in narrowing of stream channels and more frequent flooding (Graf 1978). Saltcedar has been blamed for lowering water tables because of its high evapotranspiration rate, and, on a regional scale, dense saltcedar groves use far more water than native riparian plant associations (Sala *et al.* 1996).

Soil salinities increase as a result of inputs of salt from glands on saltcedar leaves. The dome-shaped glands consist of at least two cells embedded in the epidermal pits (Decker 1961). Increased salinity inhibits growth and germination of native riparian species (Anderson 1996). Leaf litter from drought-deciduous saltcedar increases the frequency of fire. Saltcedar is capable of resprouting vigorously following fire and, coupled with changes in soil salinity, ultimately dominates riparian plant communities (Busch 1995).

Although saltcedar provides habitat and nest sites for some wildlife (e.g., white-winged dove, *Zenaida asiatica*), most authors have concluded that it has little value to most native amphibians, reptiles, birds, and mammals (Lovich and de Gouvenain 1998).

HOW DOES IT GROW AND REPRODUCE?

Saltcedar can reproduce both vegetatively and by seed. Plants can regenerate from cuttings that fall on moist soil. Plants can flower by the end of the first year of growth (DiTomaso 1996). Studies in Arizona demonstrated that dense saltcedar stands can generate 100 seeds per square inch. Seed production occurs over a 5.5-month period, with one major and one minor peak (Warren and Turner 1975). The minute seeds of the closely related (Baum 1978) *Tamarix gallica* are about 0.007 inch (0.17 mm) in diameter and about 0.018 inch (0.45 mm) long. Small hairs on the apex of the seed coat facilitate dispersal by wind. Germination can occur within twenty-four hours in warm, moist soil (Merkel and Hopkins 1957).

Following germination and establishment, the primary root grows with little branching until it reaches the water table, at which point secondary root branching is profuse (Brotherson and Winkel 1986). Under favorable conditions, saltcedar shoots reportedly grow to heights of 3-4 meters in one growing season (DiTomaso 1996). Brotherson *et al.* (1984) examined the relationship between stem diameter and age of saltcedar

Tamarix sp.

Tamarix sp.

plants. Assuming that observed growth rings of stems are annual, saltcedars in Utah require 7.68 years for a 0.39 inch (1 cm) increase in stem diameter and 2.36 years in Arizona. Germination of saltcedar seeds is not greatly affected by increased salinity under experimental conditions (Shafroth *et al.* 1995). Saltcedar can form dense thickets.

HOW CAN I GET RID OF IT?

Like most invasive species, saltcedar is easily spread but difficult to eliminate. Early detection and control are critical, as saltcedar achieves dominance rapidly under favorable conditions. Efforts should be made to prevent site disturbances that contribute to its success (fire, increased soil salinity, ground disturbance, etc.). Monitoring is essential following any control effort, as some saltcedar is capable of resprouting following treatment. In addition, seedlings will continue to establish as long as saltcedar infestations persist upwind or upstream of the target area.

Physical Control

Manual/mechanical methods: Saltcedar is difficult to kill with mechanical methods, as it is able to resprout vigorously following cutting or burning. Root plowing and cutting are effective ways of clearing heavy infestations initially, but these methods are successful only when combined with follow-up treatment with herbicide. Seedlings and small plants can be uprooted by hand.

Prescribed burning: Fire does not kill saltcedar roots, and plants return quickly after fire if untreated by other methods. Fire is valuable primarily for thinning heavy infestations prior to follow-up application of herbicide. The consequences of fire for native plants and soil chemistry must be recognized.

Flooding: Flooding thickets for one to two years can kill most saltcedar plants in a thicket.

Biological Control

Insects and fungi: The USDA is currently using an international team of researchers to test thirteen species of natural enemies to control saltcedar. Of these, two have been recommended for field release in the United States, including a mealybug (*Trabutina mannipara*) from Israel and a leaf beetle (*Diorhabda elongata*) from China. Two other species are being tested in quarantine, including a psyllid (*Colposcenia aliena*) and a gelechiid leaf tier (*Ornativalva grisea*) from China. A gall midge (*Psectorsema*) from France has been approved for quarantine testing. Overseas testing has been completed for a foliage-feeding weevil (*Coniatus tamarisci*) from France, and for a pterophorid moth (*Agdistis tamaricis*), and a foliage-feeding weevil (*Cryptocephalus sinaita* subsp. *moricei*) from Israel (DeLoach 1997).

Grazing: Cattle have been shown to graze significant amounts of sprout growth (Gary 1960).

Chemical Control

Heavy infestations may require stand thinning through controlled burns or mechanical removal with heavy equipment prior to treatment with herbicides. Six herbicides are commonly used to combat saltcedar, including; imazapyr, triclopyr, and glyphosate (Jackson 1996).

Several proven methods exist for removing tamarisk. Perhaps the best method is to apply an imazapyr marketed as Arsenal® to the foliage. This technique is especially effective when a tank mix is used with a glyphosate herbicide such as Rodeo® or RoundupPro®. The most frequently used method in California is to cut the shrub off near the ground and apply triclopyr, either as Garlon 4® or Garlon 3A®. This technique usually results in better than a 90 percent kill rate. Triclopyr (as Pathfinder II®) can even be applied directly to the basal bark of stems less than about four inches in diameter without cutting the stem (the bark must be wetted completely around the base of each stem).

Garlon 4®or Pathfinder II® have no timing restrictions, but Garlon 3A® should be applied during the growing season. Resprouts can be treated with foliar applications of herbicide. Foliar applications of glyphosate or imazapyr achieve best results when applied in late spring to early fall during good growing conditions. Triclopyr can be diluted with diesel or natural oils, a dilution of 3 parts water to 1 part of Garlon 4® has proven effective (Barrows 1993, Lovich *et al.* 1994). Application rates for these herbicides are reviewed in Jackson (1996). Only Rodeo® has an aquatic registration, making it a legal choice for application over or around water.

Ulex europaea L.

Common names: gorse, common gorse

Synonymous scientific names: *Ulex europaeus*

Closely related California natives: 0

Closely related California non-natives: 0

Listed: CalEPPC A-1; CDFA B

by Marc C. Hoshovsky

HOW DO I RECOGNIZE IT?

Distinctive Features

Common gorse (*Ulex europaea*) is a prickly evergreen shrub less than ten feet tall, with a profusion of yellow pea-like flowers from March to May. By May plants are covered with half-inch- to one-inch-long brown pods. The short, stout branches are densely packed and may appear leafless. Spines, approximately half an inch long, are located at base of leaves. The somewhat similar species, Scotch broom (*Cytisus scoparius*), is not prickly.

Description

Fabaceae. Woody leguminous shrub, heavily armed but not gland-dotted. Stems: to <10 ft (3 m) high, much-branched from base, stiffly spreading, striate; twigs hairy when young, becoming stiff, thorn-like, intricately intertwined. Leaves: simple, alternate; juvenile (on seedlings, young shoots near ground), linear; adult awl-like, stiff, becoming spines. Inflorescence: general cluster, axillary near twig tips, few-flowered. Flowers: pea-like, calyx 2-lipped, membranous, yellow, persistent, 6 in (15 mm); petals +/- equal, yellow, persistent, less than 20 mm. Fruit: <1 in (1-2 cm) pod, densely hairy, +/- exserted from calyx, ovoid or oblong, explosively dehiscent. Seed with small basal outgrowth (Hickman 1993).

WHERE WOULD I FIND IT?

In California gorse can be found in all coastal counties and in the northern Sierra Nevada foothills. It invades infertile or disturbed sites, sand dunes, gravel bars, fence rows, overgrazed pastures,

logged areas, and burned-over areas. It will grow on most soil types, from good silt to plain boulders. It has been recorded as growing on serpentine soils and, rarely, on highly calcareous soils.

Gorse is more tolerant of soil acidity than most legumes, and it readily invades soils of poor fertility. The only restrictions to soil quality seem to be adequate nutrition and availability of trace elements. It grows best in moist soils and on shaded slopes. Gorse can thrive in well drained soils and in areas with a high water table. It is intolerant of heavy shade, where it produces coarse foliage and few flowers.

WHERE DID IT COME FROM AND HOW DOES IT SPREAD?

Gorse is native to central and western Europe, where it has long been cultivated as hedgerows.

It has naturalized in Australia and New Zealand, where considerable research has been done to control its spread. It has established along the Atlantic coast of North America, from Virginia to Massachusetts.

Gorse was introduced to the West Coast before 1894, and has been established in Mendocino County for 100 years. By the 1950s gorse had spread throughout western Washington and Oregon and northern California. It has been reported in every coastal county in California from Santa Cruz to Del Norte, and sparingly in southern California (Pryor and Dana 1952).

Gorse seeds are too heavy to be dispersed by wind, and usually fall within six feet of the parent plant. Seeds may be spread by ants, quail, water, and human activity.

WHAT PROBLEMS DOES IT CAUSE?

Gorse may be slow in spreading and becoming established, but where it gains a hold, there are few other plants that will so completely dominate an area. Besides becoming a significant fire hazard, it can successfully outcompete native plants in part because of its association with nitrogen-fixing bacteria, which facilitate its colonization of nitrogen-poor soils. Gorse leaf litter acidifies and lowers the cation exchange capacity of moderately fertile soils by immobilizing the bases, making it more difficult for native species to establish. On San Bruno Mountain, San Mateo County, gorse is considered the most difficult exotic species to control, and it has caused considerable loss of valuable grassland habitat (Reid 1985).

HOW DOES IT GROW AND REPRODUCE?

Gorse reproduces by resprouting from stumps and by seed. Seed reproduction is far more troublesome to control. Seeds are very small, averaging 60,000 seeds per pound. Seed counts in the upper inch of soil may be up to 2,000 per square foot (Zabkiewicz and Gaskin 1978a).

Seeds are impermeable to water, preventing immediate germination. They may remain dormant yet viable in the soil up to thirty years, with reports of up to seventy years (Zabkiewicz 1976). Seed germination may occur under suitable conditions at any time of year. Light is not essential, but few seeds germinate in the shade of established gorse. When dense gorse cover is removed, there is a flush of germination, because of either increased light or increased temperature. Heat stimulates germination, particularly at temperatures reached just below the soil surface during fire.

Gorse plants grow quickly, producing considerable dry matter. Year-old stands may contain 1,100 lbs/acre, with older stands producing 3,300 lbs/acre per year. Nitrogen in soils occupied by gorse can accumulate at an annual rate of 20-30 lbs/acre, surpassing the production of some well managed, fertilized pastures (Egunjobi 1971). Much of this production, with its high nitrogen content, ends up as litter, accumulating faster than any other temperate plant species.

Plants grow outward, forming a central area of dry, dead vegetation. A single plant can be up to thirty feet in diameter. Plants are typically medium-sized shrubs, but when exposed to constant wind they may be mat-like or cushion-like.

Roots tend to grow in the top few inches of soil, with only the tap root extending to greater depths. Extensive lateral roots are supplemented by a fine mat of adventitious roots that descend from the lower branches.

Gorse is a successful invasive plant because it grows on a variety of soil types, fixes nitrogen, and may impoverish soil of phophorus. It produces copious amounts of heat-tolerant seeds with long-term viability, and regenerates rapidly from seeds and stumps after disturbances such as brush clearing or fires.

HOW CAN I GET RID OF IT?

Because of the longevity of buried seeds, gorse control efforts must be long-term to be successful.

Physical Control

Manual/ mechanical methods: Gorse seedlings and young plants less than five feet tall may be hand pulled, especially after rain has loosened the soil. It is important to remove the root system, which may resprout if left in the ground. Any piece of root left in the soil may produce a new plant. Hoeing is effective when plants are small. This method either cuts off the tops or exposes seedlings to the drying action of the sun. A claw-mattock is effective in pulling out large plants and their root systems.

Cutting of above-ground plant parts is only marginally effective, but it is a useful technique to prepare for other removal methods. Repeated cuttings may help to exhaust the reserve food supply in roots. Cutting is most effective when plants begin to flower. At this stage the reserve food supply is nearly exhausted, and new seeds have not yet been produced. After cutting or chopping, gorse will resprout in greater density if not treated with herbicides.

In 1983 the San Mateo County Department of Parks and Recreation manually removed dense gorse from San Bruno Mountain, a task requiring approximately 350 person-hours per acre. Gorse also was removed by chaining by bulldozers and with the use of a bulldozer-mounted rototiller. Herbicides were used as a follow-up treatment (Reid 1985).

Prescribed burning: Fire has frequently been used to eliminate gorse thickets, although burns may easily get out of control because of the high flammability of the plants. Fire may stimulate germination of buried seeds, so repeated burns may be necessary to exhaust the seedbank (Amme 1983).

Biological Control

Insects and fungi: There are no USDA approved insects for biocontrol of gorse. The gorse weevil (*Apion ulicis*) was accidentally introduced into the United States in 1953 from France, and by 1982 it had become established in California and Oregon. The weevil grub eats the seeds in the unopened legume. When the pods open, adult weevils are released to feed on spines and flowers, sometimes defoliating large plants. In California the weevil has been only partially successful in controlling gorse (Amme 1983). Plants often have enough food reserves to recover rapidly after serious injury. Additionally, the climate on the northern California coast is cool, delaying dehiscing of the pods and leaving the weevil larvae to die in the pod.

Other potential insect enemies of gorse exist but have not been tested for controlling gorse in the United States (Julien 1982).

Vegetative competition: Reseeding with native perennials after initial burning or chemical treatment of stumps may be productive. Once established, these species may displace gorse by competing for water or nutrients or by shading out lower-growing gorse plants. Amme (1983) has experimented with reseeding with native grasses in Jughandle State Reserve in Mendocino County.

Grazing: Goat grazing is effective in controlling gorse (Hartley *et al.* 1980, Hill 1955), as goats prefer woody vegetation to most grasses and herbaceous plants. Goats are less costly to use than mechanical and chemical methods. They are most cost-effective when used to clear or suppress young regrowth rather than to do the initial clearing of mature stands. A period of at least two years of goat grazing is required before there is any significant reduction in gorse (Hill 1955).

Chemical Control

Chemical control of gorse has been well researched in New Zealand. The most effective chemical treatment was a combination of picloram (as Tordon®), which is not registered for use in California, and 2,4,5-T® which is not legal anywhere in the U.S. (Ivens 1979). Good results were obtained with picloram applied during summer months. Larger plants needed retreatment, and burned stumps showed a high degree of recovery (Ivens 1979). Both chemicals have distinct disadvantages, including persistence in soils, difficulty in being leached out of organic and clay soils, and damage to other plant species. Check with your county agricultural agent or a certified pesticide applicator to determine which herbicides are currently registered for use on gorse in California.

Gorse is difficult to eradicate with a single application of herbicide (Balneaves 1980). Greater success is possible with a combination of methods, including crushing, cutting, or burning.

Glyphosate (as Roundup®) is most effective with gorse seedlings in early summer. Plants began to die the following fall and winter as herbicide was carried to the roots.

Verbascum thapsus L.

Common names: common mullein, wooly mullein, great mullein, mullein, Jacob's staff, flannel leaf, velvet plant, candlewick plant, lung wort, felt wort

Synonymous scientific names: none known

Closely related California natives: 0

Closely related California non-natives: 3

Listed: CalEPPC B; CDFA nl

by Michael J. Pitcairn

HOW DO I RECOGNIZE IT?

Distinctive Features

Common mullein (*Verbascum thapsus*) is a biennial or annual herb that sends up a large stalk, three to six feet tall, topped with yellow flowers closely attached to the stalk. It spends the first year as a rosette close to the ground. The leaves are large, six to twelve inches, densely woolly, and soft to the touch. Leaves are largest at the base and gradually become smaller up the stalk. Leaves on the stalk occur alternately, and the base of each leaf extends a short way down the stalk. Yellow flowers, three-quarters to one and a half inches in diameter, consist of five circular petals, and occur on the uppermost portion of the stalk.

Description

Scrophulariaceae. Biennial or annual herb. Stems: 1.5-6 ft (50-200 cm) tall when mature. Leaves: first-year plant is a rosette of large, woolly, gray-green leaves. Leaves 2-4 times longer than wide, 2-16 in (5-40 cm) long, alternate, the largest leaves at the base of the plant, smallest near the top; leaf pedicle short with leaf base extending a short way down the stalk to form wings. Plants usually bolt in second year and have a single stem covered with overlapping, woolly leaves from base to inflorescence. Inflorescence: a spike extending upward from top of stalk. Usually 1 inflorescence is produced, but occasionally a second or third can form with branches occurring where inflorescences begin. Flowers: densely packed along inflorescence, youngest near top; bracts 0.5-0.7 in (12-18 mm), flower pedicels short, <0.08 in (<2 mm), generally fused to stalk. Calyx deeply 5-lobed; 0.3-0.4 in (7-9 mm) long, several times longer than wide. Corolla (petals) 0.6-1.0 in (15-25 mm) wide, 5-lobed, circular, nearly regular, sulphur-yellow. Seeds: held in a 2-celled capsule 0.24 in (6 mm) diameter, covered with short, branched hairs. Seeds brown, irregular, oblong, 0.02-0.03 in (0.5-0.7 mm) long, with wavy ridges alternating with deep grooves. Each capsule holds numerous seeds. Flowering occurs from June through October (from Munz 1959, Gross and Werner 1978, and Hickman 1993).

WHERE WOULD I FIND IT?

Common mullein occurs throughout California, but is particularly abundant in dry valleys on the eastern side of the Sierra Nevada. High population densities have been observed in moist meadows and creek drainages near Mono Lake and Owens Valley. It prefers disturbed habitats with little other vegetation, especially on dry, gravelly soils. It is common along roadsides, rights-of-way, and river banks and in forest cuts, meadows, pastures, and waste areas (Gross and Werner 1978). It is an early colonizer and may be the first plant to colonize bare soil. It is found in all forty-eight contiguous states and in Hawaii. In Canada it is reported to grow abundantly in soils with a pH range of 6.5-7.8 (Gross and Werner 1978). It is found from sea level to 8,000 feet (2,440 m) elevation.

WHERE DID IT COME FROM AND HOW DOES IT SPREAD?

Common mullein is native to Asia, but it probably was introduced to the United States from Europe. It was valued for its medicinal properties and has been carried with immigrants throughout the world. It has been used as a remedy for coughs and lung diseases, diarrhea, burns, and earaches (Mitich 1989). It probably was introduced several times into North America as a medicinal herb as well as accidentally. The earliest recorded intentional introduction was in the 1700s in the Blue Ridge Mountains of Virginia (Gross and Werner 1978). It apparently naturalized and spread rapidly, for it was erroneously described as a native by Eaton (1818) and was present as far west as Michigan in 1839 (Gross and Werner 1978). It was first recorded in California in 1880 as being widely naturalized in old fields in Siskyou County (Watson 1880). It spreads by prodigious seed production and maintains its presence by long-lived seeds in the soil. Its seeds have no specialized structures for long-distance dispersal by wind or animals. Movement of soil for highway and building construction may have assisted in its dispersal.

WHAT PROBLEMS DOES IT CAUSE?

Common mullein is not a weed of agricultural crops, as it cannot tolerate cultivation. It is, however, thought to serve as a host for insects that are themselves economic pests, such as the mullein leaf bug, a pest of apples and pears in the eastern United States and Canada (Maw 1980). Common mullein is not often a significant weed of most wildlands and natural areas, as it is easily crowded out by grasses or other competing vegetation. It is a problem, however, in the sparsely vegetated soils of the eastern Sierra Nevada. In moist meadows and drainages near Mono Lake and Owens Valley, common mullein can become abundant and has invaded pristine meadows with undisturbed soils, displacing native herbs and grasses. It has also been observed to rapidly establish following forest fires in the western Sierra Nevada. High densites of rosettes appear to prevent the reinvasion of native herbs and grasses in burned areas, but eventually these give way to a developing shrub canopy. In this situation, mullein appears to disrupt the normal sequence of ecological succession.

HOW DOES IT GROW AND REPRODUCE?

Common mullein reproduces solely by seed. Flowering occurs in summer, usually July through October in the eastern Sierra and June to August in the Sacramento and San Joaquin valleys. Flowers are borne on an inflorescence that occurs as an upward extension of the stalk. Individual flowers are short-lived, opening before dawn and closing by mid-afternoon the same day. Flowers are visited by a diverse array of insects, but only the short- and long-tongued bees are effective pollinators. Flowers are also self-fertile, with self-fertilization occurring when the flower closes at the end of the day if cross-pollination has not occurred (Gross and Werner 1978).

Seeds are contained in a capsule with two cells. Field studies show that single plants produce 200 to 300 capsules with 500 to 800 seeds per capsule. Thus, seed production can be 100,000 to 240,000 seeds per plant. When dry, the capsule splits open and releases the seeds. Seeds are not adapted to dispersal by wind or animals and usually fall to the ground. Field studies report that seeds will disperse as far as eleven meters, but 75 percent fall within one meter of the parent plant (Gross and Werner 1978).

Common mullein seeds do not appear to undergo dormancy or require a period of after-ripening. They germinate rapidly under appropriate environmental conditions (Baskin and Baskin 1981). Seed germination can occur in continuous darkness (e.g., when buried) or in light. High germination rates in darkness are restricted to relatively high temperatures (>30 degrees C). In contrast, high germination rates were observed at 0 degrees C in darkness alternating with 35-40 degrees C in light (Semenza *et al.* 1978). This indicates that germination is possible on soil surfaces where extreme diurnal fluctuations occur. Despite observations of seed germination in darkness in the laboratory, field studies of buried seeds show low germination rates (<15 percent) suggesting that factors other than darkness may play a role in preventing germination of buried seeds (Baskin and Baskin 1981).

Generally, only those seeds at or near the soil surface will germinate (Semenza *et al.* 1978). Mullein seeds can survive and remain viable for thirty-five to 100 years when buried (Gross and Werner 1978, Baskin and Baskin 1981). The presence of mullein plants immediately following soil disturbance is likely a result of the presence of a seedbank rather than dispersed seeds. In California common mullein seeds usually germinate in spring following snowmelt and in fall with the onset of rains.

Common mullein is a usually a biennial, forming a taproot and a rosette in the first year and a flowering stalk in the second year. Rosettes consist of a whorl of leaves from the root crown clustered at the soil surface. In the eastern Sierra Nevada, however, it can grow as a biennial or as a winter annual. If seed germinates in spring, the plant will remain a rosette through the first growing season and the following winter. It will then bolt in spring and flower in summer. If seed germinates in fall, the plant will enter winter as a rosette and bolt the following spring (Semenza *et al.* 1978). Regardless of its flowering pattern, common mullein spends the first half of its life as a rosette, producing a deep taproot before sending up a three- to six-foot stalk and producing flowers. Flowers may be produced until the first frost or snowfall in late fall.

HOW CAN I GET RID OF IT?

The best method for controlling common mullein depends on the size of the infestation, the topography of the site, and the resources available. Timing is critical for efficient control, and follow-up is essential.

Physical Control

Manual/mechanical methods: Perhaps the most effective method of controlling common mullein is to cut plants with a weed hoe. Plants will not resprout if cut through the root crown below the lowest leaves (Gross and Werner 1978). Removing rosettes with a hand hoe can be easily accomplished by workers trained to recognize the plant. Hand hoeing can be selective and effective, and two workers may clear up to twenty acres of mullein in a few hours. Bolted plants can also be removed with a weed hoe. Sometimes bolted plants can be pulled out of sandy soil, especially following heavy rain. If plants have begun to set seed, cut off the flowering racemes with pruning shears just below the lowest seed pods and collect them in a bag to prevent seeds from being released during the hand removal operation. A second or third weeding may be necessary.

Mowing appears to be ineffective, as plants cut above the root crown do not die. Rather, the basal rosette will continue to enlarge, then later bolt and flower. Clipping the terminal flower stalk will not prevent flowering, but will cause increased growth of axillary branches, which will produce flowers later (Gross and Werner 1978).

Prescribed burning: Burning kills bolted plants and appears to kill rosettes, but creates open areas for reinfestation from seed germination. Individual bolted plants can be killed using a flame thrower, but its use is to be avoided during fire season.

Biological Control

Insects and fungi: No insects or diseases have been approved for introduction as biological control agents against common mullein in North America. A curculionid weevil, *Gymnaetron tetrum* Fab., was accidentally introduced into Canada from Europe (Maw 1980). This weevil is specific to common mullein and is considered one of two natural enemies significantly impacting the plant in Europe. The larvae feed on seeds and other tissues in the seed capsules. Larvae are able to destroy all seeds in a capsule when present; however, usually not all seed capsules are infested. Gross and Werner (1978) report that up to 50 percent of seeds may be destroyed by the larval feeding of this weevil. Over time, *G. tetrum* has spread into California and has been collected from mullein plants throughout northern and eastern California since 1942. While its impact has not been investigated in California, it is unlikely to have much impact on common mullein populations. Despite the rate of seed destruction, too many seeds remain for it to have much effect in controlling common mullein populations.

Grazing: Grazing animals generally will not eat mullein because of its hairy leaves (Whitson 1992).

Chemical Control

Common mullein is difficult to control with herbicides because the thick hairs on the leaves prevent the herbicide from reaching and penetrating the leaf surface. A surfactant is recommended for all liquid herbicides used to control this plant.

Zamora (1993) compared the effectiveness of several herbicides to control common mullein along a roadside in Montana. Herbicides were applied with a backpack sprayer calibrated to deliver 20 gal/acre at 42 psi. Treatments were applied to late rosette and bolting plants in late May. Plant height and number of plants surviving to flower were recorded at the end of July. Of the three compounds available in California, 2,4-D provided 66 percent kill at 1.9 lbs/acre;

height of standing plants averaged nineteen inches. Glyphosate provided 100 percent kill; height averaged nine inches. Sulfometuron (as Oust®) also provided 100 percent kill; average plant height was six inches. For comparison, in the unsprayed control areas, all plants survived to flowering (zero mortality); average plant height was thirty-eight inches (1 m).

Another control method, recently developed by a forest weed manager, is to spray each rosette with glyphosate by putting the spray nozzle into the center of the rosette (DiTomaso, pers. comm.). The applicator touches the plant with the spray nozzle and gives it one good squirt. The key is to ensure that the herbicide penetrates the region of the plant where the growing point is located. If the nozzle is off-center, this method does not work. Only seedlings and rosettes are susceptible using this method. In treating individual plants, it is recommended that a dye be used in the herbicide mixture to mark treated plants and prevent re-treatment.

Vinca major L.

Common names: periwinkle, bigleaf periwinkle

Synonymous scientific names: none known

Closely related California natives: 0

Closely related California non-natives: 0

Listed: CalEPPC B; CDFA nl

by Jennifer Drewitz

HOW DO I RECOGNIZE IT?

Distinctive Features

Periwinkle (*Vinca major*) is a spreading perennial vine with glabrous, dark green stems that contain a milky latex. The non-flowering stems grow close to the ground, rooting at the nodes and extending outward to three feet. Flowering stems grow erect to knee-high with solitary flowers developing in the leaf axil (Bean and Russo 1986). The purplish-blue flowers have five equal petals fused at the base. Five stamens attach near the top of the corolla tube, which is hairy within.

Description

Apocynaceae. Stems: non-flowering stems prostrate; flowering stems erect to 0.5-1.5 ft (15-45 cm). Leaves: to 2-3 in (5-7.5 cm) long, covered with a waxy cuticle, cordate at base, tapering to acute apex. Leaves opposite, 4-ranked; entire margins covered by ciliate hairs (Bean and Russo 1986). Petioles nearly glabrous, <1 in (2.5 cm) long. Flowers: rotate, 1-2 in (2.5-5 cm), with fused, 5-part calyx. The 5 stamens distinctively curve at the base and are arranged alternate to corolla lobes. There are 2 superior ovaries, which may be sterile, but if seeds are present they are curved and hairy (Hickman 1993).

WHERE WOULD I FIND IT?

Periwinkle's range extends from California throughout the southern United States. In California it is found up to 610 feet (200 m) elevation in most coastal counties, the Central Valley, and desert areas. It prefers a mediterranean climate and frost-free, damp, shaded soils (Stern 1973). It has been observed thriving along tree-covered drainages and creeks in coastal areas (Alvarez 1997). It has escaped from gardens and old homesteads where it has been used as an ornamental groundcover and is commonly found spreading from moist roadside locations where it has been dumped (Bean and Russo 1986).

WHERE DID IT COME FROM AND HOW DOES IT SPREAD?

Originally from southern Europe and northern Africa, periwinkle was introduced to the United States as an ornamental groundcover and medicinal herb (Schittler 1973). It spreads vegetatively and is not known to reproduce sexually in California (Bean and Russo 1986). Water can transport broken stem fragments throughout riparian zones, where the plant's ability to resprout enables it to spread rapidly. The rate of spread is not documented in the literature, but it appears to be limited by shade and moisture requirements.

WHAT PROBLEMS DOES IT CAUSE?

Once established, periwinkle forms a dense cover that prevents growth and establishment of other plant species (Stern 1973). Periwinkle lowers species diversity and disrupts native plant communities. Riparian zones are particularly sensitive. Major infestations at The Nature Conservancy's Ramsey Canyon Preserve in southern Arizona have suppressed natural erosional processes in a creek, promoting deepening and scouring of the creek bed and altering local hydrology and vegetation (McKnight 1993).

HOW DOES IT GROW AND REPRODUCE?

In California periwinkle reproduces vegetatively, not by seed. When produced, seeds rarely mature, and gardeners propagate it by cuttings. The plant spreads by sprawling stems that form a shallow root at the nodes. This creates a carpet of vegetation. The flowers begin to bloom in March and continue into July. Wet periods rapidly accelerate vegetative growth. Periwinkle will die back in a frost, but will resprout when optimal conditions return. It does not grow well in dry soil or direct sunlight, but does well in a moist microclimate with shaded areas (Stearn 1974).

HOW CAN I GET RID OF IT?

Physical Control

Manual/mechanical methods: Hand removal is labor-intensive, but yields good results if careful attention is paid to removing all root nodes and stolons. In Ramsey Canyon, Arizona, volunteers successfully cleared four acres using hand removal over a three-year period. An effective method is to work inward from the perimeter of the patch and pull the periwinkle back in on itself to prevent further spread of the weed between removal sessions (TNC 1997). Repeated

removal efforts, scheduled over a growing season, may allow natives to recolonize the area and reduce the chance that other weeds will move into the area following the disturbance caused by removal activities (Mcknight 1993).

Because periwinkle has the ability to resprout, mowing or cutting results in abundant regrowth and is not recommended (TNC 1997).

Biological Control

Biocontrol agents have not been determined or tested for *Vinca major*.

Chemical Control

Glyphosate (as Round up®) has been tested on large infestations of periwinkle at Ramsey Canyon, Arizona. Greatest success is achieved if plants are cut first and then sprayed immediately afterward. Cutting with a weed whip or brush cutter breaks through the waxy cuticle and allows better foliar penetration of the herbicide. Using the cut and spray method, a 5 percent glyphosate solution gave nearly 100 percent control. To reduce native plant death in the area, a 3 percent solution provides 70-75 percent control and yields good results if followed by spot applications (Bean and Russo 1986). A wick applicator is suggested for spot treatments, and a backpack sprayer is recommended for treating large areas. To aid chemical distribution throughout the plant, use surfactant and apply herbicide during an optimal growing period of good moisture and warm temperatures (70-80 degrees F) usually in late spring or early fall.

Monitoring is recommended. Follow-up on any removal actions is necessary, as any over-looked stem or plant fragments will quickly resprout. Following chemical removal, the popula-tion should be checked twice, in early fall and late spring. With manual removal, follow-up should be performed every three months to remove resprouts. After the patch is eradicated it should be checked twice a year in optimal growing seasons.

Herbicides mentioned in text by common name, trade name, or chemical formula

COMMON NAME	TRADE NAME	CHEMICAL FORMULA
Acrolein*	Magnacide	2-propenal
Ammonium Sulfate*	Ammate X-NI	
Atrazine*	AAtrex 80W	6-chloro-N'-(1-methylethyl)-1,3,5-triazine-2,4-diamine
Bromacil	Hyvar-X	5-bromo-6-methyl-3-(1-methylpropyl)-2,3(1H,3H)pyrimidinedione
Chlorsulfuron*	Glean	2-chloro-N-[[(4-methoxy-6-methyl-1,3,5-triazin-2-yl)amino]carbonyl]benzenesulfonamide
Clopyralid*	Transline	3,6-dichloro-2-pyridinecarbolic acid
Clethodim*	Select, Prism	(E,E)-(+/-)-2-[1-[[(3-chloro-2-propenyl)oxy]imino]propyl]-5-[2-(ethylthio)propyl]-3-hydroxy-2-cyclohexen-1-one
Copper Chelate*	Several Names	alkanolamine
Copper Sulfate*	Several Names	cupric sulfate pentahydrate
Cynazine*	Bladex	2-[[4-chloro-6-(ethylamino)-1,3,5-triazine-2-yl]amino]-2-methylpropanenitrile
Dicamba*	Banvel	3,6-dichloro-2-methoxybenzoic acid
Diclofop*	Hoelon 3EC	(+/-)-2-[4-(2,4-diclorophenoxy)phenoxy)propanoic acid
Diquat*	Reward, Diquat	6,7-dihydrodipyrido[1,2-a:2',1'-c]pyrazinedium ion
Diuron*	Karmex	N'-(3,4-dichlorophenyl)-N,N-dimethylurea
Endothall*	Endothal	7-oxabicyclo[2,2,2]heptane-2,3-dicarboxylic acid
Fluazifop-p*	Fusilade	(R)-2-[4-[[5[(trifluoromethyl)-2-pyridinyl]oxy]phenoxy]propanoic acid
Fluridone*	Sonar	1-methyl-3-phenyl-5-[3-(trifluoromethyl)phenyl]-4(1H)-pyridinone
Fosamine*	Krenite	ethyl hydrogen (aminocarbonyl)phosphonate
Glyphosate*	Roundup, Rodeo, Accord	N-(phosphonomethyl)glycine
Hexazinone*	Velpar	3-cyclohexyl-6-(dimethylamino)-1-methyl-1,3,5-triazine-2,4(1H,3H)-dione
Imazapyr	Arsenal, Chopper, Contain	(+/-)-2-[4,5-dihydro-4-methyl-4-(1-methylethyl)-5-oxo-1H-imidazol-2-yl]-3-pyridinecarboxylic acid
Linuron*	Lorox	N'-(3,4-dichlorophenyl)-N-methoxy-N-methylurea
Metham*	Vapam	methylcarbamodithioic acid
Metolachlor*	Dual, Bicep	2-chloro-N(2-ethyl-6-methylphenyl)-N-(2-methoxy-1-methylethyl)acetamide
Metribuzin*	Lexone, Sencor	4-amino-6-(1,1-dimethylethyl)-3-(methylthio)-1,2,4-triazin-5(4H)-one
Metsulfuron	Ally, Escort	methyl 2-[[[[(4-methoxy-6-methyl-1,3,5-triazin-2-yl)amino]carbonyl]amino]sulfonyl]benzoate
Napropamide*	Devrinol 50 WP	N,N-diethyl-2-(1-naphthalenyoxy)propanamide
Oryzalin*	Surflan	4-(dipropylamino)-3,5-dinitrobenzenesulfonamide
Paraquat*	Gramoxone, Cyclone	1,1'-dimethyl-4,4'-bipyridinium
Picloram*	Tordon	4-amino-3,5,6-trichloro-2-pyridinecarboxylic acid
Sethoxydim*	Poast	2-[1[(ethoxyimino)butyl]-5[2-(ethylthio)propyl]-3-hydroxy-2-cyclohexen-1-one
Simazine*	Princep	6-chloro-N,N'-diethyl-1,3,5-triazine-2,4-diamine
Sodium Chlorate*	DeFol, Harvestaid	sodium chlorate
Sulfometuron*	Oust	methyl 2-[[[[(4,6-dimethyl-2-pyrimidinyl)amino]carbonyl]amino]sulfonyl]benzoate
Terbacil	Sinbar	5-chloro-3-(1,1-dimethyl)-6-methyl-2,4-(1H,3H)-pyrimidinedione
Thifensulfuron	Pinnacle	methyl 3-[[[[(4-methoxy-6-methyl-1,3,5-triazin-2-yl)amino]carbonyl]amino]sulfonyl]-2-thiophenecarboxylate
Triasulfuron	Amber	2-(2-chloroethoxy)-N-[[(4-methoxy-6-methyl-1,3,5-triazin-2-yl)amino]carbonyl]benzenesulfonamide
Tribenuron	Express	methyl 2-[[[[(4-methoxy-6-methyl-1,3,5-triazin-2-yl),ethylamino]carbonyl]amino]sulfonyl]benzoate
Triclopyr*	Garlon	[(3,5,6-trichloro-2-pyridinyl)oxy]acetic acid
2,4-D or MCPA*	Several Names	phenoxy carboxylic acid

**Registered in California

Authors

Albert, Marc, Presidio Natural Resources, Bldg. 201,
Fort Mason, San Francisco, CA 94123

Alvarez, Maria, GGNRA, 75 Henry St., Cotati, CA
94931 <henry75@sonic.net>

Alverson, Edward, The Nature Conservancy, c/o
Eugene Public Works Dept., Engineering
Division, 858 Pearl St., Eugene, OR 97401
<ed.r.alverson@ci.eugene.or.us>

Anderson, Lars, Weed Science Program, Dept. of Veg
Crops, 1 Shields Av. University of CA, Davis, CA
95616-8730 <landerson@ucdavis.edu>

Apteker, Rachel, University of CA, Davis, 146 Ipanema
Pl., Davis, CA 95616

Bayer, David, Weed Science Section of Plant Biology,
Robbins Hall, University of CA, Davis, 95616
<dbayer@ucdavis.edu>

Benefield, Carri, CA Dept. of Food and Agri.,
Integrated Pest Control Branch, 1220 N. Street
Room A357, Sacramento, CA 95814
<cbenefield@cdfa.ca.gov>

Bossard, Carla, Dept. of Biology, Saint Mary's College
of California, P.O. Box 4507, Moraga, CA 94575
<cbossard@stmarys-ca.edu> or
<bossard3@pacbell.net>

Boyd, David, CA Dept. of Parks & Rec., 7665
Redwood Blvd., Suite 150, Novato, CA 94945

Brooks, Matthew, U.S. Dept. of the Interior, USGS,
41704 S. Fork Dr., Three Rivers, CA 93271

Brown, Allison, Rider University, 2083 Lawrenceville
Rd., Lawrence, NJ 08648-3099
<abrown@voicenet.com>

Burrascano, Cindy, California Native Plant Society, San
Diego, 771 Lori Lane, Chula Vista, CA 91910

Chipping, David, California Native Plant Society, 1530
Bayview Hts. Dr., Los Osos, CA 93402
<dchipping@polymail.cpunix.calpoly.edu>

Daehler, Curtis, Dept. of Botany, Univ. of Hawaii, 3190
Maile Way, Honolulu, HI 96822-2279

Deiter, Laurie, City of Boulder, Land Management
Division., 7315 Red Deer Rd., Boulder, CO
80301

Dewey, Steve, Utah State University, Dept. of Plants,
Soils and Biometeorology, 4400 Old Main Hill,
Logan, UT 84322-4900 <steved@ext.usu.edu>

D'Antonio, Carla, Dept. of Integrative Biology, Univ.
of CA, Berkeley, CA 94720-3140
<cdantonio@socrates.berkeley.edu>

DiTomaso, Joseph, Weed Science Program, Dept. of
Veg Crops, University of CA, Davis, CA 95616
<jmditomaso@ucdavis.edu>

Drewitz, Jennifer, Section of Plant Biology, Univ. of
CA, Davis, CA 95616 <jdrewitz@ucdavis.edu>

Dudley, Thomas, Dept. of Integrative Biology, Univ. of
CA, Berkeley, 209 Mulford Hall, Berkeley, CA
94720 <tdudley@socrates.berkeley.edu>

Faber, Phyllis, 212 Del Casa, Mill Valley, CA 94941
<pmfaber@aol.com>

Gerlach, John, Agronomy and Range Science, 1 Shields
Ave., Univ. of California, Davis, CA 95616

Godfrey, Kris, CA Dept. of Food and Agriculture,
Biocontrol Program, 3288 Meadowview Rd.,
Sacramento, CA 95832

Harrington, Kerry, Practical Teaching Complex,
Institute of Natural Resources, Massey University,
Private Bay 11-222, Palmerston, North New

Zealand <k.harrington@massey.ac.nz>

Harris, Steve, CalEPPC, P.O. Box 341, Arcata, CA
95518-0341 <sharris@igc.apc.org>

Hoban, Gavin, Pt. Reyes National Seashore
<hoban@earthling.net>

Hoshovsky, Marc, Biological Conservation Planner, CA
Dept. of Fish & Game, Natural Heritage
Division, 1416 9th Street, Sacramento, CA 95814
<mhoshovs@kirk.dfg.ca.gov>

Howald, Ann, CalEPPC, California Native Plant
Society, 210 Chestnut Ave, Sonoma, CA 95476-
3472 <annhowald@vom.com>

Hunter, John, State Univ. of NY, College of Brockport,
350 New Campus Dr., Brockport, NY 14220-
2973; <jhunter@brockport.edu>

Jacobsen, Edie, Naval Facilities, Engineering
Command, SW Division, 1220 Pacific Highway,
San Diego, CA 93132-5190

Kan, Tamara, The Nature Conservancy, 16060 Skyline
Blvd. Woodside, CA 94062

Kelly, Michael, CalEPPC, 11591 Polaris Dr., San
Diego, CA 92126-1507 <mkellysd@aol.com>

Kitz, Jo, Mountains Restoration Trust, 6223 Lubao
Ave., Woodland Hills, CA 91367
<mtnsrt@aol.com>

Klinger, Robert, Div. of Biological Sciences, Dept. of
Evolution and Ecology, Univ. of CA, Davis, 1
Shields Ave., Davis, CA 95616-8755
rckscip@fcdarwin.org.ec

Kreps, Butch, Associate Agricultural Biologist,
California Dept. of Food and Agriculture,
Integrated Pest Control, 20235 Charlanne Dr.,
Redding, CA 96002

Kyser, Guy, Weed Science Program, Dept. of Veg
Crops and Weed Science, Univ. of CA, Davis, CA
95616

Lanini, Thomas, Weed Science Program, 1 Shields
Av., Dept. of Veg. Crops and Weed Science,
Univ. of CA, Davis, CA <lanini@ucdavis.edu>

Lichti, Richard, Creighton University Medical School,
Omaha, Nebraska < rlichti@uswest.net>

Lovich, Jeffrey, Dept. of the Interior, U.S. Geological
Survey, Univ. of CA, Bio. Dept., Riverside, CA
92521-0427 <jeffrey_lovich@usgs.gov>

Minnich, Richard, Dept. of Earth Science, Univ. of CA,
Riverside, CA 92521-0423
<richard.minnich@ucr.edu>

Nilsen, Erik Tallak, Virginia Tech. University, Dept. of
Biology, Blacksburg, VA 24061-0406

O'Connell, Ross, CA Dept. of Food and Agriculture,
Integrated Pest Control Branch, 1220 N Street,
P.O. Box 942871, Sacramento, CA 94271

Palmquist, Debra, USDA, 920 Valley Rd., Reno, NV
89512

Pickart, Andrea, Humboldt Bay National Wildlife
Refuge, Lanphere Dunes Unit, 6800 Lanphere
Rd, Arcata, CA 95521
<andrea_pickart@mail.fws.gov>

Pitcairn, Michael, CA Dept. of Food and Agriculture,
Integrated Pest Control Branch, Biological
Control Program, 3288 Meadowview Road,
Sacramento, CA 95832; CalEPPC
<mpitcairn@cdfa.ca.gov>

Pollak, Oren, deceased

Randall, John, The Nature Conservancy, Wildland

Invasive Species Program, Dept. of Veg. Crops and Weed Science,124 Robbins Hall, Univ. of CA, Davis, CA 95616 <jarandall@ucdavis.edu>

Randall, Jonathan, Graduate Group in Ecology, Univ. of CA, Davis, CA 95616 <jjrandall@ucdavis.edu>

Reichard, Sarah, Center for Urban Horticulture, Box 354115, Univ. of Washington, Seattle, WA 98195-4115 <reichard@homer09.u.washington.edu>

Sanders, Andrew, Herbarium, Dept. of Botany and Earth Sciences, Univ. of CA, Riverside, CA 92521

Sigg, Jake, California Native Plant Society, CalEPPC, 338 Ortega St., San Francisco, CA 94122 <jsigg@pacbell.net>

Warner, Peter, GGNRA, 555 Magnolia Ave., Petaluma, CA 94952 <peter_warner@ggnra.org> or <peterwarner@earthlink.net>

Young, Jim, USDA, Conservation Biology of Range-lands Lab, 920 Valley Rd., Reno, NV 89512

Photograph Sources: Closeup / Scenic

Ageratina adenophora, Tony Bomkamp; CalEPPC
Ailanthus altissima, Carla Bossard; Carla Bossard
Alhagi pseudalhagi, Bill Heveron, Bill Heveron
Ammophila arenaria, John Randall; John Randall
Aptenia cordifolia, Carla Bossard; Carla Bossard
Arctotheca calendula, John Randall; Greg Archbald
Arundo donax, Tom Dudley; John Randall
Atriplex semibaccata, Dale Smith; Dale Smith
Bassia hyssopifolia, CDFA; CDFA
Bellardia trixago, Tom Echols; Tom Echols
Brassica tournefortii, Matt Brooks; Matt Brooks
Bromus madritensis, John Randall; John Randall
Bromus tectorum, John Randall; John Randall
Cardaria chalepensis, John Randall; John Randall
Cardaria draba, Jerry Asher; Clyde Elmore
Carduus pycnocephalus, CalEPPC; Carla Bossard
Carpobrotus edulis, John Randall; John Randall
Centaurea calcitrapa, John Randall; John Randall
Centaurea melitensis, John Randall; John Randall
Centaurea solstitialis, Joe DiTomaso; Joe DiTomaso
Cirsium arvense, John Randall; John Randall
Cirsium vulgare, CalEPPC, John Randall
Conicosia pugioniformis, Sharon Farrell; Sharon Farrell
Conium maculatum, John Randall; John Randall
Cortaderia jubata, John Randall; Greg Gaar
Cortaderia selloana, Joe DiTomaso; Joe DiTomaso
Cotoneaster spp., Greg Gaar; Maria Alvarez
Crataegus monogyna, Dale Smith; Dale Smith
Cynara cardunculus, Mike Kelly; Mike Kelly
Cytisus scoparius, Carla Bossard, Carla Bossard
Cytisus striatus, Maria Alvarez; Maria Alvarez
Delairea odorata, Carla Bossard, John Randall
Digitalis purpurea, John Randall; John Randall
Egeria densa, Ross O'Connell; Lars Anderson
Ehrharta spp., Dale Smith; Dale Smith
Eichhornia crassipes, Kris Godfrey; Kris Godfrey
Elaeagnus angustifolia, John Randall; John Randall
Erechtites glomerata, Gavin Hoban; Gavin Hoban
Erechtites minima, Gavin Hoban; Gavin Hoban

Eucalyptus globulus, Dave Boyd; Dave Boyd
Euphorbia esula, John Randall; John Randall
Ficus carica, John Randall; John Randall
Foeniculum vulgare, Carla Bossard; John Randall
Genista monspessulana, Carla Bossard; Greg Archbald
Halogeton glomeratus, Steve Dewey; Steve Dewey
Hedera helix, Carla Bossard; John Randall
Helichrysum petiolare, John Randall; John Randall
Hydrilla verticillata, Kris Godfrey; Kris Godfrey
Lepidium latifolium, Joe DiTomaso; John Randall
Leucanthemum vulgare, John Randall; Maria Alvarez
Lupinus arboreus, Joe DiTomaso; Joe DiTomaso
Lythrum salicaria, Carri Benefield; John Randall
Mentha pulegium, Peter Warner; Maria Alvarez
Mesembryanthemum crystallinum, John Randall; John Randall
Myoporum laetum, John Randall; Jo Kitz
Myriophyllum aquaticum, John Randall; John Randall
Myriophyllum spicatum, CA IPM Project; Lars Anderson
Pennisetum setaceum, Jeff Lovich; Jeff Lovich
Phalaris aquatica, Tom Lanini
Retama monosperma, Edie Jacobson; Carla Bossard
Ricinus communis, John Randall; John Randall
Robinia pseudoacacia, Joe DiTomaso; John Hunter
Rubus discolor, John Randall; John Randall
Schinus terebinthifolius, John Randall; John Randall
Schismus sp., Matt Brooks; Matt Brooks
Senecio jacobaea, Clyde Elmore; Tim Butler
Spartina alterniflora, CalEPPC; John Randall
Spartina anglica, Debra Ayres
Spartina densiflora, Phyllis Faber
Spartina patens, CalEPPC
Spartium junceum, Eric Nilson; Marcel Rejmanek
Taeniatherum caput-medusae, Oren Pollak; Oren Pollak
Tamarix sp., Carla Bossard; Carla Bossard
Ulex europaea, Michael Pitcairn; Michael Pitcairn
Verbascum thapsus, Michael Pitcairn; John Randall
Vinca major, Dale Smith; Alison Fisher

Illustration Sources

Ageratina adenophora: Lesley Randall
Ailanthus altissima: *Jepson Manual of the Higher Plants of California*, ed. Hickman, 1993, P. 1069.
Alhagi pseudalhagi: *Jepson Manual of the Higher Plants of California*, ed. Hickman, 1993, P. 593.
Ammophila arenaria: *Jepson Manual of the Higher Plants of California*, ed. Hickman, 1993, P. 1234.
Aptenia cordifolia: Lesley Randall

Arctotheca calendula: Lesley Randall

Arundo donax: *Jepson Manual of the Higher Plants of California*, ed. Hickman, 1993, P. 1241.

Atriplex semibaccata: Lesley Randall

Bassia hyssopifolia: *Jepson Manual of the Higher Plants of California*, ed. Hickman, 1993, P. 507.

Bellardia trixago: *Weeds of California*, Robbins, Bellue and Ball, CA Department of Agriculture, 1970, P. 395.

Brassica tournefortii: *Jepson Manual of the Higher Plants of California*, ed. Hickman, 1993, P. 409.

Bromus madritensis: *Jepson Manual of the Higher Plants of California*, ed. Hickman, 1993, P. 1245.

Bromus tectorum: Lesley Randall

Cardaria chalepensis: Lesley Randall

Cardaria draba: *Weeds of California*. Robbins, Bellue, and Ball, CA. Department of Agriculture, 1970, P. 220.

Carduus pycnocephalus: *Jepson Manual of the Higher Plants of California*. ed. Hickman, 1993, P. 227.

Carpobrotus edulis: *Jepson Manual of the Higher Plants of California*. ed. Hickman, 1993, P. 133.

Centaurea calcitrapa: *Weeds of California*. Robbins, Bellue, and Ball, CA Department of Agriculture, 1970, P. 437.

Centaurea melitensis: Lesley Randall

Centaurea solstitialis: *Jepson Manual of the Higher Plants of California*. ed. Hickman, 1993, P. 227.

Cirsium arvense: *Jepson Manual of the Higher Plants of California*. ed. Hickman, 1993, P. 237.

Cirsium vulgare: *Jepson Manual of the Higher Plants of California*. ed. Hickman, 1993, P. 243.

Conicosia pugioniformis: *Jepson Manual of the Higher Plants of California*. ed. Hickman, 1993, P. 133.

Conium maculatum: *Jepson Manual of the Higher Plants of California*. ed. Hickman, 1993, P. 149.

Cortaderia jubata: *Jepson Manual of the Higher Plants of California*. ed. Hickman, 1993, P. 1251.

Cortaderia selloana: Lesley Randall

Cotoneaster lacteus, C. monogyna, C. pannosa: Lesley Randall

Crataegus monogyna: Lesley Randall

Cynara cardunculus: *Jepson Manual of the Higher Plants of California*. ed. Hickman, 1993, P. 247.

Cytisus scoparius: *Jepson Manual of the Higher Plants of California*. ed. Hickman, 1993, P. 611.

Cytisus striatus: Lesley Randall

Delairea odorata: Lesley Randall

Digitalis purpurea: Dr. Rudolf Becking

Egeria densa: *Common Weeds of the United States*, USDA, 1971, P. 31.

Ehrharta calycina: *Jepson Manual of the Higher Plants of California*. ed. Hickman, 1993, P. 255.

Ehrharta erecta: Lesley Randall

Ehrharta longiflora: Lesley Randall

Eichhornia crassipes: *Jepson Manual of the Higher Plants of California*. ed. Hickman, 1993, P. 1317.

Elaeagnus angustifolia: Lesley Randall

Erechtites glomerata: *Jepson Manual of the Higher Plants of California*. ed. Hickman, 1993, P. 259.

Eucalyptus globulus: *Jepson Manual of the Higher Plants of California*. ed. Hickman, 1993, P. 773.

Euphorbia esula: *Weeds of the North Central States*. University of Illinois Experiment Station. 1954, P. 104.

Ficus carica: Lesley Randall

Foeniculum vulgare: *Jepson Manual of the Higher Plants of California*. ed. Hickman, 1993, P. 149.

Genista monspessulana: *Jepson Manual of the Higher Plants of California*. ed. Hickman, 1993, P. 611.

Halogeton glomeratus: *Weeds of California*. Robbins, Bellue, and Ball, CA Department of Agriculture, 1970, P. 156.

Hedera helix: Lesley Randall

Helichrysum petiolare: Lesley Randall

Hydrilla verticillata: *Jepson Manual of the Higher Plants of California*. ed. Hickman, 1993, P. 1161.

Lepidium latifolium: Lesley Randall

Leucanthemum vulgare: *Jepson Manual of the Higher Plants of California*. ed. Hickman, 1993, P. 311.

Lupinus arboreus: *Jepson Manual of the Higher Plants of California*. ed. Hickman, 1993, P. 649.

Lythrum salicaria: Lesley Randall

Mentha pulegium: Lesley Randall

Mesembryanthemum crystallinum: Lesley Randall

Myoporum laetum: Lesley Randall

Myriophyllum aquaticum: Lesley Randall

Myriophyllum spicatum: *Common Weeds of the United States*. USDA, Dover Publications, 1971, P. 276.

Pennisetum setaceum: Lesley Randall

Phalaris aquatica: *Grass Weeds 2*, Hafliger and Scholz, CIBA-Geigy, Basle, 1981, P. 111.

Retama monosperma: Lesley Randall

Ricinus communis: *Jepson Manual of the Higher Plants of California*. ed. Hickman, 1993, Univ. of CA Press, P. 579.

Robinia pseudoacacia: *Jepson Manual of the Higher Plants of California*. ed. Hickman, 1993, P. 651.

Rubus discolor: Lesley Randall

Schinus terebinthifolius: Lesley Randall

Schismus arabicus: *Jepson Manual of the Higher Plants of California*. ed. Hickman, 1993, P. 1295.

Schismus barbatus: Lesley Randall

Senecio jacobaea: Lesley Randall

Spartina alterniflora: Lesley Randall

Spartina anglica: Lesley Randall

Spartina densiflora: Lesley Randall

Spartina patens: Lesley Randall

Spartium junceum: *Jepson Manual of the Higher Plants of California*. ed. Hickman, 1993, P. 651.

Taeniatherum caput-medusae: *Jepson Manual of the Higher Plants of California*. ed. Hickman, 1993, P. 1301.

Tamarix ramosissima, T. parviflora: *Jepson Manual of the Higher Plants of California*. ed. Hickman, 1993, P. 1087.

Tamarix gallica: *British Wildflowers*, Vol. 1. Hitchison, 1957, P. 116.

Ulex europaeus: *Jepson Manual of the Higher Plants of California*. ed. Hickman, 1993, P. 659.

Verbascum thapsus: *Weeds of the North Central States*. University of Illinois Experiment Station, 1954, P. 185.

Vinca major: Lesley Randall

References

Aberle, B. 1993. The biology and control of introduced *Spartina* (cordgrass) worldwide and recommendations for its control in Washington. Masters thesis, Evergreen State College, Olympia, WA.

Adam, P. 1990. Salt Marsh Ecology. Cambridge University Press, Cambridge, UK.

Ahart, L. 1985, California Native Plant Society, Mount Lassen Chapter, personal communication, March 1985.

Ahmed, M., A. Jabbar, and K. Samad. 1977. Ecology and behavior of *Zyginidia guyumi* (Typhlocyloinae: Cicadellidae) in Pakistan. Pakistan J. of Zoology. 9(1):79-85.

Ahrens, W.H. 1994. Herbicide Handbook. 7th ed. Weed Science Society of America, Champaign, IL.

Aiken, S.G., P.R. Newroth, and I. Wile. 1979. The biology of Canadian weeds. Canadian J. Plant Science. 59:201-15.

Alex, J.F. 1982. Canada. In: Holzner, W. and N. Numata (eds.). Biology and Ecology of Weeds. Dr. W. Junk Publishers, The Hague, The Netherlands.

Allan, H.H. 1982. Flora of New Zealand. DSIR Publishing, Auckland, NZ.

Allen, L. 1997, US National Park Service, Point Reyes National Seashore, personal communication, July 15, 1997.

Allen, L. 1999, US National Park Service, Point Reyes National Seashore, personal communication, February 24, 1999.

Alvarez, M. 1995a. The impact of German ivy (*Senecio mikanioides*) on the richness and composition of three native plant communities. Masters thesis, Sonoma State University, Rohnert Park, CA.

Alvarez, M. 1995b. The growth of German ivy (*Senecio mikanioides*) to increased shading levels and decreased resource availability. Internal report. Golden Gate National Recreation Area, Sausalito, CA.

Alvarez, M., director, Habitat Restoration Team, Golden Gate National Recreation Area, Sausalito, CA, interview, 1997.

Amme, D. 1983. Gorse control at Jughandle State Reserve: resource restoration and development. California Department of Parks and Recreation, unpublished report.

Amme, D. 1988. Poison Hemlock Control at Elkhorn Slough, 1987—Final Report.

Amor, R.L. 1972. A study of the ecology and control of blackberry (*Rubus fruticosus* L. agg.). J. Australian Institute of Agricultural Science. 38(4):294.

Amor, R.L. 1974. Ecology and control of blackberry (*Rubus fruticosus* L. agg.): II. Reproduction. Weed Research. 14:231-38.

Anderson, B.W. 1996. Salt cedar, revegetation and riparian ecosystems in the Southwest. In: Lovich, J. E., J. Randall, and M. Kelly (eds.). Proceedings of the California Exotic Pest Plant Council Symposium '95, Pacific Grove, CA. California Exotic Pest Plant Council, Davis, CA. Pp. 32-41.

Anderson, E. Bureau of Land Management, personal communication, 1996.

Anderson, L.W. 1981. Effect of light on the phototoxicity of fluridone in American pondweed (*Potamogeton nodosus*) and sago pondweed (*P. pectinatus*). Weed Science. 29(6):7723-28.

Anderson, L.W. 1987. Hydrilla (*Hydrilla verticillata*). Exotic Pest Profile No. 11. California Dept. of Food and Agriculture.

Anderson, L.W. 1990. Aquatic weed problems and management in the western United States and Canada. In: Pieterse, A. and K. Murphy (eds.). Aquatic Weeds: the Ecology and Management of Aquatic Nuisance Vegetation. Oxford University Press, London, UK.

Anderson, L.W. 1996. Annual Report on Aquatic Weed Control Investigations. USDA, ARS, AWCRL, Davis, CA. Effect of long-term Sonar® treatment on the growth and pigment content of *Egeria densa*. Pp. 44-47. Effect of Sonar® followed with Komeen® treatments on the growth of *Egeria densa*. Pp. 48-51.

Anderson, L.W. 1998. Report to California Department of Boating and Waterways on Fragment Production and Mechanical Control.

Anderson, L.W., S. Perry, and S. Fellows. 1992. Survival and germination of buried dioecious and monoecious hydrilla tubers. In: Anderson, L. (ed.). Annual report on Aquatic Weed Control Investigations. USDA, ARS, AWCRL, Davis, CA. Pp. 41-42.

Anderson, N.O. 1987. Reclassification of the genus *Chrysanthemum* L. Hortscience. 22:313.

Aplet, G.H. 1990. Alteration of earthworm community biomass by the alien *Myrica faya* in Hawaii. Oecologia. 82:414-16.

Aptekar, R. 1999. The ecology and control of European beachgrass (*Ammophila arenaria*). Ph.D. dissertation, University of California, Davis, CA.

Archbald, G. 1995. Biology and control of German ivy. Report to California Department of Fish and Game, Pesticide Applications Seminar.

Archbald, G. 1996. A French broom control method. CalEPPC News. Summer/Fall: 4-6.

Archbald, G., GGNRA, personal communication, July 1998.

Arnold, W.J. and L.E. Warren. 1966. Dowpon C® Grass Killer: a new product for controlling perennial grasses such as Johnson grass and Bermuda grass. Down to Earth. 21:14-16.

Ashton, F.M. and T.J. Monaco. 1991. Weed Science: Principles and Practices. John Wiley & Sons, New York, NY.

Austin D.F. 1978. Exotic plants and their effects in southeastern Florida. Environmental Conservation. 5:25-34.

Avault, J.W. 1965. Biological weed control with herbivorous fish. In: Proceedings of the Eighteenth Southern Weed Control Conference. Pp. 590-91.

Ayres, D.R., D. Garcia-Rossi, H.G. Davis, and D.R. Strong. In press. Extent and degree of hybridization between exotic (*Spartina alterniflora*) and native (*S. foliosa*) cordgrass (Poaceae) in California, USA, determined by RAPDs. Molecular Ecology.

Bailey, L. 1945. The genus *Rubus* in North America. Gentes Herbarium. 5(1):851-54.

Bailey, L.H. 1949. Manual of Cultivated Plants. Macmillan Publishing, New York, NY.

Bailey, L.H. 1930. Standard Cyclopedia of Horticulture. Macmillan Publishing, New York, NY.

Bailey, L.H. 1933. Standard Cyclopedia of Horticuture. Vol. II. P. 1322. Macmillan Publishing, New York, NY.

Bailey, L.H. 1949. Manual of Cultivated Plants. Macmillan Publishing, New York, NY.

Bailey, L.H. 1964. Manual of Cultivated Plants. Macmillan Publishing, New York, NY.

Baker, H.G. 1965. Characteristics and modes of origin of weeds. In: Baker, H.G. and G.L. Stebbins (eds.). The Genetics of Colonizing Species, Academic Press, New York, NY.

Balciunas, J. 1985. Final report on the overseas surveys (1981-1983) for insects to control hydrilla. Tech. Rpt. A-85-4. US Army Eng. Waterways Experimental Station, Vicksburg, MS.

Balciunas, J. and B. Villegas. 1999. Two new seed head flies attack yellow starthistle. California Agriculture. 53(2):8-11.

Balneaves, J.M. 1980. A programme for gorse control in forestry using a double kill spray regime. Proceedings of the 33rd New Zealand Weed and Pest Control Conference. Pp. 170-73.

Balogh, G.R. 1986. Ecology, distribution, and control of purple loosestrife in Northwestern Ohio. Master's thesis, Ohio State University, Columbus, OH.

Barbe, G.D. 1990. Noxious Weeds of California: Detection Manual. California Department of Food and Agriculture, Sacramento, CA.

Barbe, G.D., CDFA, BioControl, personal communication, 1996.

Barbour, M.G. 1970. The flora and plant communities of Bodega Head, California. Madroño. 20(6): 289-336.

Barbour, M.G. and A.F. Johnson. 1988. Beach and dune. In: Barbour, M.G. and J. Major (eds.). Terrestrial Vegetation of California. John Wiley & Sons, New York, NY.

Barbour, M.G., T.M. De Jong, and A.F. Johnson. 1976. Synecology of beach vegetation along the Pacific Coast of the United States of America: a first approximation. J. of Biogeography. 3:55-69.

Barnhart, R.A., J.B. Milton, and J.E. Pequegnat. 1992. The Ecology of Humboldt Bay, California: an Estuarine Profile. Biological Report 1. US Fish and Wildlife Service. Washington, DC.

Baron, D. 1994. Limitations on spread of introduced cordgrass (*Spartina*) in South San Francisco Bay. Masters thesis, California State University, Hayward, Hayward, CA.

Barrows, C.W. 1993. Tamarisk control II: a success story. Restoration and Management Notes. 11: 35-38.

Barthell, J., University of Central Oklahoma, personal communication, 1996.

Baruah, N.C., J. Sarma, S. Sarma, and R. Sharma. 1993. Seed germination and growth inhibitory cadinenes from *Eupatorium adenophorum* Spreng. J. of Chemical Ecology. 20:8.

Baskin, C.C. and J.M. Baskin. 1996. Role of temperature and light in the germination ecology of buried seeds of weedy species of disturbed forests. II. *Erechtites hieracifolia*. Canadian J. of Botany. 74:12, 2002-05.

Baskin, J.M. and C.C. Baskin. 1981. Seasonal changes in germination responses of buried seeds of *Verbascum thapsus* and *V. blattaria* and ecological implications. Canadian J. of Botany. 59:1769-75.

Baskin, J.M. and C.C. Baskin. 1990. Seed germination ecology of poison-hemlock. Canadian J. Botany. 68:2018-24.

Batra, S.W.T., J.R. Coulson, P.H. Dunn, and P.E. Boldt. 1981. Insects and fungi associated with *Carduus* thistles. United States Department of Agriculture, Technical Bulletin No. 1616. Washington, DC.

Bauder, E., University of California, Davis, unpublished data.

Baum, B.R. 1978. The genus *Tamarix*. Israel Academy of Sciences and Humanities. Tel Aviv, Israel.

Bazzaz, F.A. 1986. Life history of colonizing plants: some demographic, genetic, and physiological features. In: Mooney, H.A. and J.A. Drake (eds.). Ecology of Biological Invasions of North America and Hawaii. Springer-Verlag, New York, NY.

Bean, D. and M.J. Russo. 1986. Element Stewardship Abstract for Periwinkle, Myrtle (*Vinca major*). The Nature Conservancy, San Francisco, CA.

Beatty, S.W. 1991. The interaction of grazing, soil disturbance, and invasion success of fennel on Santa Cruz Island, CA. Report to The Nature Conservancy, Santa Barbara, CA.

Beatty, S.W. and D.L. Licari. 1992. Invasion of fennel into shrub communities on Santa Cruz Island, California. Madroño. 39:54-66.

Beauchamp, R.M. 1986. A Flora of San Diego County, California. Sweetwater River Press, National City, CA.

Beck, N.G. and E.M. Lord. 1988. Breeding system in *Ficus carica*, the common fig. II. pollination events. American J. of Botany. 75:1913-22.

Bedell, T.E., R.B. Hawkes, and R.E. Whitesides. 1995. Pasture management for control of tansy ragwort. Pacific Northwest Cooperative Extension Publication No. 210. Portland, OR.

Bell, G. 1994. Biology and growth habits of giant reed (*Arundo donax*). In: Jackson, N.E. *et al. Arundo donax* workshop. California Exotic Pest Plant Council. San Diego, CA. Pp. 1-6.

Bell, G. 1998. Ecology and management of *Arundo donax* and approaches to riparian habitat restoration in southern California. In: Brock, J.H., M. Wade, P. Pysek, and D. Green (eds.). Plant Invasions. Backhuys Publ., Leiden, The Netherlands.

Bellue, M.K. 1936. *Lepidium latifolium* L., a new perennial peppergrass. California Department of Agriculture Bulletin. 25:359.

Bellue, M.K. 1951. Current status of halogeton in California. California Department of Agriculture Bulletin. 40:1.

Bendall, G.M. 1973a. The control of slender thistle, *Carduus pycnocephalus* L. and *C. tenuiflorus* Curt. (Compositae), in pasture by grazing management. Australian J. of Agricultural Research. 24:831-37.

Bendall, G.M. 1975. Some aspects of the biology, ecology, and control of slender thistle, *Carduus pycnocephalus* L. and *C. tenuiflorus* Curt. (Compositae), in Tasmania. J. Australian Institute of Agricultural Science. 41:52-53.

Bender, J. 1987. *Lythrum salicaria*. The Nature Conservancy, Element Stewardship Abstract.

Benecke, R. 1990. Investigation of seedling emergence and seed bank of *Ammophila arenaria*. The Nature Conservancy. Internal report. Lanphere-Christensen Dunes Preserve, Arcata, CA.

Benefield, C.B., J.M. Tomaso, G.B. Kyser, S.B. Orloff, K.R. Churches, D.B. Marcum, and G.A. Nader. 1999. Success of mowing to control yellow starthistle depends on timing and plants branching form. California Agriculture. 53(2):17-21.

Benham, P.E.M. 1990. *Spartina anglica*: a research review. Research Publication No. 2, Huntingdon Institute of Terrestrial Ecology, UK.

Bennett, F.D. 1988. Brazilian peppertree: prospects for biological control. Proceedings of the Seventh International Symposium on Biological Control of Weeds, March, 6-11, 1988, Rome, Italy.

Benton, N. 1998. Fountain grass. The Nature Conservancy Native Plant Conservation Initiative, Alien Plant Working Group (August 1997). Updated 8/13/98. http://www.nps.gov/plants/alien/fact/pese1.htm

Berenbaum, M.R. and T.L. Harrison 1994. *Agonopterix alstroemeriana* (Oecophoridae) and other Lepidopteran associates of poison hemlock (*Conium maculatum*) in east central Illinois. The Great Lakes Entomologist. 27(1):1-5.

Berenbaum, M.R. and S. Passoa. 1983. Notes on the biology of *Agonopterix alstroemeriana* (Clerck), with descriptions of the immature stages (Oecophoridae). J. of the Lepidopterist Society. 37:38-45.

Billings, W. D. 1990. Grasslands. In: Woodwell, G.M. (ed.). The Earth in Transition: Patterns and Processes of Biotic Impoverishment. Cambridge University Press, NY.

Bily, P., Dow Blanco, personal communication, 1997.

Birdsall, J., C. Quimby, T. Svejcar, and J. Young. 1997.

Potential for biocontrol of perennial pepperweed (*Lepidium latifolium* L.). In: Management of Perennial Pepperweed (tall whitetop). Special Report 972. USDA, Agricultural Research Service and Agricultural Experimental Station, Oregon State University, Corvallis, OR.

Black, C.C, T.M. Chen, and R.H. Brown. 1969. Biochemical basis of plant competition. Weed Science. 17:338-44.

Blackburn, W., R.W. Knight, and J. Schuster. 1982. Saltcedar influence on sedimentation in the Brazos River. J. Soil Water Conservation. 37:298-301.

Bleck, J., University of California, Santa Cruz, personal communication, 1997.

Blossey, B. and D. Schroeder. 1995. Host specificity of three potential biological weed control agents attacking flowers and seeds of *Lythrum salicaria*. Biological Control. 5:47-53.

Blumin, L. and R. Peterson. 1997. Manual removal of German ivy: pilot project in Volunteer Canyon, Bolinas Lagoon Preserve, Audubon Canyon Ranch. CalEPPC News.

Bohmont, D.W., A.A. Beetle, and F.L. Rauchfuss. 1955. Halogeton: what should we do about it? Circular 48. Wyoming Agricultural Experimental Station.

Boland, D.J. 1984. Forest Trees of Australia. Nelson Wadsworth, Melbourne, Australia.

Bor, N.L. 1968a. *Bromus*. In: Townsend C.C., E. Guest, and A. Al-Rawi (eds.). Flora of Iraq. Volume 9. Ministry of Agriculture of the Republic of Iraq, Baghdad, Iraq.

Bor, N.L. 1968b. *Schismus*. In: Townsend, C.C., E. Guest, and A. Al-Rawi. Flora of Iraq. Volume 9. Ministry of Agriculture of the Republic of Iraq. Baghdad, Iraq.

Borell, A.E. 1971. Russian-olive for wildlife and other conservation uses. Leaflet 292. US Department of Agriculture, Washington, DC.

Borman, M.M., D.E. Johnson, and W.C. Krueger. 1992. Soil moisture extraction by vegetation in Mediterranean/maritime climatic region. Agronomy J. 84:897-904.

Bossard, C. 1990a. Tracing of ant-dispersed seeds: a new technique. Ecology. 71:2370-72.

Bossard, C. 1990b. Secrets of an ecological interloper: ecological studies on *Cytisus scoparius* (Scotch broom) in California. Ph.D. dissertation, University of California, Davis, CA.

Bossard, C. 1991a. Establishment and dispersal of *Cytisus scoparius* (Scotch broom) in California Sierra Nevada foothill and northern coastal habitats. In: Proceedings of the International Symposium on Exotic Pest Plants, Miami, Florida, November 3-6, 1988. University of Miami Press, Miami, FL.

Bossard, C. 1991b. The role of habitat disturbance, seed predation, and ant dispersal on establishment of the exotic shrub *Cytisus scoparius* in California. American Midland Naturalist. 126:1-13.

Bossard, C. 1993. Seed germination in the exotic shrub *Cytisus scoparius* in California. Madroño. 40:47-61.

Bossard, C. 1998. Effects of floating Cape ivy (*Senecio mikanioides*) foliage on golden shiners and crayfish. Report to Golden Gate National Park Association, January 1998, San Francisco, CA.

Bossard, C. and C. Benefield. 1995. The war on German ivy: good news from the front. Proceedings of CalEPPC Symposium 95, Monterey, CA. CalEPPC, Sacramento, CA.

Bossard, C. and M. Rejmánek, 1994. Herbivory, seed production, growth, and resprouting of *Cytisus scoparius* in California. Biological Conservation. 67:193-200.

Bossard, C., M. Alvarez, G. Archbald, R. Gibson, D. Glusernkamp, E. (Grotkopf) Kuo, S. Jones, L. Nelson, and D. Smith. 1995. A French broom control project. Proceedings of CalEPPC Symposium 95, Monterey, CA. CalEPPC, Sacramento, CA.

Bottell, A.E. 1933. Introduction and control of camelthorn. California State Department of Agriculture Bulletin. 22:261-63.

Bovey, R.W. 1965. Control of Russian olive by aerial application of herbicides. J. of Range Management. 18(4):194-95.

Bowen, B. 1995. Exotic plant species profile: Eurasian watermilfoil. TN-EPPC Newsletter. Pp.4-5.

Boyd, D. 1992. The influence of *Ammophila arenaria* on foredune plant microdistributions at Point Reyes National Seashore, CA. Madroño. 39(1): 67-76.

Boyd, D. 1994. Prescribed burning of French broom. Proceedings of CalEPPC Symposium 94, Monterey, CA. CalEPPC, San Diego, CA.

Bradley, J. 1988. Bringing Back the Bush—the Bradley Method of Bush Regeneration. Landowne Press, Sydney, Australia.

Bravo, L.M. 1985. We are losing the war against broom. Fremontia. 12:27-29.

Breckon, G.J. and M.G. Barbour. 1974. Review of North American Pacific Coast vegetation. Madroño. (22):333-60.

Breidenbach, R.W., N.L. Wade, and J.M. Lyons. 1974. Effects of chilling temperatures on the activities of glyoxysomal and mitochondrial enzymes from castor bean seedlings. Plant Physiology. 54:324-27.

Brenton, B. and R.C. Klinger. 1994. Modeling the expansion and control of fennel (*Foeniculum vulgare*) on the Channel Islands. In: Halvorson, W. and G. Maender (eds.). Fourth California Islands Symposium: Update on the Status of Resources. Santa Barbara Museum of Natural History, Santa Barbara, CA.

Brey, C. 1996. What? Another *Ehrharta*? CalEPPC News. 4(2):4-5.

Brinkman, K.A. 1974. *Rubus*. In: Schopmeyer, C. (ed.). Seeds of Woody Plants in the US. Agriculture Handbook No. 450. US Govt. Printing Office, Washington, DC.

Brooks, M.L. 1998. Ecology of a biological invasion: alien annual plants in the Mojave Desert. Dissertation, University of California, Riverside, CA.

Brooks, M.L. In press. Alien annual grasses and fire in the Mojave Desert. Madroño.

Broom, D.M. and G.W. Arnold. 1987. Selection by grazing sheep of pasture plants at low herbage availability and responses of the plants to grazing. Australian J. of Agriculture Research. 37:(5) 527-38.

Brotherson, J.D., J.G. Carman, and L.A. Szyska. 1984. Stem-diameter age relationships of *Tamarix ramosissima* in central Utah. J. of Range Management. 37:362-64.

Brotherson, J.D. and D. Field. 1987. *Tamarix*: impacts of a successful weed. Rangelands. 9:10-19.

Brotherson, J.D. and V. Winkel. 1986. Habitat relationships of saltcedar (*Tamarix ramosissima*) in central Utah. Great Basin Naturalist. 46:535-41.

Bruckart, W. 1991. Phenotypic comparison of *Puccinia carduorum* from *Carduus thoermieri*, *C. tenuiflorus*, and *C. pycnocephalus*. Phytopathology. 81:192-97.

Bruegmann, M.M. 1996. Hawaii's dry forests. Endangered Species Bulletin. 21:26-27.

Bruns, V.F. 1965. The effects of fresh water storage on the germination of certain weed seeds. Weeds. 13(1):38-40.

Buell, A.C., A. Pickart and J.D. Stuart. 1995. Introduction, history and invasion patterns of *Ammophila arenaria* on the north coast of California. Conservation Biology. 9:1587-93.

Bureau of Land Management. 1996. Partners Against Weeds: Final Action Plan for the Bureau of Land Management. 27 pp. Washington, DC.

Burge, L.M. 1950. *Halogeton glomeratus*, a poisonous plant in Nevada. Nevada State Department of Agriculture Bulletin, Reno, NV.

Burgess, T.L., J.E. Bowers, and R.M. Turner. 1991. Exotic plants at the desert laboratory, Tucson, Arizona. Madroño. 38:96-114.

Burtt, B.L. 1977. Aspects of diversification in the capitulum. In: Heywood, V.H., J.B. Harborne, and B.L. Turner (eds.). The Biology and Chemistry of the Compositae. Academic Press, London, UK.

Busch, D.E. 1995. Effects of fire on southwestern riparian plant community structure. Southwestern Naturalist. 40:259-67.

Bush, Lisa & Associates, Environmental Consultants, personal communication.

Butterfield, C. and J. Stubbendieck. 1999. *Euphorbia esula*. Species Abstracts of Highly Disruptive Exotic Plants at Pipestone National Monument. US Geological Survey, Biological Resources Division, Northern Prairie Research Center. http://www.npwrc.usgs.gov/resource/othrdata/exoticab/pipeeuph.htm

Butterfield, H.M. 1964. Dates of introductions of trees and shrubs to California. University of California Press, Berkeley, CA.

Cade, J.W. 1980. Perennial veldt grass. Order no. 1019/80. F.D. Atkinson, Government Printer, Melbourne, Australia.

California Department of Fish and Game, Natural

Heritage Division, Sacramento, CA.

California Department of Food and Agriculture. 1971. Plant Detection Manual. Sacramento, CA.

California Department of Food and Agriculture. 1991. Plant Detection Manual: Hydrilla Review. Sacramento, CA.

California Exotic Pest Plant Council. 1996. Exotic pest plants of greatest ecological concern in California as of August 1996.

Callaway, J.C. and M.N. Josselyn. 1992. The introduction and spread of smooth cordgrass (*Spartina alterniflora*) in south San Francisco Bay. Estuaries. 15:218-26.

Callihan, R.H., T.S. Prather, and F.E. Northam. 1993. Longevity of yellow starthistle (*Centaurea solstitialis*) achenes in soil. Weed Technology. 7:3-35.

Cameron, E. 1935. A study of the natural control of ragwort (*Senecio jacobaea* L.). J. Ecology. 23:265-322.

Carafa, A.M., G. Carratu, and G.F. Tucci. 1980. Ecologia delle Rhinantheae parassite. Osservazioni sulla fisiologia nutrizionale di *Bellardia trixago* L. All. coltivata in vitro. Annali della Facolta di Scienze Agrarie della Universita degli Studi di Napoli Portici. 14(1):25-31.

Carmin, J. 1950. Plants in Israel: their biology, diseases, and cryptogamic inhabitants. Bulletin of the Independent Biological Laboratories, Kefar Malal, P.O. Ramatayim, Israel. 6(2):1-6.

Carrithers, V.F. 1997. Using Transline herbicide to control invasive plants. In: Kelly, M.E., E. Wagner, and P. Warner (eds.). Proceedings of CalEPPC Symposium 1997, San Diego, CA.

Carrithers, V.F., Dow Elanco, personal communication, 1998.

Carson, E. 1998. Overview of Scotch broom in British Columbia. Invasive Plants of Canada Project. http://infoweb.magi.com/~ehaber/bc_broom.html.

Carter, G.A. and A.H. Teramura. 1988. Vine photosynthesis and relationships to climbing mechanisms in a forest understory. American J. of Botany. 75:1011-18.

Catalano, S., S. Lusch, G. Flamini, P.L. Cioni, E.M. Neri, and I. Morelli. 1996. Chemistry of *Senecio mikanioides.* Phytochemistry. 42:1605-07.

Center, T.A. and N. Spencer. 1981. The phenology and growth of water hyacinth (*Eichhornia crassipes* (Mart.) Solms) in a eutrophic north-central Florida lake. Aquatic Botany. 10:1-32.

Center, T.A., Cofrancesco, and J. Balciunas. 1989. Biological control of aquatic and wetland weeds in the southeastern United States. In: Delfosse, E. (ed.). Proceedings of the VII International Symposium on Biological Control of Weeds, Rome, Italy. Pp. 239-62.

Chabreck, R.H. and R.E. Condrey. 1979. Common vascular plants of the Louisiana marsh. Sea Grant Publication No. LSU-T-79-003. Louisiana State University, Center for Wetland Resources, Baton Rouge, LA.

Challoner, K.R. and M.M. McCarron. 1990. Castor bean intoxication. Annals of Emergency Medicine. 19(10):1177-83.

Chamberlain, S.J. 1995. Comparison of methods to control *Spartina alterniflora* in San Francisco Bay. Masters thesis, San Francisco State University, San Francisco, CA.

Charlton, J.F.L., J.G. Hampton, and D.J. Scott. 1986. Temperature effects on germination of New Zealand herbage grasses. Proceedings of the New Zealand Grassland Association. 47:165-72.

Cheeke, P.R. (ed.). 1979. Proceedings of the Symposium on Pyrrolizidine (*Senecio*) Alkaloids: toxicity, metabolism, and poisonous plant control measures, Oregon State University, Corvallis, OR. Pp. 42-54.

Chemical & Pharmaceutical Press. 1997. Crop protection reference. Chemical & Pharmaceutical Press, New York, NY.

Cheng, Sheauchi. In press. Genetics, habitat, and French broom (*Genista monspessulanus*) control. Weed Science.

Chesnut, J.W. 1999a. A review of weed threats to the Nipomo Dunes. Report prepared for the Land Conservancy of San Luis Obispo, CA.

Chesnut, J.W. 1999b. Assessing the impacts of grazing on community composition in veldt grass infested dune scrub, Guadalupe Dunes, San Luis Obispo. Prepared for Land Conservancy of San Luis Obispo, CA.

Chippendale, G.M. 1988. Flora of Australia, Volume 19, Eucalyptus, Angophora. Australian Government Publishing Service, Canberra, Australia.

Chipping, D. 1992. Hoary cress. Proceedings of the CalEPPC Symposium 1992. Weed Note. CalEPPC, San Diego, CA.

Chipping, D. 1993. German ivy infestation in San Luis Obispo. CalEPPC News. 1(1):15-16.

Chipping, D., California Native Plant Society, personal communication, November 5, 1996, January 20, 1997, and March 3, 1997.

Christensen, E.M. 1963. Naturalization of Russian olive (*Elaeagnus angustifolia* L.) in Utah. American Midland Naturalist. 70:133-37.

Cicero, V., California Department of Parks and Recreation, San Luis Obispo Coast District, personal communication, November 4, 1996.

Clarke, O. University of California, Riverside, personal communication, 1998.

Clausen, C.P. (ed.). 1982. Introduced Parasites and Predators of Arthropod Pests and Weeds: A World Review. US Dept. of Agriculture Handbook 480. Washington, DC.

Collingwood, G.H. 1937. Knowing Your Trees. American Forestry Association. Washington, DC.

Collins, S.L. 1987. Interaction of disturbances in tallgrass prairie: a field experiment. Ecology. 68:1243-50.

Collins, S.L. and W.H. Blackwell, Jr. 1979. *Bassia* (Chenopodiaceae) in North America. SIDA. 8(1):57-64.

Colvin, W.I. 1996. Fennel (*Foeniculum vulgare*) removal

from Santa Cruz Island, California: managing successional processes to favor native over non-native species. Senior thesis, Board of Environmental Studies, University of California, Santa Cruz, CA.

Conert, H.J. and A.M. Turpe. 1974. Revision der gattung *Schismus* (Poaceae: Arundinaceae: Danthonaceae). Abhendlungender Senckenbelgischen. Naturforschenden Gesellschaft. 532:1-81.

Connor, H.E. 1973. Breeding systems in *Cortaderia* (Gramineae). Evolution. 27:663-78.

Connor, H.E. and D. Charlesworth. 1989. Genetics of male-sterility in gynodioecious *Cortaderia* (Gramineae). Heredity. 63:373-82.

Conway, K. 1976. Evaluation of *Cercospora rodmanii* as a biological control of water hyacinths. Phytopathology. 66:914-17.

Cook, C.W. 1965. Grass seedling response to halogeton competition. J. of Range Management. 18: 317-21.

Cook, C.W. and L.A. Stoddart. 1953. The halogeton problem in Utah. Utah State Agricultural Experimental Station, Bulletin 364.

Cook, M.J. 1987. Hoary cress in California parklands. Report by California Department of Food and Agriculture for California Department of Parks and Recreation, Sacramento, CA.

Cooper, M.R. and A.W. Johnson. 1984. Poisonous Plants in Britain and Their Effects on Animals and Man. Her Majesty's Stationery Office, London, UK.

Cooper, W.S. 1967. Coastal sand dunes of California. Geological Society of America. 104:1-131.

Corliss, J. 1993. Tall whitetop's crowding out the natives. Agricultural Research. May 1993:16.

Costas-Lippman, M. 1977. More on the weedy "pampas grass" in California. Fremontia. 4:25-27.

Costello, L.R. 1986. Control of ornamentals gone wild: pampas grass, bamboo, English and Algerian ivy. California Weed Conference. 38:162-65.

Couch, R. and E. Nelson. 1986. *Myriophyllum spicatum*. In: Proceedings of the First International Symposium on Watermilfoil (*Myriophyllum spicatum*) and Related Haloragaceae, Aquatic Plant Management Society, Vicksburg, MS. Pp. 8-18.

Cowan, B.D. 1976. The menace of pampas grass. Fremontia. 4:14-16.

Cowell, W.S. 1973. Wildflowers of the US Vol 5: The Northwestern States. Cowell LTD, Butter Market, Ipswich, UK.

Cox, C.S. and P.B. McEvoy. 1983. Effect of summer moisture stress on the capacity of tansy ragwort (*Senecio jacobaea* L.) to compensate for defoliation by the cinnabar moth (*Tyria jacobaeae* L.) J. Applied Ecology. 20:225-34.

Cox, T. 1997. Perennial pepperweed in Idaho. In: Management of Perennial Pepperweed (tall whitetop). Special Report 972. USDA, Agricultural Research Service and Agricultural Experimental Station, Oregon State University, Corvallis, OR.

Cozzo, D. 1972. Initial behavior of *Ailanthus altissima* in experimental plantations. Revista Forestal Argentina. 16(2):47-52 (Spanish).

Cronin, E.H. 1965. Ecological and physiological factors influencing chemical control of *Halogeton glomeratus*. Technical Bulletin No. 1325. US Department of Agriculture in cooperation with Utah State Agricultural Experimental Station, Washington, DC.

Cronin, E.H. and M.C. Williams. 1966. Principles for managing ranges infested with halogeton. J. of Range Management. 19:226-27.

Cronk, Q.C. and J.L Fuller. 1995. Plant Invaders: the Threat to Natural Ecosystems. Chapman and Hall. London, UK.

Crooks, K. 1994. Comparative ecology of the island spotted skunk and the island fox of Santa Cruz Island, California. Masters thesis, University of California, Davis, CA.

Crouchley, G. 1980. Regrowth control by goats—plus useful meat returns. New Zealand J. Agriculture. 141(5):9-14.

Cuddihy, L.W., C.P. Stone, and J.T. Tunison. 1988. Alien plants and their management in Hawaii Volcanoes National Park. Transactions of the Western Section of the Wildlife Society. 24:42-46.

Cudney, D.W. and D. Hodel. 1986. German ivy in the coastal sage scrub regions of southern California. Research Progress Reports of the Western Society of Weed Science. P. 11.

Dale, R., director, Sonoma Ecological Center, personal communication, 1997.

Danin, A., S. Rae, M. Barbour, N. Jurjavcic, P. Connors, and E. Uhlinger. 1998. Early primary succession on dunes at Bodega Head, California. Madroño. 45(2):101-09.

Daar, S. 1983. Using goats for brush control. The IPM Practitioner. 5(4):4-6.

Daehler, C.C. 1996a. Seed set variability, inbreeding depression, and effects of herbivory in introduced smooth cordgrass (*Spartina alterniflora*) invading Pacific estuaries. Ph.D. dissertation, University of California, Davis, CA.

Daehler, C.C. 1996b. Spartina invasions in Pacific estuaries: biology, impact, and management. In: Sytsma, M.D. (ed.). Proceedings of a Symposium on Non-indigenous Species in Western Aquatic Ecosystems, Portland State University. Lakes and Reservoirs Program Publication No. 96-8, Portland, OR.

Daehler, C.C. and D.R. Strong. 1994. Variable reproductive output among clones of *Spartina alterniflora* (Poaceae) invading San Francisco Bay, California: the influence of herbivory, pollination, and establishment site. American J. of Botany. 81:307-13.

Daehler, C.C. and D.R. Strong. 1996. Status, prediction, and prevention of introduced cordgrass (*Spartina* spp.) invasions in Pacific estuaries, USA. Biological Conservation. 78: 51-58.

Daehler, C.C. and D.R. Strong. 1997a. Reduced herbivore resistance in introduced smooth cordgrass (*Spartina alterniflora*) after a century of herbivore-free growth. Oecologia. 110:99-108.

Daehler, C.C. and D.R. Strong. 1997b. Hybridization between introduced smooth cordgrass (*Spartina alterniflora*; Poaceae) and native California cordgrass (*S. foliosa*) in San Francisco Bay, California, USA. American J. of Botany. 84:607-11.

Dahl, B.E. and E.W. Tisdale. 1975. Environmental factors related to medusahead distribution. J. Range Management. 28:463-68.

Dai, Z., G. Edwards, and M. Ku. 1996. Control of photosynthesis and stomatal conductance in *Ricinus communis* L. (castor bean) by leaf to air vapor pressure deficit. Plant Physiology. 99(4): 1426-33.

D'Antonio, C.M. 1990a. Invasion of coastal plant communities in California by the introduced iceplant, *Carpobrotus edulis* (Aizoaceae). Ph.D dissertation, University of California, Santa Barbara, CA.

D'Antonio, C.M. 1990b. Seed production and dispersal in the non-native, invasive succulent *Carpobrotus edulis* in coastal strand communities of central California. J. Applied Ecology. 27:693-702.

D'Antonio, C.M. 1993. Mechanisms controlling invasion of coastal plant communities by the alien succulent *Carpobrotus edulis*. Ecology. 74(1):83-95.

D'Antonio, C.M. and B. Mahall. 1991. Root profiles and competition between the invasive exotic perennial *Carpobrotus edulis* and two native shrub species in California coastal scrub. American J. Botany. 78:885-94.

D'Antonio, C.M. and P.M. Vitousek. 1992. Biological invasions by exotic grasses, the grass/fire cycle, and global change. Annual Review of Ecology and Systematics. 23:63-87.

Damjanic, A. and B. Akacic. 1974. Forocoumarins in *Ficus carica*. Planta Medica 26:119-23.

Darling, C., agricultural inspector/biologist, Dept. of Agriculture/Measurement Standards, San Luis Obispo County, CA, personal communication.

Darwin, C. 1989. Voyage of the Beagle. Penguin Classics, London, UK.

Dash, B.A. and S.R. Gliessman. 1994. Nonnative species eradication and native species enhancement: fennel on Santa Cruz Island. In: Halvorson, W. and G. Maender (eds.). Fourth California Islands Symposium: Update on the Status of Resources. Santa Barbara Museum of Natural History, Santa Barbara, CA. Pp. 505-12.

Davidson, A. 1907. The change in our weeds. Bulletin of the Southern California Academy of Sciences. 6:11-12.

Davidson, E.D. and M.G. Barbour. 1977. Germination, establishment and demography of coastal bush lupine (*Lupinus arboreus*) at Bodega Head, California. Ecology. 58:592-600.

Davis, P.H., R.R. Mill, and K. Tan. 1985. Flora of Turkey and the East Aegean Islands. University Press, Edinburgh, UK.

Davy, B.J. 1902. Stock ranges of northwestern California. In: US Department of Agriculture, Bureau of Plant Industry Bulletin 12. Washington, DC. Pp. 1-81.

Dayton, W.A. 1951. Historical sketch of barilla (*Halogeton glomeratus*). J. of Range Management. 4:375-81.

de Jong, T.J. and P.G. Klinkhamer. 1988a. Population ecology of the biennials *Cirsium vulgare* and *Cynoglossum officinale* in a coastal sand-dune area. J. of Ecology. 76:366-82.

de Jong, T.J. and P.G.Klinkhamer. 1988b. Seedling establishment of the biennials *Cirsium vulgare* and *Cynoglossum officinale* in a sand-dune area: the importance of water for differential survival and growth. J. of Ecology. 76:393-402.

de Winton, M.D. and J.S. Clayton. 1996. The impact of invasive submerged weed species on seedbanks in lake sediments. Aquatic Botany. 53(1-2):31-45.

De Villiers, A., M. Van Royen, G. Theron, and A. Claasens. 1995. The effect of leaching and irrigation on the growth of *Atriplex semibaccata*. Land Degradation and Rehabilitation 6:125-31.

Decker, J.P. 1961. Salt secretion by *Tamarix pentandra*. Forest Science. 7:214-17.

del Pardo, C.F. and R.G. Encina. 1977. Control quimico de *Bellardia trixago*. Control de Malezas: Resultados de la Investigacion y Nuevos Herbicidas, 1976. 77:43-46. (Sociedad Chilena de Control de Malezas, Santiago, Chile).

DeLoach, C.J. 1997. Biological control of weeds in the United States and Canada. In: Luken, J.O and J.W. Thieret (eds.). Assessment and Management of Plant Invasions. Springer-Verlag, New York, NY.

Derr, J.F. 1992. Weed control in nursery crops. In: Pest Management Guide for Horticulture and Forest Crops. Virginia Cooperative Extension Publication 456-017. Blacksburg, VA. Pp. 105-16.

Derr, J.F. 1993. English ivy (*Hedera helix*) response to postemergence herbicide. J. of Environmental Horticulture. 11:45-48.

Detmers, F. 1927. Canada thistle (*Cirsium arvense* Tourn.), field thistle, creeping thistle. Ohio Agricultural Experiment Station Bulletin 414.

di Castri, F. 1990. On invading species and invaded ecosystems: the interplay of historical chance and biological necessity. In: di Castri, F., A.J. Hansen, and M. Debussche (eds.). Biological Invasions in Europe and the Mediterranean Basin. Pp. 3-16. Kluwer Academic Publishers, Dordrecht, The Netherlands.

Dick, M.A. 1994. Blight of *Lupinus arboreus* in New Zealand. New Zealand J. of Forestry Science. 24: 51-68.

Dirr, M. and C.W. Heuser. 1987. The Reference Manual of Woody Plant Propagation. Varsity Press, Athens, GA.

DiTomaso, J.M. 1996. Identification, biology and ecology of salt cedar. In: Proceedings of the Saltcedar Management Workshop. University of California, Cooperative Extension, Imperial County,

University of California, Davis, and the California Exotic Pest Plant Council, Davis, CA.

DiTomaso, J.M., E. Healy, C.E. Bell, J. Drewitz, and A. Tscholl. 1999. Pampas grass and jubata grass threaten California coastal habitats. Leaflet #98-1. University of California Weed RIC, Davis, CA.

DiTomaso, J.M., G.B. Kyser, and M.S. Hastings. 1999a. Prescribed burning for control of yellow starthistle (*Centaurea solstitialis*) and enhanced native plant diversity. Weed Science. 47:233-42.

DiTomaso. J.M., W.T. Lanini, C.D. Thomsen, T.S. Prather, C.E. Turner, M.J. Smith, C.L. Elmore, M.P. Vayssieres, and W.A. Williams. 1998. Yellow starthistle. University of Calfornia, DANR. Pest Notes 3:1-4.

Dodd, A.P. 1961. Biological control of *Eupatorium adenophorum* in Queensland. Australian J. of Science. 23:356-65.

Doidge, E.M. 1948. South African rust fungi. Bothalia. 4:895-937.

Donald, W.W. 1990. Management and control of Canada thistle (*Cirsium arvense*). Reviews of Weed Science. 5:193-250.

Doody, J.P. (ed.). 1984. *Spartina anglica* in Great Britain: report of a meeting held at Liverpool University, November 10, 1982. Huntingdon Nature Conservancy Council. (Focus on Nature Conservation, No. 5). Liverpool, UK.

Doren, R.F., L.D. Whiteaker, G. Molnar, and D. Sylvia. 1990. Restoration of former wetlands within the Hole-in-the-Donut in Everglades National Park. In: Webb, F.J. (ed.). Proceedings of the Seventh Annual Conference on Wetlands Restoration and Creation, Hillsborough Community College, Tampa, FL.

Douce, R. 1994. The biological pollution of *Arundo donax* in river estuaries and beaches. In: Jackson, N.E. *et al. Arundo donax* workshop. California Exotic Pest Plant Council. San Diego, CA. Pp. 11-13.

Douthit, S. 1994. *Arundo donax* in the Santa Ana River Basin. In: Jackson, N. *et al. Arundo donax* workshop. California Exotic Pest Plant Council. San Diego, CA. Pp. 7-10.

Dudley, T. and B. Collins. 1995. Biological Invasions in California Wetlands: The Impacts and Control of Non-indigenous Species in Natural Areas. Pacific Institute for Studies in Development, Environment, and Security, Oakland, CA.

Dujardin, M. and W.W. Hanna. Crossability of pearl millet with wild *Pennisetum* species. Crop Science. 29:77-80.

Duke, J.A. 1985. CRC Handbook of Medicinal Herbs. CRC, Boca Raton, FL.

Dumka, W., Rick Engineering Co., representing the Black Mountain Ranch developer, San Diego, CA, personal communication, 1997.

Egunjobi, J.K. 1971. Ecosystem processes in a stand of *Ulex europaeus*. J. Ecology. 59(1):31-38.

Eicher, A. 1987. Salt marsh vascular plant distribution in relation to tidal elevation, Humboldt Bay, California. Masters thesis, Humboldt Bay State University, Arcata, CA.

Elfers, S.C. 1995. *Schinus terebinthifolius* (Brazilian peppertree). Element Stewardship Abstract. The Nature Conservancy. Washington, DC.

Elliot, W. 1994. German ivy engulfing riparian forests and heading for the uplands. CalEPPC News. 2(1):9.

Elliott, K.J., L.R. Boring, W.T. Swank, and B.R. Haines. 1997. Successional changes in plant species diversity and composition after clearcutting a southern Appalachian watershed. Forest Ecology and Management. 92:1-3, 67-85.

Else, J.A., D. Lawson, and V. Vartanian. 1996. Removal of *Arundo donax* on the Santa Margarita River: effectiveness of different removal treatments and recovery of native species. Abstract, CalEPPC Symposium 96, San Diego, CA.

Erasmus, D.J., P.H. Bennett, and J. van Staden. 1991. The effect of galls induced by the gall fly *Procecidochares utilis* on vegetative growth and reproductive potential of Crofton weed, *Ageratina adenophora*. Annals of Applied Biology. 120:173-81.

Erickson, L.C., E.W. Tisdale, H.L. Morton, and G. Zappettini. 1951. Halogeton: intermountain range menace. University of Idaho Agricultural Experiment Station, and Forest, Wildlife, and Range Experimental Station, Circular No. 117.

Espigares, T. and B. Peco. 1995. Mediterranean annual pasture dynamics: impact of autumn drought. J. of Ecology. 83:135-42.

Evans, F.J. and R. Schmidt. 1980. Plants and plant products that induce contact dermatitis. Planta-Medica. 38:289-316.

Evans, R.A., J.A. Young, and R. Hawkes. 1979. Germination characteristics of Italian thistle, *Carduus pycnocephalus*, and slenderflower thistle, *Carduus tenuiflorus*. Weed Science. 27:327-32.

Everist, S.L. 1959. Strangers within the gates (plants naturalised in Queensland). Queensland Naturalist. 16:49-60.

Ewel, J.J. 1979. Ecology of *Schinus*. In: Workman, R. (ed.). Technical Proceedings of Techniques for Control of *Schinus* in South Florida: Workshop for Natural Area Managers. The Sanibel-Captiva Conservation Foundation, Sanibel, FL.

Ewel, J.J. 1986. Invasibility: lessons from south Florida. In: Mooney, H.A. and J.A. Drake (eds.). Ecology of Biological Invasions of North America and Hawaii. Springer-Verlag, New York, NY. Pp. 214-30.

Ewel, J.J., D.S. Ojima, D.A. Karl, and W.F. DeBusk. 1982. *Schinus* in Successional Ecosystems of Everglades National Park. Report T-676. South Florida Research Center, Everglades National Park, FL.

Fagg, P.C. 1989. Control of *Delairea odorata* (Cape ivy) in native forest with the herbicide clopyralid. Plant Protection Quarterly. 4(2):107-10.

Faruqi, S.A. 1981. Studies on the Libyan grasses. VII. Additional notes on *Schismus arabicus* and *Schismus barbatus*. Pakistan J. of Botany. 13:225.

Faruqi, S.A. and H.B. Quraish. 1979. Studies on the Libyan grasses. V. Population variability and distribution of *Schismus arabicus* and *Schismus barbatus* in Libya. Pakistan J. of Botany. 11:167-72.

Fawcett, R.S. and J.E. Nelson. 1981. Weed Control—Biennial Thistles: Bull, Musk, Tall. Iowa State University Cooperative Extension. Pm-772. Ames, IA.

Featherstone, C.I. 1957. The progress of chemical weed control in Hawke's Bay. In: Proceedings of the Tenth New Zealand Weed and Pest Control Conference. Pp. 7-12. Auckland, NZ.

Feijoo, C.S., F.R. Momo, C.A. Bonetto, and N.M. Tur. 1996. Factors influencing biomass and nutrient content of the submerged macrophyte *Egeria densa* in a pampasic stream. Hydrobiologia. 341(1):21-26.

Feliz, D., assistant manager, Grizzly Island Wildlife Area, California Dept. of Fish and Game, personal communication, 1999.

Fenley, J.M. 1952. How to live with halogeton while limiting its spread and reducing losses. Bulletin 106. University of Nevada Agricultural Extension Service.

Fenster, C.R. and G.A. Wicks. 1978. Know and control downy brome. NebGuide G 78-422. University of Nebraska, Lincoln, NB.

Ferguson, L., T.J. Michailides, and H.H. Shorey. 1990. The California fig industry. In: Janick, J. (ed.). Horticultural Reviews, volume 12. Timber Press, Portland, OR.

Fischer, B.B., A.H. Lange, and J. McCaskell. 1979. Five-hook bassia: *Bassia hyssopifolia*. In: Growers Weed Identification Handbook. Publication 4030. University of California Cooperative Extension Service, Berkeley, CA.

Fisher, R. and T. Esque. National Biological Survey, unpublished data.

Fish and Wildlife Information Exchange, Species Information Library. 1995. CD-ROM maintained by Virginia Polytechnical Institute.

Foster, R., E. Knake, R.H. McCarty, J.J. Mortvedt, and L. Murphy (eds.). 1994. Weed Control Manual. Meister Publishing Company, Willoughby, OH.

Fox, M.D. 1990. Mediterranean weeds: exchanges of invasive plants between the five Mediterranean regions of the world. In: di Castri, F., A.J. Hansen, and M. Debussche (eds.). Biological Invasions in Europe and the Mediterranean Basin. Kluwer Academic Publishers, Dordrecht, The Netherlands.

Frandsen, P. and N. Jackson. 1994. The impact of *Arundo donax* on flood control and endangered species. In: Jackson, N. *et al. Arundo donax* workshop, California Exotic Pest Plant Council, San Diego, CA. Pp. 13-16.

Franklin, B.B. 1996. Eradication/control of the exotic pest plants tamarisk and *Arundo* in the Santa Ynez River drainage. USDA-FS-PSW, Washington, DC.

French, W.J. 1972. *Cristulariella pyramidalis* in Florida: an extension of range and new hosts. Plant Disease Report. 56(2):135-38.

Frenkel, R.E. 1970. Ruderal Vegetation Along Some California Roadsides. University of California Publications in Geography, Volume 20, University of California Press, Berkeley, CA.

Frenkel, R.E. 1977. Ruderal Vegetation Along Some California Roadsides. University of California Press, Berkeley, CA.

Frenkel, R.E. 1987. Introduction and spread of cordgrass (*Spartina*) into the Pacific Northwest. North West Environmental J. 3:152-54.

Frenkel, R.E. and T.R. Boss. 1988. Introduction, establishment and spread of *Spartina patens* on Cox Island, Siuslaw Estuary, Oregon. Wetlands. 8:33-45.

Frey H. 1984. Encounter with *Arctotheca calendula*, a composite plant from the Cape of Good Hope, South Africa, is invading the southern European coast. Dissertationes Botanicae. 72:453-58.

Fryor, J.D. and R.S. Makepeace (eds.). 1978. Weed Control Handbook: Recommendations. Vol.2. 8th ed. Blackwell Scientific Publications, Oxford, UK.

Fuller, T. C. 1976. Pampas grass: its history as a weed. Fremontia. 4:16.

Fuller, T.C. and G.D. Barbe. 1985. The Bradley method of eliminating exotic plants from natural reserves. Fremontia. 13(2):24-25.

Fuller, T.C. and E. McClintock. 1986. Poisonous Plants of California. University of California Press, Berkeley, CA.

Furbush, P. 1953. Control of medusa-head on California ranges. J. Forestry. 51:118-21.

Gadgil, R.L., A.L. Knowles, and J.A. Zabkiewicz. 1984. Pampas: a new forest weed problem. Proceedings of New Zealand Weed and Pest Control Conference. 37:187-90.

Gadgil, R.L., A.M. Sandberg, P. Allen, and S.S. Gallagher. 1990. Partial suppression of pampas grass by other species at the early seedling stage. In: Bassett, C., L.J. Whitehouse, and J.A. Zabkiewicz (eds.). Alternatives to the Chemical Control of Weeds, Ministry of Forestry, FRI Bulletin 155. Pp. 120-27.

Gaffney, K. and H. Cushman. 1998. Transformation of a riparian plant community by grass invasion. Abstract. Soc. for Conserv. Biol., 12th Annual Meeting, Sydney, Australia, May 1998.

Galil, J. and G. Neeman. 1977. Pollen transfer and pollination in the common fig (*Ficus carica* L.). New Phytologist. 79:163-71.

Garland, S. 1979. The Complete Book of Herbs and Spices. Viking press, New York, NY.

Gary, H.L. 1960. Utilization of five-stamen tamarisk by cattle. Rocky Mountain Forest and Range Experiment Station. Research Notes. 51:1-3.

Gautier, C. 1982. The effects of ryegrass on erosion and natural vegetation recovery after fire. In: USDA Forest Service. Dynamics and Management of Mediterranean-Type Systems. General Technical Report PSW-58. Pacific SW Forest and Range Experiment Station, Berkeley, CA.

Geicky, H.M. 1957. Weeds of the Pacific Northwest. Oregon Press, Portland, OR.

Gemmell, A.R., P. Greig-Smith, and C.H. Gimingham. 1953. A note on the behavior of *Ammophila arenaria* Link in relation to sand dune formation. Translation and Proceedings of the Botanical Society of Edinburgh. 36:132-36.

Gerlach, J.D. 1998. How the West was lost: reconstructing the invasion dynamics of yellow starthistle and other plant invaders of Western rangelands and natural areas. In: M. Kelly, E. Wagner, and P. Warner (eds.). Proceedings of the California Exotic Pest Plant Council Symposium. 3:1997. Pp. 67-72.

Gibbs, R. and E.R. Robinson. 1983. Speciation environments and centres of species diversity in southern Africa: 2. Case studies. Bothalia. 14:1007-12.

Gilbertson, R.L. and M. Blackwell. 1988. Some new or unusual corticoid fungi from the Gulf Coast region. Mycotaxon. 33:375-86.

Gilkey, H. 1957. Weeds of the Pacific Northwest. Oregon State College, Eugene, OR.

Gill, N.T. 1938. The viability of weed seeds at various stages of maturity. Annals of Applied Biology. 25:447-56.

Glusenkamp, D.A. 1999. Relative abundance of two thistle species: evaluating non-exclusive hypotheses for the success of an introduced invader. Abstracts of the Ecological Society of America 84th Annual Meeting, Spokane, WA, August 1999.

Glusenkamp, D., University of California, Berkeley, personal communication, 1999.

Goeden, R.D. 1974. Comparative survey of the *Phytophagous* insect faunas of Italian thistle, *Carduus pycnocephalus*, in southern California and southern Europe relative to biological weed control. Environmental Entomology. 3:464-74.

Goeden, R.D. and D.W. Ricker. 1982. Poison hemlock (*Conium maculatum*) in southern California: an alien weed attacked by few insects. Annals of the Entomological Society of America. 75:173-76.

Goeden, R.D. and D.W. Ricker. 1986a. Phytophagous insect faunas of two introduced *Cirsium* thistles, *C. ochrocentrum* and *C. vulgare*, in southern California. Annals of the Entomological Society of America. 79:945-52.

Goeden, R.D. and D.W. Ricker. 1986b. Phytophagous insect faunas of the two most common native *Cirsium* thistles, *C. californicum* and *C. proteanum*, in southern California. Annals of the Entomological Society of America. 79:953-62.

Gogue, G.J., C.J. Hurst, and L. Bancroft. 1974. Growth inhibition by *Schinus terebinthifolius*. American Society of Horticultural Science. 9:45.

Golden Gate National Recreation Area. 1989. Poison Hemlock and Thistle Management Plan.

Goldman, C., DES, University of California, Davis, personal communication, 1997.

Goodwin, P.J. and W.T. Williams. 1961. Investigations into dieback in *Spartina townsendii* agg. III. Physiological correlates of dieback. J. of Ecology. 49:391-98.

Gouin, F.R. 1979. Controlling brambles in established Christmas tree plantations with glyphosate. HortScience. 14(2):189-90.

Grace, J.B. and R.G. Wetzel. 1978. The production biology of Eurasian watermilfoil (*Myriophyllum spicatum*): a review. J. Aquatic Plant Management. 16:1-10.

Graf, W.L. 1978. Fluvial adjustments to the spread of tamarisk in the Colorado Plateau region. Geological Society of America Bulletin. 89:1491-1501.

Granath, T. 1992. Fennel on Santa Cruz Island. Year II. Senior thesis, Board of Environmental Studies, University of California, Santa Cruz, CA.

Granval, P. 1993. The impact of agricultural management on earthworms (Lumbricidae), common snipe (*Gallinago gallinago*) and the environmental value of grasslands in the Dives marshes (Calvados). Gibier Faune Sauvage. 10:59-73.

Gray, K., California State Parks, Monterey, personal communication, 1998.

Green, H.E. 1937. Dispersal of *Senecio jacobaea*. J. of Ecology. 25:569.

Grey, W.E., P.C. Quimby, Jr., D.E. Mathre, and J.A. Young. 1995. Potential for biological control of downy brome (*Bromus tectorum*) and medusahead (*Taeniatherum caput-medusae*) with crown and root rot fungi. Weed Technology. 9:362-65.

Grieve, M. 1959. A Modern Herbal. Hafner, New York, NY.

Griffiths, A. and E. McClintock. 1971. Transient horticultural taxa: the need for documentation, as exemplified in *Myoporum*. California Horticultural J. 32(1):24-26.

Grime, J.P. 1965. Shade tolerance in flowering plants. Nature. 208:161-63.

Grime, J.P., J.G. Hodgson, and R. Hunt. 1988. Comparative Plant Ecology: a Functional Approach to Common British Species. Unwin Hyman, London, UK.

Griswold, G.B. 1985. Population biology of ox-eye daisy (*Chrysanthemum leucanthemum*) in different habitats. Ph.D. thesis, University of Kansas, Department of Botany.

Grobbelaar, E., O. Nesar, and S. Nesar. A survey of the potential insect biological control agents of *Delairea odorata* Lemaire in South Africa. Report to the USDA Agricultural Research Service, July, 1999. Agricultural Research Council, Landbounavorsingsraad, South Africa.

Groenendaal, G.M. 1983. Eucalyptus helped solve a timber problem: 1853-1880. In: Proceedings of a Workshop on Eucalyptus in California. General Technical Report PSW-69. USDA, Pacific Southwest Forest and Range Experiment Station,

Forest Service, Berkeley, CA.

Gross, K.L. and P.A. Werner. 1978. The biology of Canadian weeds. 28. *Verbascum thapsus* L. and *V. blattaria* L. Canadian J. of Plant Science. 58:410-13.

Grotkopf, E., University of California, Davis, personal communication, 1999.

Groves, R.H. 1991. Status of environmental weed control in Australia. Plant Protection Quarterly. 6:95-98.

Gutterman, Y. 1993. Seed germination in desert plants. Springer-Verlag, Berlin, Germany.

Gutterman, Y. 1994. Strategies of seed dispersal and germination in plants inhabiting deserts. The Botanical Review. 60:373-425.

Gutterman, Y. and M. Evanari. 1994. The influences of amounts and distribution of irrigation during the hot and dry season on emergence and survival of some desert winter annual plants in the Negev Desert of Israel. Israel J. of Plant Science. 42:1-14.

Haderlie, L., S. Dewey, and D. Kidder. 1987. Canada Thistle Biology and Control. Bulletin No. 666. University of Idaho, Cooperative Extension Service.

Hafliger, E. and H. Scholz. 1981. Grass Weeds 2. Documenta, Ciba-Geigy Ltd., Basle, Switzerland.

Haller, W., J. Miller, and L. Garrard. 1976. Seasonal production and germination of hydrilla vegetative propagules. Hyacinth Control J. 14:26-29.

Halvorson, W.L. 1992. Alien plants at Channel Islands National Park. In: Stone, C.P., C.W. Smith, and J.T. Tunison (eds.). Alien Plant Invasions in Native Ecosystems of Hawaii: Management and Research. University of Hawaii Cooperative National Park Studies Unit, Honolulu, HI. Pp. 64-96.

Halvorson, W.L., D. Fenn, and W. Allardice. 1988. Soils and vegetation of Santa Barbara Island, Channel Islands National Park, California. Environmental Management. 12:109-18.

Hammer, R. 1996. Fountain grass: turn off the spigot! Newsletter of the Florida Exotic Pest Plant Council. 6:1.

Hanson, B. 1996. Tools and techniques: chemical-free controls. In: Randall, J. M. and J. Marinelli (eds.). Invasive Plants: Weeds of the Global Garden. Brooklyn Botanic Garden, Brooklyn, NY.

Hardt, R.A. and R.T. Forman. 1989. Boundary form effects on woody colonization of reclaimed surface mines. Ecology. 70(5):1252-60.

Harlow, W.M, E.S. Harrar, and F.M. White. 1979. Textbook of Dendrology. McGraw-Hill, New York, NY.

Harper, J.L. 1958. The ecology of ragwort with special references to control. Commonwealth Burl Pastures & Field crops. Herbage Abstracts. 28(3): 151-57.

Harper, J.L. 1965. Establishment, aggression and cohabitation in weedy species. In: Baker, H.G. and G.L. Stebbins (eds.). The Genetics of Coloniz-

ing Species. Academic Press, New York, NY. Pp. 243-68.

Harper, J.L. and W.A. Wood. 1957. Biological flora of the British Isles: *Senecio jacobaea* L. J. of Ecology. 45:617-37.

Harradine, A.R. 1985. Dispersal and establishment of slender thistle, *Carduus pycnocephalus*, as effected by ground cover. Australian J. of Agricultural Research. 36:791-97.

Harradine, A.R. 1991. The impact of pampas grass as weeds in southern Australia. Plant Protection Quarterly. 6:111-15.

Harris, G.A. 1967. Some competitive relationships between *Agropyron spicatum* and *Bromus tectorum*. Ecological Monographs. 37:89-111.

Harris, P. and A.T. Wilkinson. 1984. *Cirsium vulgare* (Savi) Ten., bull thistle (Compositae). In: Kelleher, J.S. and M.A. Hulme (eds.). Biological Control Programmes Against Insects and Weeds in Canada 1969-1980. Commonwealth Agricultural Bureaux, Farnham Royal, Slough, England. Pp. 147-53.

Hartley, M.J. 1983. Effect of Scotch thistles on sheep growth rates. Proceedings of the New Zealand Weed and Pest Control Conference. 36:86-89.

Hartley, M.J., D.K. Edmonds, H.T. Phung, A.I. Popay, and P. Sanders. 1980. The survival of gorse seedlings under grazing, treading, and mowing. Proceedings of the 33rd New Zealand Weed and Pest Control Conference. Pp. 161-64.

Hatzios, K.K. 1998. Herbicide Handbook. Supplement to the 7th edition. Weed Science Society of America. Champaign, IL.

He, X. 1991. A preliminary study in the ecology of the lesser bamboo rat (*Cannomys badius*) in China. Zoological Research. 12:41-46.

Heidorn, R. 1990. Vegetation management guideline: purple loosestrife. Illinois Nature Preserves Commission, Vegetation Management Manual. 1(17): 1-5.

Hellmers, H. and W.C. Ashby. 1958. Growth of native and exotic plants under controlled temperatures and in the San Gabriel Mountains, California. Ecology. 39(3):416-28.

Hendry, G.W. 1931. The adobe brick as a historical source. Agricultural History. 5(3):110-27.

Hendry, G. W. and M. P. Kelley. 1925. The plant content of adobe bricks. California Historical Society Quarterly. 4:361-73.

Herre, H. 1971. The genera of the *Mesembryanthemum*. University of California Press, Berkeley, CA.

Hewett, R., former preserve manager, The Nature Conservancy, Kern River Preserve, personal communication, January 31, 1985.

Hickman, J. (ed.). 1993. The Jepson Manual: Higher Plants of California. University of California Press, Berkeley, CA.

Higaki, T. 1973. Chemical weed control in *Anthurium*. Research Report, Hawaii Agricultural Experiment Station, No. 212. Honolulu, HI.

Hight, S.D. and J. Drea. 1991. Prospects for a classical

biological control project against purple loosestrife. Natural Areas J. 11(3):151-57.

Hilken, T.O. and R.F. Miller. 1980. Medusahead (*Taeniatherum asperum* Nevski): a review and annotated bibliography. Station Bulletin 664. Agricultural Experimental Station, Oregon State University, Corvallis, OR.

Hill, D.D. 1955. Gorse control. Bulletin 553. Oregon State College Agricultural Experimental Station, Corvallis, OR.

Hillman, F.H. and H.H. Henry. 1928. The incidental seeds found in commercial seed of alfalfa and red clover. Proceedings of the International Seed Testing Association. 6:1-19.

Hitchcock, A.S. 1935. Manual of the Grasses of the United States. USDA Misc. Publ. No. 200. Washingtion, DC.

Hitchcock, A.S. 1950. Manual of the Grasses of the United States. USDA Misc. Publ. No. 200, Washington, DC (revised edition by Agnes Chase).

Hodgson, J.M. 1964. Variations in ecotypes of Canada thistle. Weeds. 12:167-71.

Hodgson, J.M. 1968. The Nature, Ecology, and Control of Canada Thistle. USDA, Technical Bulletin 1386. Washington, DC.

Hoffman, J. and W. Kearns. 1997. Eurasian watermilfoil (*Myriophyllum spicatum*). In: Wisconsin Manual of Control Recommendations for Ecologically Invasive Plants. State of Wisconsin, Madison, WI.

Hogan, E.L. (ed.). 1992. Sunset Western Garden Book. Sunset Publishing, Menlo Park, CA.

Holden, L. 1989. Biological control of tansy ragwort (*Senecio jacobaea* L.) on a disturbed coastal prairie in Del Norte County, California. Masters thesis, Humboldt State University, Arcata, CA.

Holm, L., J.V. Pancho, J.P. Herberer, and D.L. Poucknett. 1979. A Geographical Atlas of World Weeds. John Wiley & Sons, New York, NY.

Horng, L.C. and L.S. Leu. 1979. Control of five upland perennial weeds with herbicides. Proceedings 7th Asian-Pacific Weed Science Conference. Pp. 165-67.

Hoshovsky, M. 1988. Element stewardship abstract: *Arundo donax*. The Nature Conservancy, San Francisco, CA.

Hoshovsky, M. 1995. Element stewardship abstract for brooms. The Nature Conservancy, Washington, DC.

Hosking, J.R. 1994. The impact of seed and pod feeding insects on *Cytisus scoparius* Link. In: Delfosse, E.S. and R.R. Scott (eds.). Proceedings of the 8th International Symposium on Biological Control of Weeds, 2-7, 1992, Lincoln University, Canterbury, New Zealand. DSIR/CSIRO, Melbourne, Australia.

Howe, W.H. and F.L. Knopf. 1991. On the imminent decline of Rio Grande cottonwoods in central New Mexico. Southwestern Naturalist. 36:218-24.

Howell, J.T. 1949. Marin Flora. University of California Press, Berkeley, CA.

Howell, J.T. 1969. Marin Flora Supplement. University of California Press, Berkeley, CA.

Howell, J.T. 1979. Marin Flora. University of California Press, Berkeley, CA.

Hrusa, F., CDFA Herbarium, personal communication, 1998.

Hu, S.Y. 1979. *Ailanthus*. Arnoldia. 39(2):29-50.

Huikes, A.H. 1977. The natural establishment of *Ammophila arenaria* from seed. Oikos. 29:133-36.

Huikes, A.H. 1979. Biological flora of the British Isles: *Ammophila arenaria* (L.) Link. J. of Ecology. 67:363-82.

Hujik, P., Dye Creek Preserve, Los Molinos, CA, personal communication, 1999.

Hulbert, L.C. 1955. Ecological studies of *Bromus tectorum* and other annual brome grasses. Ecological Monographs. 25:181-213.

Hull, A.C., Jr., and G. Stewart. 1948. Replacing cheatgrass by reseeding with perennial grass on southern Idaho ranges. Agronomy J. 40:694-703.

Hull, A.C., Jr., and W.T. Hansen, Jr. 1974. Delayed germination of cheatgrass seed. J. of Range Management. 27:366-68.

Humphries, S.E., R.H. Groves, and D.S. Mitchell. 1991a. Plant invasions of Australian ecosystems: a status review of management directions. Australian National Parks and Wildlife Service, Canberra.

Humphries, S.E., R.H. Groves, and D.S. Mitchell (eds.). 1991b. Plant Invasions: the Incidence of Environmental Weeds in Australia. Australian National Parks Wildlife Service, Canberra.

Hunter, J.C. 1995. *Ailanthus altissima*: its biology and recent history. CalEPPC News. 3(4):4-5.

Hunter, R. 1991. *Bromus* invasion on the Nevada test site: present status of *B. rubens* and *B. tectorum* with notes on their relationship to disturbance and altitude. Great Basin Naturalist. 51:176-82.

Huntley, J.C. 1990. *Robinia pseudoacacia* L., black locust. In: Burns, R. M. and B. H. Honkala. Silvics of North America, Volume 2, Hardwoods. Agricultural Handbook 654. US Department of Agriculture, Washington, DC.

Ivens, G.W. 1979. Effects of sprays on gorse regrowth at different growth stages. Proceedings of the 32nd New Zealand Weed and Pest Control Conference. Pp. 303-06.

Jackson, L.E. 1985. Ecological origins of California's Mediterranean grasses. J. of Biogeography. 12:349-61.

Jackson, N.E. 1994. Control of *Arundo donax*: techniques and pilot project. In: Jackson, N.E. *et al. Arundo donax* workshop, California Exotic Pest Plant Council, San Diego, CA. Pp. 27-34.

Jackson, N.E. 1996. Chemical control of saltcedar (*Tamarix ramosissima*). In: Proceedings of the Saltcedar Management Workshop. University of California, Cooperative Extension, Imperial County, University of California, Davis, and California Exotic Pest Plant Council, Davis, CA.

Jackson, N.E., P. Frandsen, and S. Douthit. 1994. Proceedings of the *Arundo donax* workshop, Nov. 1993, Ontario, CA. Calif. Exotic Pest Plant Council, Riverside, CA.

Jacot-Guillarmod, A. 1979. Water weeds in southern Africa. Aquatic Botany. 6:377-91.

James, L.F., R.F. Keeler, A.E. Johnson, M.C. Williams, E.H. Cronin, and J.D. Olsen. 1980. Plants poisonous to livestock in the western United States. USDA Information Bulletin No. 415. Washington, DC.

James, L.F., M.C. Williams, and T.T. Bleak. 1976. Toxicity of *Bassia hyssopifolia* to sheep. J. of Range Management. 29(4):284-85.

Jeffery, L.S. and L.R. Robinson. 1990. Poison-hemlock (*Conium maculatum*) control in alfalfa (*Midicaga sativa*). Weed Technology. 4:585-87.

Jensen, W.I. and J.P. Allen. 1981. Naturally occurring and experimentally induced castor bean (*Ricinus communis*) poisoning in ducks. Avian Disease. 25(1):184-94.

Jepson, W.L. 1911. A Flora of Western Middle California. Cunningham, Curtis and Welch, San Francisco, CA.

Jepson, W.L. 1919. Field Book. 36:147.

Jepson, W.L. 1925. A Manual of the Flowering Plants of California. University of California Press, Berkeley, CA.

Johnson, P.N. 1982. Naturalized plants in southwest South Island, New Zealand. New Zealand J. Botany. 20:131-42.

Joley, D.B. 1995. Yellow starthistle, *Centaurea solstitialis* L., seed germination study: effect of red and far-red light. In: Bezark, L.G. (ed.). Biological Control Program Annual Summary, 1994. California Department of Food and Agriculture, Division of Plant Industry, Sacramento, CA.

Julien, M.H. 1982. Biological Control of Weeds: A World Catalogue of Agents and Their Target Weeds. Commonwealth Institute of Biological Control, Silwood Park, UK.

Julien, M.H. (ed.). 1987. Biological Control of Weeds: a World Catalogue of Agents and Their Target Weeds. 2nd ed. CAB International, Wallingford, UK.

Junak, S., T. Ayers, R. Scott, D. Wilken, and D. Young. 1995. A Flora of Santa Cruz Island. Santa Barbara Botanic Garden and California Native Plant Society, Santa Barbara and Sacramento, CA.

Kane, M., E. Gilman, and M. Jenks. 1991. Regenerative capacity of *Myriophyllum aquaticum* tissues cultured in vitro. J. of Aquatic Plant Management. 29:102-09.

Kassas, M. 1952a. On the distribution of *Alhagi maurorum* in Egypt. Proceedings of the Egyptian Academy of Sciences. 8:140-51.

Kassas, M. 1952b. On the reproductive capacity and life cycle of *Alhagi maurorum*. Proceedings of the Egyptian Academy of Sciences. 8:114-22.

Katibah, E.F., N.E. Nedeff, and K.J. Dummer. 1984. Summary of riparian vegetation aerial and linear extent measurements from the Central Valley

Riparian Mapping Project. In: Warner, R.E. and K.M. Hendrix (eds.). Proceedings of the conference California Riparian Systems: Ecology, Conservation, and Productive Management, September 17-19, 1981, Davis, CA. University of California Press, Berkeley, CA.

Katovich, S.E., R.L. Becker, and B.D. Kinkaid. 1996. Influence of non-target neighbors and spray volume on retention and efficacy of triclopyr in purple loosestrife. Weed Science. 44:143-47.

Kay, B.L. 1969. Harding grass and annual legume production in the Sierra foothills. J. Range Management. 22(3):174-77.

Kearney, T.H., R.H. Peebles, J.T. Howell, and E. McClintock. 1960. Arizona Flora. 2d ed. University of California Press, Berkeley, CA.

Keeley, S.C., J. Keeley, S.M. Hutchinson, and A.W. Johnson. 1981. Postfire succession of the herbaceous flora in southern California chaparral. Ecology. 62:1608-21.

Kelly, D. 1988. Demography of *Carduus pycnocephalus* and *C. tenuiflorus*. New Zealand Natural Sciences. 15:17-24.

Kelly, M. 1996. *Cynara cardunculus*. In: Randall, J.M. and J. Marinelli (eds.). Invasive Plants: Weeds of the Global Garden. Brooklyn Botanic Garden Pub., Brooklyn, NY. P. 75.

Kerpez, T.A. and N.S. Smith. 1987. Saltcedar control for wildlife habitat improvement in the southwestern United States. Resource Publication 169. US Fish and Wildlife Service, Washington, DC.

Kerr, H.D. 1963. A study of the establishment and control of *Alhagi camelorum* Fisch. Ph.D. dissertation, Washington State University.

Kerr, H.D., W.C. Robocker, T.J. Muzik. 1965. Characteristics and control of camelthorn. Weeds 13(2):156-63.

Kingsbury, J.M. 1964. Poisonous Plants of the United States and Canada. Prentice-Hall, Englewood Cliffs, NJ.

Kjellberg, F., P.H. Gouyon, M. Ibrahim, M. Raymond, and G. Valdeyron. 1987. The stability of the symbiosis between dioecious figs and their pollinators: a study of *Ficus carica* L. and *Blastophaga psenes* L. Evolution. 41:693-704.

Klinkhamer, P.G. and T.J. de Jong. 1988. The importance of small-scale disturbance for seedling establishment in *Cirsium vulgare* and *Cynoglossum officinale*. Journal of Ecology. 76:383-92.

Klinkhamer, P.G., T.J. de Jong, and E. Meelis. 1987a. Delay of flowering in the "biennial" *Cirsium vulgare*: size effects and devernalization. Oikos. 49:303-08.

Klinkhamer, P.G., T.J. de Jong, and E. Meelis. 1987b. Life-history and the control of flowering in short-lived monocarps. Oikos. 49:309-14.

Klinkhamer, P.G., T.J. de Jong, and E. van der Meijden. 1988. Production, dispersal and predation of seeds in the biennial *Cirsium vulgare*. J. of Ecology. 76:403-14.

Kloot, P.M. 1983. The role of common iceplant (*Mes-

embryanthemum crystallinum) in the deterioration of mesic pastures. Australian J. Ecology. 8:301-6.

Kluge, R.L. 1991. Biological control of Crofton weed, *Ageratina adenophora* (Asteraceae), in South Africa. Agriculture, Ecosystems and Environment. 37:187-91.

Knopf, F.L. and T.E. Olson. 1984. Naturalization of Russian-olive: implications for Rocky Mountain wildlife. Wildlife Society Bulletin. 12:289-98.

Koehler, J.W., M.R. Pryor, C.E. Pratt. 1956. Camelthorn: new approaches to eradication. Bulletin of the California State Department of Agriculture. 45:229-32.

Koske, R.E. 1988. Vesicular-arbuscular mycorrhizae of some Hawaiian dune plants. Pacific Science. 42:217-29.

Kowarik, I. 1983. On the naturalization and plant-geographic control of the tree-of-heaven (*Ailanthus altissima* (Mill.) Swingle) in the French Mediterranean area (Bas-Languedoc). Phytocoenologia. 11:385-405 (German with English summary).

Kowarik, I. 1995. Clonal growth in *Ailanthus altissima* on a natural site in West Virginia. J. of Vegetation Science. 6:853-56.

Kranz, J., H. Schmutterer, and W. Koch (eds.). 1977. Diseases, Pests and Weeds in Tropical Crops. John Wiley & Sons, New York, NY.

Krugman, S.L. 1974. Eucalyptus L'Herit eucalyptus. In: Schopmeyer, C.S. (ed.). Seeds of Woody Plants in the United States. Agriculture Handbook No. 450. US Department of Agriculture, Forest Service, Washington, DC.

Kunkel, G. 1978. Flowering trees in subtropical gardens. Dr. W. Junk Publishers, The Hague, The Netherlands.

Lacey, J.R., C.B. Marlow, and J.R. Lane. 1989. Influence of spotted knapweed (*Centaurea maculosa*) on surface runoff and sediment yield. Weed Technology. 3:627-31.

Lalonde, R.G. and B.D. Roitberg. 1994. Mating system, life-history, and reproduction in Canada thistle (*Cirsium arvense*; Asteraceae). American J. of Botany. 81:21-28.

Lambrechtsen, N.C. 1992. What grass is that? New Zealand Department of Scientific and Industrial Research Information Series No. 87, Wellington, NZ.

Land Conservancy of San Luis Obispo. 1999. Veldt grass suppression demonstration project final report. San Luis Obispo, CA.

Lanford, J. and L. Nelson. 1992. Arthropod populations at three Golden Gate habitats compared. Park Science, Spring issue, National Park Service, Washington DC.

Langeland, K. 1990. Hydrilla: a continuing problem in Florida waters. Florida Cooperative Extension Circular No. 884. Miami, FL.

Langer, R.H.M. 1990. Pasture plants. In: Langer, R.H.M. (ed.). Pastures: Their Ecology and Management. Oxford University Press. Oxford, UK.

LaRoche, F.B. (ed.). 1994. Melaleuca Management Plan for Florida: Recommendations from the Melaleuca Task Force. 2nd ed. Exotic Pest Plant Council, West Palm Beach, FL.

Lasca Leaves. 1968. This month's cover plant. 18:22.

Lasca Leaves. 1973. *Arctotheca calendula* (L.) Levins, Cape weed, Cape dandelion. 23:59.

Lawrence, J.G. 1991. The ecological impact of allelopathy in *Ailanthus altissima*. American J. of Botany. 78:948-58.

Liegel, K., R. Marty, and J. Lyon. 1984. Black locust control with several herbicides, techniques tested. Restoration and Management Notes. 2(2):87-88.

Lisci, M. and E. Pascini. 1994. Germination ecology of drupelets of the fig (*Ficus carica* L.). Botanical J. of the Linnean Society. 114:133-46.

Little, S. 1974. *Ailanthus altissima* (Mill.) Swingle, ailanthus. In: Schopmeyer, C.S. (ed.). Seeds of Woody Plants in the United States. Agricultural Handbook No. 450. United States Department of Agriculture, Washington, DC. Pp. 201-02.

Lodi, A., S. Leuchi, L. Mancini, G. Chiarelli, and C. Crosti. 1992. Allergy to castor oil and colophony in a wart remover. Contact Dermatitis. 26(4):266-67.

Loope, L.L. and V.L. Dunevitz. 1981. Impact of Fire Exclusion and Invasion on *Schinus terebinthifolius* on Limestone Rockland Pine Forests of Southeastern Florida. Report T-645. South Florida Research Center, Everglades National Park, Homestead, FL.

Loria, M. and I. Noy-Meir. 1979-1980. Dynamics of some annual populations in a desert loess plain. Israel J. of Botany. 28:211-25.

Louda, S.M. 1998. Population growth of *Rhinocyllus conicus* (Coleoptera: Cucurliondiae) on two species of native thistles in prairie. Environmental Entomology. 27(4):834-41.

Louda, S.M., D. Kendall, J. Connor, and D. Simberloff. 1997. Ecological effects of an insect introduced for the biological control of weeds. Science. 277:1088-90.

Love, R. and M. Feigen. 1978. Interspecific hybridization between native and naturalized *Crataegus* (Rosaceae) in western Oregon. Madroño. 25:211-17.

Love, R.M. 1948. Eight new plants developed for California ranges. California Agriculture. 2:7.

Lovich, J.E. and R.G. de Gouvenain. In press. Saltcedar invasion in desert wetlands of the southwestern United States: ecological and political implications. In: Majumdar, S.K. (ed.). Ecology of Wetlands and Associated Systems. Pennsylvania Academy of Science.

Lovich, J.E., T.B. Egan, and R.C. de Gouvenain. 1994. Tamarisk control on public lands in the desert of southern California: two case studies. In: California Weed Science Society. Proceedings of the Forty-Sixth Annual California Weed Conference. Pp. 166-77.

Luken, J.O. 1992. Bark girdling by herbivores as a potential biological control of black locust (*Robinia pseudoacacia*) in power-line coridors. Transactions of the Kentucky Academy of Science. 53(1):26.

Lusk, W.C., M.B. Jones, P.J. Torell, and C.M. McKell. 1961. Medusahead palatability. J. Range Management. 14:248-51.

Lym, R.G. and R.K. Zollinger. 1995. Integrated Management of Leafy Spurge. Bulletin W-866. North Dakota State University Extension Service, Fargo, ND.

Mack, R.N. 1981. Invasion of *Bromus tectorum* into western North America: an ecological chronicle. Agroecosystems. 7:145-65.

Mack, R.N. and D.A. Pyke. 1984. The demography of *Bromus tectorum*: the role of microclimate, grazing, and disease. J. of Ecology. 72:731-48.

Maddox, D.M. 1981. Introduction, phenology, and density of yellow starthistle in coastal, intercoastal, and Central Valley situations of California. USDA-ARS. Agricultural Research Reports ARR-W-20:1-33.

Maddox, D.M. and A. Mayfield. 1985. Yellow starthistle infestations on the rise. California Agriculture. 39(6):10-12.

Maddox, D.M., D.B. Joley, D.M. Supkoff, and A. Mayfield. 1996. Pollination biology of yellow starthistle (*Centaurea solstitialis*) in California. Canadian J. of Botany. 74:262-67.

Maddox, D.M., A. Mayfield, and N.H. Poritz. 1985. Distribution of yellow starthistle (*Centaurea solstitialis*) and Russian knapweed (*Centaurea repens*). Weed Science. 33:315-27.

Madison, J. 1992. Pampas grasses: one a weed and one a garden queen. Pacific Horticulture. 53:48-52.

Magnani, G. 1975. A weak parasite of *Ailanthus altissima*. Publ. Centro. Sperimentazione. Agricultural Forestale. 12(1):79-83 (Italian).

Maron, J.L. and P.G. Connors. 1996. A native nitrogen-fixing shrub facilitates weed invasion. Oecologia. 105:302-12.

Marquis, L.Y., R.D. Comes, and C.P. Yang. 1981. Absorption and translocation of fluridone and glyphosate (herbicides) in submersed vascular plants. Weed Science. 29(2):229-36.

Martin, R.W. and J.H. Popenoe. 1984. Buried, viable seed in the forest floor in Redwood National Park. Second Biennial Conference of Research in California's National Parks, 5-7 September 1984, University of California, Davis, CA.

Mastroguiseppe, R.J., N.T. Blair, and D.J. Vezie. 1982. Artificial and biological control of tansy ragwort (*Senecio jacobaea* L.) in Redwood National Park. In: Proceedings of the Second Biennial Conference of Research in National Parks. University of California, Davis, CA.

Mathias, M.E. (ed.). 1982. Flowering Plants in the Landscape. University of California Press, Berkeley, CA.

Maw, M.G. 1976. An annotated list of insects associated with Canada thistle (*Cirsium arvense*) in Canada. Canadian Entomologist. 108:235-44.

Maw, M.G. 1980. *Cucullia verbasci*, an agent for the biological control of common mullein (*Verbascum thapsus*). Weed Science. 28(1):27-30.

May, M. 1995. *Lepidium latifolium* L. in the San Francisco Estuary. Dept. of Geography, University of California, Berkeley. Unpublished report.

McCavish, W.J. 1980. Herbicides for woody weed control by foliar application. In: Proceedings of the 1980 British Crop Protection Conference on Weeds.

McClintock, E. 1981. Trees of Golden Gate Park: tree-of-heaven, *Ailanthus altissima*. Pacific Horticulture. 42(3):16-18.

McClintock, E. 1985. Status reports on invasive weeds: brooms. Fremontia. 12(4):17-18.

McDonald, P.M. and J.C. Tappeiner II. 1986. Weeds: the cycles suggest controls. J. of Forestry. 84:33-37.

McDonald, R., North Carolina Department of Agriculture Plant Industry Division. Oct. 1994. Biological control of bull thistle. http://ipm_www.ncsu.edu/ncda/bcbull.html.

McKell, C.M., A.M. Wilson, and B.L. Kay. 1962. Effective burning of rangelands infested with medusahead. Weeds. 10:125-31.

McKinnon, D. 1984. Pampas problem may surpass gorse. New Zealand Farmer. 105:20-21.

McKnight, B.N. (ed.). 1993. Biological Pollution: The Control and Impact of Invasive Exotic Species. Indiana Academy of Science, Indianapolis, IN.

McNabb, T. and L.W. Anderson. 1989. Aquatic weed control. In: California Weed Conference (eds.). Principles of Weed Control in California. 2nd ed.

Mendes, E. 1980. Asthma provoked by castor-bean dust. In: Occupational Asthma. Nostrand Reinholt Company, New York, NY. Pp. 272-82.

Mensing, S.A. and R. Byrne. 1998. Pre-mission invasion of *Erodium cicutarium* in California. J. of Biogeography. 25(4):757-62.

Mergen, F. 1959. A toxic principle in the leaves of ailanthus. Botanical Gazette. 121:32-36.

Merkel, D.L. and H.H. Hopkins. 1957. Life history of salt cedar (*Tamarix gallica* L.). Transactions of the Kansas Academy of Science. 60:360-69.

Michailides, T.J., D.P. Morgan, and K.V. Subbarao. 1996. Fig endosepsis: an old disease still a dilemma for California growers. Plant Disease. 80:828-41.

Milberg, P. and B.B. Lamont. 1995. Fire enhances weed invasion of roadside vegetation in southwestern Australia. Biological Conservation. 73:45-49.

Miller, G.K., J.A. Young, and R.A. Evans. 1986. Germination of seeds of perennial pepperweed (*Lepidium latifolium*). Weed Science. 34:252-55.

Miller, J.H. 1990. *Ailanthus altissima* (Mill.) Swingle, ailanthus. In: Burns, R.M. and B.H. Honkala. Silvics of North America, Vol. 2, Hardwoods. Agricultural Handbook 654. US Department of Agriculture, Washington, DC.

Miller, L.M. 1988. How yellow bush lupine came to Humboldt Bay. Fremontia. 16(3):6-7.

Miller, T. 1996. Use pesticides safely. In: William, R.D.,

D. Ball, T.L. Miller, R. Parker, K. Al-Khatib, R.H. Callihan, C. Eberlein, and D.W. Morishita. 1996. Pacific Northwest Weed Control Handbook. Cooperative Extension, Washington State University, Pullman, WA.

Misra, R.M. 1978. A mermithid parasite of *Attera fabricella*. Indian Forester. 104(2):133-34.

Mitich, L.W. 1989. Common mullein: the roadside torch parade. Weed Technology. 3:704-05.

Monsanto Corp. 1992. Native habitat restoration: Controlling *Arundo donax*. Monsanto Co. Application Guide Circular No. 170-92-06.

Monsen, S.B. and S.G. Kitchen. 1994. Proceedings. Ecology and Management of Annual Rangelands (on cheatgrass). Gen. Tech. Report 313. USDA, Forest Service, Ogden, UT.

Montllor, C.B., E.A. Bernays, and R.V. Barbehenn. 1990. J. Chemical Ecology 16(6):1853-65.

Moody, M.E. and R.N. Mack. 1988. Controlling the spread of plant invasions: the importance of nascent foci. J. of Applied Ecology 25:1009-21.

Mooney, H.A., S.P. Hamburg, and J.A. Drake. 1986. The invasions of plants and animals into California. In: Mooney, H.A. and J.A. Drake (eds.). Ecology of Biological Invasions of North America and Hawaii. Springer-Verlag, New York, NY.

Moore, K. 1994. Pulling pampas: controlling *Cortaderia* by hand with a volunteer program. CalEPPC News. 2:7-8.

Moore, R.J. 1975. The biology of Canadian weeds. 13. *Cirsium arvense* (L.) Scop. Canadian J. of Plant Science. 55:1033-48.

Moore, R.J. and C. Frankton. 1974. The Thistles of Canada. Research Branch, Canada Department of Agriculture, Monograph No. 10.

Morris, M.J. 1989. Host specificity studies of a leaf spot fungus, *Phaeoramularia* sp., for the biological control of Crofton weed (*Ageratina adenophora*) in South Africa. Phytophylactica. 21:281-83.

Morrow, L.A. and P.W. Stahlman. 1984. The history and distribution of downy brome (*Bromus tectorum*) in North America. Weed Science. 32:Supplement 1:2-6.

Moss, T.K. 1994. Ice plant eradication and native landscape restoration. Proceedings of the California Weed Conference, Fremont, CA, 1994. Pp. 155-56.

Moss, T., California State Parks, Monterey, personal communication, 1998.

Mountain, W.L. 1994. Purple loosestrife, *Lythrum salicaria*. Regulatory Horticulture. 20:27-34.

Mountjoy, J.H. 1979. Broom: a threat to native plants. Fremontia. 6:11-15.

Muenscher, W. 1955. Weeds. 2nd ed. MacMillan, New York, NY.

Muldavin, E.H., J.M. Lenihan, W.S. Lennox, and S.D. Veirs, Jr. 1981. Vegetation succession in the first ten years following logging of coast redwood forests. Technical Report 6. Redwood National Park, Arcata, CA.

Mulligan G.A. and J.N. Findlay. 1974. The biology of Canadian weeds. 3. *Cardaria draba, C. chalepensis* and *C. pubescens*. Canadian J. of Plant Science. 54: 149-60.

Mulligan, G.A. and C. Frankton. 1962. Taxonomy of the genus *Cardaria* with particular reference to the species introduced into North America. Canadian J. Botany. 40:1413-25.

Mulroy, T., biologicial consultant, Santa Barbara, CA. personal communication, 1998.

Mulroy, T.W., M.L. Dungan, R.E. Rich, and B.C. Mayerle. 1992. Wildland weed control in sensitive natural communities: Vandenberg Air Force Base, California. In: Proceedings of the 44th California Weed Conference, January 22, 1992, Fremont, CA. Pp. 166-80.

Muniappan, R. and C.A. Viraktamath. 1993. Invasive alien weeds in the western Ghats. Current Science. 64:8.

Munz, P.A. 1959. A California Flora. University of California Press, Berkeley, CA.

Munz, P.A. 1974. A Flora of Southern California. University of California Press, Berkeley, CA.

Munz, P.A. and D.D. Keck. 1963. A California Flora. University of California Press, Berkeley, CA.

Munz, P.A. and D.D. Keck. 1968. A California Flora. University of California Press, Berkeley, CA.

Munz, P.A. and D.D. Keck. 1973. A California Flora and Supplement. University of California Press, Berkeley, CA.

Murphy, A.H. and W.C. Lusk. 1961. Timing medusahead burns to destroy more seed and save good grasses. California Agriculture. 15:6-7.

National Academy of Science. 1975.

Neal, J.C. and A.F. Senesac. 1991. Preemergent herbicide safety in container-grown ornamental grasses. HortScience. 26:157-59.

Neal, J.C. and W.A. Skroch. 1985. Effects of timing and rate of glyphosate application on toxicity to selected woody ornamentals. J. of the American Society of Horticultural Scientists. 110:860-64.

Neill, W. 1983. The tamarisk invasion of desert riparian areas. Educational Bulletin #83-84. Education Foundation of the Desert Protective Council, Spring Valley, CA.

Nelson, L. 1994. French broom control in mission blue butterfly habitat in the Golden Gate National Recreation Area. Restoration and Management Notes. NPS, San Francisco, CA.

Nelson, L.S. and K.D. Getsinger. 1994. Battling purple loosestrife brings some success. The Wetlands Research Program Bulletin, Vol. 4, No. 1, pp. 5-8. Washington, DC.

Newman, O.W. 1917. Yellow star thistle (*Centaurea solstitialis*). Monthly Bulletin of the California State Commission of Horticulture. 6:27-29.

Nilsen, E.T. and W.H. Muller. 1980a. A comparison of the relative naturalization ability of two *Schinus* species in southern California. I. Seed germination. Bulletin of the Torrey Botanical Club. 107:51-56.

Nilsen, E.T. and W.H. Muller. 1980b. A comparison

of the relative naturalization ability of two *Schinus* species (Anacardiaceae) in southern California. II. Seedling establishment. Bulletin of the Torrey Botanical Club. 107(2):232-37.

Nilsen, E.T. and W.H. Muller. 1980c. A comparison of the relative naturalization ability of two *Schinus* species in southern California. I. Seed germination. Bulletin of the Torrey Botanical Club. 107(1):51-56.

Nilsen, E.T. and Y. Bao. 1990. The influence of water stress on stem and leaf photosynthesis in Glycine max and *Spartium junceum* (Leguminosae). American J. of Botany. 77(8):1007-15.

Nilsen, E.T. and D. Karpa. 1994. Seasonal acclimation of stem photosynthesis in two invasive, naturalized legumes from coastal habitats of California. Photosynthetica. 30:77-90.

Nilsen, E.T., D. Karpa, H.A. Mooney, and C. Field. 1993. Patterns of stem assimilation in two species of invasive legumes in coastal California. American J. of Botany. 80(10):1126-36.

North Dakota State University Extension Service, 1998. Weed-Pro 1998: North Dakota Agricultural Weed Control Guide. http://www.ext.nodak.edu/extnews/weedpro/.

Northcroft, E.G. 1927. The blackberry pest. I. Biology of the plant. New Zealand. J. Agriculture. 34:376-88.

Noss, R. 1990. Indicators for monitoring biodiversity: a hierarchical approach. Conservation Biology. 4(4):355-64.

O'Conner, B. 1996. Novachem Manual 1996: New Zealand Guide to Agrichemicals for Plant Protection. Novachem, Palmerston, New Zealand.

Odion, D., D.E. Hickson, and C.M. D'Antonio. 1992. Central Coast Maritime Chaparral on Vandenberg Air Force Base: an Inventory and Analysis of Management Needs for a Threatened Vegetation Type. The Nature Conservancy, San Francisco, CA.

Olaifa, J.I., F. Matsumura, J.A.D. Zeevaart, C.A. Mullin, and P. Charalambus. 1991. Lethal amounts of ricinine in green peach aphids (Myzuspersicae suzler) fed on castor bean plants. Plant Science. 73(2):253-56.

Olivieri, I. 1984. Effect of *Puccinia cardui-pycnocephali* on slender thistles (*Carduus pycnocephalus* and *C. tenuiflorus*). Weed Science. 32:508-10.

Olivieri, I. 1985. Comparative electrophoretic studies of *Carduus pycnocephalus, C. tenuiflorus*, and their hybrids. American J. Botany 72:715-18.

Olivieri, I., J. Swan, and P. Gouyon. 1983. Reproductive system and colonizing strategy of two species of *Carduus*. Oecologia. 60:114-17.

Olson, D.F. 1974. *Robinia pseudoacacia* L. locust. In: Schopmeyer, C. S. (ed.). Seeds of Woody Plants in the United States. Agricultural Handbook No. 450. US Dept. of Agriculture, Washington, DC.

Olson, D.F., Jr. 1974. *Elaeagnus* L. elaeagnus. In: Schopmeyer, C.S., technical coordinator. Seeds

of Woody Plants in the United States. Agricultural Handbook 450. US Department of Agriculture, Forest Service: 376-79. Washington, DC.

Olson, T.E. and F.L. Knopf. 1986. Naturalization of Russian-olive in the western United States. Western J. of Applied Forestry. 1:65-69.

Olson, W.W. 1975. Effects of Controlled Burning on Grassland within the Tewaukon National Wildlife Refuge. Masters thesis, North Dakota State University, Fargo, ND.

Orchard, A. 1981. A revision of South American *Myriophyllum* (Haloragaceae) and its repercussions on some Australian and North American species. Brunonia. 4:27-65.

Orr, B. and V. Resh. 1989. Experimental test of the influence of aquatic macrophyte cover on the survival of *Anopheles* larvae. J. of the American Mosquito Control Association. 5:579-85.

Ortho. 1977. All about groundcovers, western edition. Ortho Books, San Francisco, CA.

Panetta, F.D. 1993. A system of assessing proposed plant introductions for weed potential. Plant Protection Quarterly. 8:10-14.

Pannill, P. 1995. Tree-of-Heaven Control. Stewardship Bulletin, Maryland Department of Natural Resources, Hagerstown, MD.

Parish, S.B. 1920. The immigrant plants of southern California. Bulletin of Southern California Academy of Science. 19:3-30.

Parker, K.F. 1972. An Illustrated Guide to Arizona Weeds. University of Arizona Press, Tucson, AZ.

Parker, V.T. and R. Kersnar. 1989. Regeneration potential in French broom, *Cytisus monspessulanus*, and its possible management. Report to the Land Management Division of the Marin Municipal Water District, Corte Madera, CA.

Parsons, J.M. 1992. (ed.) Australian Weed Control Handbook. Inkata Press, Melbourne, Australia.

Parsons, W.T. 1973. Noxious weeds of Victoria. Inkata Press, Melbourne, Australia.

Parsons, W.T. 1977. The ecology and physiology of two species of *Carduus* as weeds in pastures in Victoria. Ph.D. thesis, University of Melbourne, Australia.

Parsons, W.T. 1992. Noxious Weeds of Australia. Inkata Press, Melbourne, Australia.

Parsons, W.T. and R.L. Amor. 1968. Comparison of herbicides and times of spraying for the control of blackberry (*Rubus fruticosus*). Australian J. Experimental Agriculture and Animal Husbandry. 8:238-43.

Parsons, W.T. and E. Cuthbertson. 1992. Noxious Weeds of Australia. Inkata Press, Melbourne, Australia. Pp. 204-08.

Partridge, T.R. 1989. Soil seedbanks of vegetation on the Port Hills and Banks Peninsula, Canterbury, New Zealand, and their role in succession. New Zealand J. of Botany. 27:421-36.

Pattens, T., California State Parks, Benicia, personal communication, 1998.

Pavlik, B.M. 1983a. Nutrient and productivity relations

of the dune grasses *Ammophila arenaria* and *Elymus mollis* 1. Blade photosynthesis and nitrogen use efficency in the laboratory and field. Oecologia. 57:227-37.

Pavlik, B.M. 1983b. Nutrient and productivity relations of the dune grasses *Ammophila arenaria* and *Elymus mollis* 2. Growth patterns of dry matter and nitrogen allocation as influenced by nitrogen supply. Oecologia. 57:233-38.

Pavlik, B.M. 1983c. Nutrient and productivity relations of the dune grasses *Ammophila arenaria* and *Elymus mollis* 3. Spatial aspects of clonal expansion with reference to rhizome growth and the dispersal of buds. Bulletin of the Torrey Botanical Club. 110:271-79.

Paynter, Q.E. 1997. European Work for the USA Broom Biocontrol Programme. Report June 1997. Institute of International Biological Control. CSIRO European laboratory, Montpellier, France.

Peck, M. E. 1993. A Manual of the Higher Plants of Oregon. 2nd ed. Binford and Morton, Portland, OR.

Peeper, T.F. 1984. Chemical and biological control of downy brome (*Bromus tectorum*) in wheat and alfalfa in North America. Weed Science. 32:Supplement 1:18-25.

Pemberton, R.W. 1986. The distribution of halogeton in North America. J. of Range Management. 39:281-82.

Penfound, W. and T. Earle. 1948. The biology of the water hyacinth. Ecological Monographs. 18:447-72.

Penrose, D. 1994. Artichoke fly, new state records, CPPDR July-September 1994. In: Proceedings, Thistle Management in California Conference 1995, San Luis Obispo. County Dept. of Agriculture, San Luis Obispo, CA.

Pepper, A. and M. Kelly. 1994. The Ecology and Management of the Wild Artichoke *Cynara cardunculus*. CalEPPC News. Winter 94:4-6.

Perdue, R.E. 1958. *Arundo donax*—source of musical reeds and industrial cellulose. Economic Botany. 12:368-404.

Perry, B. 1989. Trees and Shrubs for Dry California Landscapes. Land Design Publishing, Claremont, CA.

Perrine, P., California Department of Fish and Game, personal communication, 1997.

Peterlee, S. and Associates. 1990. An assessment of broom control methods. Marin County Open Space District, Fairfax, CA.

Pickart, A.J. 1988. Dune restoration in California: a beginning. Restoration and Management Notes. 6(1):8-12.

Pickart, A.J., L.M. Miller, and T.E. Duebendorfer. 1998a. Yellow bush lupine invasion in northern California coastal dunes: I. Ecology and manual restoration techniques. Restoration Ecology. 6: 59-68.

Pickart, A.J. and J.O. Sawyer. 1998. Removal of invasive plants from coastal dunes. Ecology and Restoration of Northern California Coastal Dunes, Chapter 4. California Native Plant Society, Sacramento, CA.

Pickart, A.J. and J.O. Sawyer. 1998. Ecology and restoration of northern California coastal dunes. California Native Plant Society, Sacramento, CA.

Pickart, A.J., K.C. Theiss, H.B. Stauffer, and G.T. Olsen. Yellow bush lupine invasion in northern California coastal dunes: II. Mechanical restoration techniques. Restoration Ecology. 6:69-74.

Pierce, S.M. and R.M. Cowling. 1991. Disturbance regimes as determinants of seedbanks in coastal dune vegetation of the southeastern Cape. J. of Vegetation Science. 2:403-12.

Pimm, S.L. and M.E. Gilpin. 1989. Theoretical issues in conservation biology. In: Roughgarden, J., R. May, and S.A. Levin (eds.). Perspectives in Ecological Theory. Princeton University Press, Princeton, NJ. Pp. 287-305.

Pitcairn, M.J. 1997a. Yellow starthistle control methods: biological control. In: Lovich, J.E., J. Randall, and M.D. Kelly (eds.). Proceedings of the California Exotic Pest Plant Council, Symposium 96. CalEPPC, San Diego, CA.

Pitcairn, M.J. 1997b. Biological control of wildland weeds. Fremontia, in press.

Pitcher, D. 1986. Element Stewardship Abstract for Poison Hemlock (*Conium maculatum*). The Nature Conservancy, San Francisco, CA.

Pitcher, D. and M. Russo. 1988. *Carduus pynocephalus*. Element Stewardship Abstract. The Nature Conservancy, California Field Office, San Francisco, CA.

Plant Advisor. 1997. Southwestern Deserts Edition. http://www.plantadvisor.com/plants/atrisemi.htm

Plass, W.T. 1975. An Evaluation of Trees and Shrubs for Planting Surface Mine Spoils. Research Paper 317. USDA Forest Service, Northeast Forest Experimental Station, Washington, DC.

Pollak, O. and T. Kan. In press. The use of prescribed fire to control invasive exotic weeds at Jepson Prairie Preserve. June 1996 Vernal Pool Conference Symposium, University of California, Davis, CA.

Polunin, O. 1969. Flowers of Europe: a Field Guide. Oxford University Press, London, UK.

Poole, A.L. 1990. Trees and Shrubs of New Zealand. DSIR Publishing, Auckland, NZ.

Poole, A.L. and D. Cairns. 1940. Botanical aspects of ragwort (*Senecio jacobaea* L.) control. New Zealand Department of Science and Industrial Research Bulletin 82:1-66. Government printer, Wellington, NZ.

Popenoe, J.H., US National Park Service, Redwood National Park, personal communication, February 24, 1999.

Powles, S.B., E.S. Tucker, and T.R. Morgan. 1989. A capeweed (*Arctotheca calendula*) biotype in Australia resistant to bipyridyl herbicides. Weed Science. 37:60-62.

Pryor, M. 1959. Hoary cress: new control findings. California Department of Agriculture Monthly Bulletin. 48(1):11-14.

Pryor, M.R. and R.H. Dana. 1952. Gorse control. California Department of Agriculture Bulletin. 41(1):43-45.

Pullman, G.D. 1992. The management of Eurasian watermilfoil in Michigan. Issues in Aquatic Management Series. Midwest Aquatic Plant Management Society.

Putz, F.E. and H.A. Mooney. 1991. The Biology of Vines. Cambridge University Press, Cambridge, MA.

Pyke, D.A. 1987. Demographic responses of *Bromus tectorum* and seedlings of *Agropyron spicatum* to grazing by small mammals: The influence of grazing frequency and plant age. J. of Ecology. 75:825-35.

Pylar, D.B. and T.E. Proseus. 1996. A comparison of the seed dormancy characteristics of *Spartina patens* and *Spartina alterniflora* (Poaceae). American J. of Botany. 83:11-14.

Randall, J.M. 1991. The ecology of an invasive biennial, *Cirsium vulgare*, in California. Dissertation, Ecology Graduate Group, University of California, Davis, CA.

Randall, J.M. 1996a. Weed control for the preservation of biological diversity. Weed Technology. 10:370-83.

Randall, J. M. 1996b. Assessment of the invasive weed problem on preserves across the United States. Endangered Species Update. 12:4-6.

Randall, J.M. and M. Rejmánek. 1993. Interference of bull thistle (*Cirsium vulgare*) with growth of ponderosa pine (*Pinus ponderosa*) seedlings in a forest plantation. Canadian J. of Forest Research. 23:1507-13.

Randall, J.M., M. Rejmánek, and J.C. Hunter. 1998. Characteristics of the exotic flora of California. Fremontia. 26(4):3-12.

Rapoport, E.H. 1982. Areagraphy: Geographical Strategies of Species. Pergamon Press, Oxford, UK.

Rawinski, T.J. 1982. The ecology and management of purple loosestrife in New York. Master's thesis, Cornell University, Ithaca, NY.

Rayner C.J. and D.F. Langidge. 1985. Amino acids in bee-collected pollens from Australian indigenous and exotic plants. Australian J. of Experimental Agriculture. 25:322-26.

Reader, R.J. 1991. Relationship between seedling emergence and species frequency on a gradient of ground cover density in an abandoned pasture. Canadian J. of Botany. 69:1397-1404.

Reddy, K., M.P. Rolston, and W.R. Scott. 1996. A study of seed production in phalaris (*Phalaris aquatica* L.). J. of Applied Seed Production. 14:11-16.

Reed, P.B., Jr. 1988. National list of plant species that occur in wetlands: California (Region 0). Biological Report 88. US Fish & Wildlife Service, Washington, DC.

Rees, N.E, P.C. Quimby, G.L. Piper, E.M. Coombs, C.E. Turner, N.R. Spencer, and L.V. Knutson. 1996. Biological Control of Weeds in the West. Western Society of Weed Science, Bozeman, MO.

Reichard, S.H. 1997. Prevention of invasive plant introductions on national and local levels. In: Luken, J. O. and J. W. Thieret (eds.). Assessment and Management of Plant Invasions. Springer, New York, NY.

Reichard, S.H. and C.W. Hamilton. 1997. Predicting invasions of woody plant introductions into North America. Conservation Biology. 11:193-203.

Reid, T., Thomas Reid Associates, personal communication, May 1995.

Rejmánek, M. 1995. What makes a species invasive? In: Pysek, P., K. Prach, M. Rejmánek, and P.M. Wade (eds.). Plant Invasions. SPB Academic Publishing, The Hague, The Netherlands.

Rejmánek, M. 1996. A theory of seed plant invasiveness: the first sketch. Biological Conservation. 78:171-81.

Rejmánek, M. and J.M. Randall. 1994. Invasive alien plants in California: 1993 summary and comparison with other areas in North America. Madroño. 41:161-77.

Rejmánek, M. and D.M. Richardson. 1996. What attributes make some plant species more invasive? Ecology. 77:1655-61.

Rezk, M.R. and T.Y. Edany. 1979. Comparative responses of two reed species to water table levels. Egyptian J. of Botany 22:157-72.

Richardson, J.M. 1953. Camelthorn (*Alhagi camelorum* Fisch.). J. of the Department of Agriculture South Australia. 57:18-20, 33.

Rissing, S.W. 1988. Dietary similarity and foraging range of two seed-harvester ants during resource fluctuations. Oecologia. 75:362-66.

Robacker, C.D. and W.L. Corley. 1992. Plant regeneration of pampas grass from immature inflorescences cultured in vitro. HortScience. 27:841-43.

Robbins, W.W. 1940. Alien plants growing without cultivation in California. Bulletin of the California Agricultural Experiment Station. 637:1-128.

Robbins, W.W., M.K. Bellue, and M.S. Ball. 1941. Weeds of California. California Department of Agriculture, Sacramento, CA.

Robbins, W.W., M.K. Bellue, and W.S. Ball. 1951. Weeds of California. California Department of Agriculture, Sacramento, CA.

Robbins, W.W., M.K. Bellue, and W.S. Ball. 1970. Weeds of California. California Department of Food & Agriculture, Sacramento, CA.

Robbins, W.W., M.K. Bellue, and W.S. Ball. 1974. Weeds of California, California Department of Agriculture, Sacramento, CA.

Roberts, H.A. 1979. Periodicity of seedling emergence and seed survival in some Umbelliferae. J. of Applied Ecology. 16:195-201.

Robertson, J.H. and C.K. Pearse. 1945. Range reseeding and the closed community. Northwest Science. 19:58-66.

Robinson, E.R. 1984. Naturalized species of *Cortaderia*

(Poaceae) in southern Africa. S. Afr. Tydskr. Plantk. 3:343-46.

Robinson, T.W. 1965. Introduction, spread and areal extent of saltcedar (*Tamarix*) in the western states. Geological Survey Professional Paper No. 491-A. Washington, DC.

Robison, M., University of California, Davis, unpublished data.

Roche, C. 1992. Slenderflower thistle, Italian thistle, plumeless thistle. Weeds. Pacific Northwest Extension, Portland, OR.

Roché, B.F., Jr. 1992. Achene dispersal in yellow starthistle (*Centaurea solstitialis* L.). Northwest Science. 66:62-65.

Roché, B.F., Jr., C.T. Roché, and R.C. Chapman. 1994. Impacts of grassland habitat on yellow starthistle (*Centaurea solstitialis* L.) invasion. Northwest Science. 68:86-96.

Roché, C.T. and B.F. Roché Jr. 1988. Distribution and amount of four knapweed (*Centaurea* L.) species in eastern Washington. Northwest Science. 62: 242-53.

Rogers, C.F. 1928. Canada Thistle and Russian Knapweed and Their Control. Colorado Agricultural Experiment Station Bulletin 348. Fort Collins, CO.

Rossiter, R.C. 1947. Studies on perennial veldt grass (*Ehrharta calycina* Sm.) CSIR Bulletin 227.

Saelinger, C.B. 1990. Trafficking of Bacterial Toxins. CRC Press, Boca Raton, FL.

Sala, A., S.D. Smith, and D.A. Devitt. 1996. Water use by *Tamarix ramosissima* and associated phreatophytes in a Mojave Desert floodplain. Ecological Applications. 6:888-98.

Sales F. 1994. A reassessment of the *Bromus madritensis* complex (Poaceae): a multivariate approach. Israel J. of Plant Sciences. 42:245-55.

Sallabanks, R. 1992. Fruit fate, frugivory, and fruit characteristics: a study of the hawthorn, *Crataegus monogyna* (Rosaceae). Oecologica. 91:296-304.

Sallabanks, R. 1993. Fruiting plant attractiveness to avian seed dispersers: native vs. invasive *Crataegus* in western Oregon. Madroño. 40:108-16.

Salmon, J.T. 1986. The Native Trees of New Zealand. DSIR Publishing, Auckland, NZ.

Sanders, A., California Native Plant Society, personal communication, March 1985.

Sanders, A., Department of Botany and Plant Sciences, University of California, Riverside, personal communication, August 19, 1997.

Sawyer, J.O. and T. Keeler-Wolf. 1995. A Manual of California Vegetation. California Native Plant Society, Sacramento, CA.

Sayce, K. 1988. Introduced cordgrass, *Spartina alterniflora* Loisel. in salt marshes and tidelands of Willapa Bay, Washington. FWSI-87058(TS). US Fish and Wildlife Service, Washington, DC.

Scheerer, M. and M.T. Jackson. 1989. Experimental use of herbicides to control black locust (*Robinia pseudoacacia* L.) populations. Natural Areas J. 9: 176.

Schittler, J. 1973. Introduction to *Vinca* alkaloids. In: Taylor, W. (ed.). The Vinca Alkaloids. Mariel Dekker Inc., New York, NY.

Schlesselman, J.T., G.L. Ritenour, and M.M. Hile. 1989. Cultural and physical control methods. In: Proceedings of the California Weed Conference on Principles of Weed Control in California, Davis, CA.

Schmidl, L. 1972. Biology and control of ragwort (*Senecio jacobaea* L.) in Victoria, Australia. Weed Research. 12:37-45.

Schmitz, D.C., D. Simberloff, R.H. Hofstetter, W. Haller, and D. Sutton. 1997. The ecological impact of nonindigenous plants. In: Simberloff, D., D.C. Schmitz, and T.C. Brown (eds.). Strangers in Paradise: Impact and Management of Nonindigenous Species in Florida. Island Press, Washington, DC. Pp. 39-61.

Schultz, A.M., J.L. Launchbaugh, and H.H. Biswell. 1955. Relationship between grass density and brush seedling survival. Ecology. 36:226-38.

Schwartz, M.W. and J.M. Randall. 1995. Valuing natural areas and controlling nonindigenous plants. Natural Areas J. 15:98-100.

Schwendiman, J.L. 1977. Coastal and sand dune stabilization in the Pacific Northwest. International J. Biometeorology. 21:281-89.

Scott, G.D. 1994. Fire threat from *Arundo donax*. In: Jackson, N. *et al. Arundo donax* workshop. California Exotic Pest Plant Council, San Diego, CA. Pp. 17-18.

Scott, J.M. and D.S. Wilcove. 1998. Improving the future for endangered species. Bioscience. 48(8): 579-80.

Scott, W.T. 1997. Grecian foxglove, *Digitalis lanata* Ehrh. Kansas Department of Agriculture, Division of Plant Health, Plant Protection Section.

Sebastian, J.R. and K.G. Beck. 1993. Halogeton control with metsulfuron, dicamba, picloram, and 2,4-D in Colorado rangeland. Western Society of Weed Science Research Progress Report. 1:29-30.

Semenza, R.J., J.A. Young, and R.A. Evans. 1978. Influence of light and temperature on the germination and seedbed ecology of common mullein (*Verbascum thapsus*). Weed Science. 26(6):577-81.

Serpa, L., letter to Terri Thomas, area manager for The Nature Conservancy Preserve, Tiburon, CA, 1989.

Shafroth, P.B., G.T. Auble, and M.L. Scott. 1995. Germination and establishment of the native plains cottonwood (*Populus deltoides* Marshall subsp. *monilifera*) and the exotic Russian-olive (*Elaeagnus angustifolia* L.). Conservation Biology. 9:1160-75.

Shafroth, P.B., J.M. Friedman, and L.S. Ischinger. 1995. Effects of salinity on establishment of *Populus fremontii* (cottonwood) and *Tamarix ramosissima* (saltcedar) in southwestern United States. Great Basin Naturalist. 55:58-65.

Shaw, K.A. and E. Bruzzese. 1979. The use of fosamine for control of two *Rubus* species. In: Proceedings of the Seventh Asian-Pacific Weed Science Society Conference. Pp. 189-92.

Sheldon, J.C. and F.M. Burrows. 1973. The dispersal effectiveness of the achene-pappus units of selected Compositae in steady winds with convection. New Phytologist. 72:665-75.

Sheldon, S.P. and R.P. Creed. 1995. Use of a native insect as a biocontrol for an introduced weed. Ecological Application. 5(4):1122-32.

Shepard, A.W., J.P. Aeshlimann, J.L. Sagliocco, and J.Vitou. 1991. Natural enemies and population stability of the winter-annual *Carduus pycnocephalus* in Mediterranean Europe. Acta Ecologica. 12(6).

Shmalzer and Hinkle. 1987. Species biology and potential for controlling four exotic plants (*Ammophila arenaria, Carpobrotus edulis, Cortaderia jubata,* and *Gasoul crystallinum*) on Vandenberg Air Force Base, California. NASA technical memorandum 100980. Redwood City, CA.

Sholars, T., Biology Department, College of the Redwoods, Fort Bragg, California, personal communication, September 11, 1996.

Shreve, F. and I. Wiggins 1964. Vegetation and Flora of the Sonoran Desert. Stanford University Press, Palo Alto, CA.

Siccama, T.G., G. Weir, and K. Wallace. 1976. Ice damage in a mixed hardwood forest in Connecticut in relation to *Vitis* infestation. Bulletin of the Torrey Botanical Club. 103(4):180-83.

Sigg, J. 1996. *Ehrharta erecta*: sneak attack in the making? CalEPPC News. 4(3):8-9.

Silander, J.A. 1979. Microevolution and clone structure in *Spartina patens*. Science. 203:658-60.

Simpson, C.E. and E.C. Bashaw. 1969. Cytology and reproductive characteristics in *Pennisetum setaceum*. American J. of Botany. 56:31-36.

Sindel, B.M. 1991. A review of the ecology and control of thistles in Australia. Weed Research. 31:189-201.

Skinner, L.C., W.J. Rendall, and E.L. Fuge. 1994. Minnesota's purple loosestrife program: history, findings and management recommendations. Special Publication 145. Minnesota Department of Natural Resources, Minneapolis, MN.

Skinner, M.W. and B.M. Pavlik. 1994. Inventory of Rare and Endangered Vascular Plants of California. Special Publication No. 1. 5th ed. California Native Plant Society, Sacramento, CA.

Skolmen, R.G. 1983. Growth and yield of some eucalypts of interest in California. In: Proceedings of a Workshop on Eucalyptus in California. General Technical Report PSW-69. USDA, Pacific Southwest Forest and Range Experiment Station, Berkeley, CA.

Skolmen, R.G. and F.T. Ledig. 1990. *Eucalyptus globulus* Labill. bluegum eucalyptus. In: Burns, R.M. and B.H. Honkala (technical coordinators). Silvics of North America: Vol. 2, Hardwoods. Agriculture Handbook 654. US Department of Agriculture, Washington, DC.

Slobodchikoff, C.N. and J.T. Doyden. 1977. Effects of *Ammophila arenaria* on sand dune arthropod communities. Ecology. 58:1171-75.

Smith, C.W. and J.T. Tunison. 1992. Fire and alien plants in Hawai'i: research and management implications for native ecosystems. In: Stone, C.P., C.W. Smith, and J.T. Tunison (eds.). Alien Plant Invasions in Native Ecosystems of Hawaii: Management and Research, 1986. Hawaii Volcanoes National Park. University of Hawaii Press, Honolulu, HI.

Smith, G. and C. Wheelor. 1990. Flora of Vascular Plants of Mendocino County. University of San Francisco, San Francisco, CA.

South Australia Department of Agriculture. 1973. Weed Control Note: Hoary cress. Department of Agriculture of South Australia.

Spicher, D. and M. Josselyn. 1985. *Spartina* (Graminae) in northern California: distribution and taxonomic notes. Madroño. 32:158-67.

Stachon, W.J. and R. L Zimdahl. 1980. Allelopathic activity of Canada thistle (*Cirsium arvense*) in Colorado. Weed Science. 28:83-86.

Stanton, T.R. and E.G. Boerner. 1936. An interesting seed combination. Agronomy J. 28:329.

State Resources Agency. 1990. Annual report of the status of California state listed threatened and endangered species. California Department of Fish and Game, Sacramento, CA.

Stearn, W.T. 1973. Taxonomy and nomenclature of *Vinca*. In: Taylor, W. (ed.). The Vinca Alkaloids. Mariel Dekker Inc., New York, NY.

Stebbins, G.L. 1985. Polyploidy, hybridization, and the invasion of new habitats. Annals of the Missouri Botanical Garden. 72:824-32.

Stein, B.A. and S.R. Flack (eds.). 1996. America's Least Wanted: Alien Species. The Nature Conservancy, Arlington, VA.

Stelljes, M.E., R.B. Kelly, R.J. Molyneux, and J.N. Seiber. 1991. GC-MS determination of pyrrolizidine alkaloids in four *Senecio* species. J. of Natural Products. 54(3):759-73.

Stephens, H.A. 1973. Woody Plants of the North Central Plains. The University Press of Kansas, Lawrence, KS.

Stephenson, R.J. and J.E. Recheigl. 1991. Effects of dolomite and gypsum on weeds. Communications in Soil Science and Plant Analysis. 22:15-16, 1569-79.

Stoddart, L.A., H. Clegg, B.S. Markham, and G. Stewart. 1951. The halogeton problem on Utah rangelands. J. of Range Management. 4:223-27.

Strode, D.D. 1977. Black locust: *Robinia pseudoacacia* L. In: Woody Plants as Wildlife Food Species. SO-16. USDA, Forest Service, Southern Forest Experiment Station, Atlanta, GA. Pp. 215-16.

Strong, D.R., J.L. Maron, and P.G. Connors. 1995a. Top down from underground? The underappreciated influence of subterranean food webs on above-ground ecology. In: Polis, G.A. and K.O. Winemiller (eds.). Food Webs. Chapman and Hall, New York, NY. Pp. 170-75.

Strong, D.R., J.L. Maron, P.G. Connors, A. Whipple, S. Harrison, and R.L. Jefferies. 1995b. High mortality, fluctuation in numbers, and heavy subterranean insect herbivory in bush lupine, *Lupinus arboreus*. Oecologia. 104:85-92.

Stubbendieck, J., C.H. Butterfield, and T.R. Flessner. 1992. *Chrysanthemum leucanthemum* L. In: An Assessment of Exotic Plants of the Midwest Region. Final Report. Department of Agronomy, University of Nebraska, Lincoln, NB.

Stubbendieck, J., G.Y. Friisoe, and M.R. Bolick. 1994. Weeds of Nebraska and the Great Plains. Nebraska Department of Agriculture, Omaha, NB.

Sukasmanm, M. 1979. A short note on weed control experiments with Roundup in tea plantations (in Indonesian). Simposium herbisida Roundup-3, 1979, Medan, Indonesia.

Sunset. 1967. Western Garden Book. Lane Books, Menlo Park, CA.

Sunset. 1985. Western Garden Book. Sunset Publishing, Menlo Park, CA.

Sunset. 1996. Western Garden Book. Sunset Publishing, Menlo Park, CA.

Sutton, D. and T. Van. 1992. Growth potential of hydrilla. Aquatics. 14(4):6, 8, 10-11.

Sytsma, M. and L. Anderson. 1989. Parrotfeather impact and management. Proceedings of the 41st California Weed Conference. Pp. 137-46.

Talbot, M.W., H.H. Biswell, and A.L. Hormay. 1939. Fluctuations in the annual vegetation of California. Ecology. 20:394-402.

Tasmanian Department of Agriculture. Annual Report for 1976-77. No. 64, p.88. Hobart, Australia.

Tasmanian Department of Primary Industry and Fisheries. 1998. Capeweed. Service Sheet, 70/96. Agdex 642. Tasmania, Australia.

The Nature Conservancy. 1996. Control and management of giant reed (*Arundo donax*) and saltcedar (*Tamarix* spp.) in waters of the United States and wetlands. Report by The Nature Conservancy, Southern Calif. Projects Office, to US Army Corps of Engineers, Los Angeles.

The Nature Conservancy. 1997. Wildland Weeds Management & Research Weed Report, *Vinca major* L., Periwinkle.

Theiss, Karen and Associates. 1994. Methods for Removal of *Carpobrotus edulis* from coastal dunes, Humboldt County. Report prepared for The Nature Conservancy, Lanphere-Christensen Dunes Preserve, Arcata, CA.

Thill, D.C., K.G. Beck, and R.H. Callihan. 1984. The biology of downy brome (*Bromus tectorum*). Weed Science. 32:Supplement 1:7-12.

Thomas, L. and L. Anderson. 1984. Water hyacinth control in California. Aquatics. 6(2):11-15.

Thomas, L.K. 1980. The impact of three exotic plant species on Potomac Island. National Park Service Monograph Series. Number 13. Washington, DC.

Thomsen, C.D., G.D. Barbe, W.A. Williams, and M.R. George. 1986. Escaped artichokes are troublesome pests, California Agriculture. March-April (198):7-9.

Thomsen, C.D., W.A. Williams, W. Olkowski, and D.W. Pratt. 1996. Grazing, mowing and clover plantings control yellow starthistle. The IPM Practitioner. 18:1-4.

Thomsen, C.D., W.A. Williams, M. Vayssieres, F.L. Bell, and M.R. George. 1993. Controlled grazing on annual grassland decreases yellow starthistle. California Agriculture. 47(6):36-40.

Thomsen, C.D., W.A. Williams, M.P. Vayssieres, C.E. Turner, and W.T. Lanini. 1996. Yellow Starthistle Biology and Control. University of California, Division of Agriculture and Natural Resources, Publication 21541.333.

Tinnin, R.O. and C.H. Muller. 1972. The allelopathic potential of *Avena fatua*: the allelopathic mechanism. Bulletin of the Torrey Botanical Club. 99(6):287-92.

Tomlinson, P.B. 1980. The Biology of Trees Native to Tropical Florida. Harvard University Printing Office, Allston, MA.

Tothill, J.C. 1962. Autecological studies on *Ehrharta calycina*. Ph.D. dissertation, University of California, Davis.

Tothill, J.C. and J.B. Hacker. 1983. The Grasses of Southern Queensland. University of Queensland Press, St. Lucia, Australia.

Townsend, C.E. and E. Guest (eds.). 1980. Flora of Iraq. Vol. 4, Part 2, pp. 848-49. Ministry of Agriculture and Agrarian Reform, Iraq.

Townsend, C.H. 1896. The transplanting of eastern oysters to Willapa Bay, Washington, with notes on the native oyster industry. US Commissioner of Fish and Fisheries for 1895, Bureau of Fisheries Document 30. Pp. 193-202. Washington, DC.

Tracy, J.L. and C.J. DeLoach. 1999. Suitability of classical biological control for giant reed (*Arundo donax*) in the United States. In: Bell, C.R. (ed.). Arundo and Saltcedar: the Deadly Duo. Proceedings of the Arundo and Saltcedar Workshop, June 17, 1998, Ontario, CA. UC Cooperative Extension, Holtville, CA. Pp. 73-109.

Tremolieres, M., R. Carbiener, A. Exinger, and J.C. Tulot. 1988. Un exemple d'interaction non competitive entre especes ligneuses: le cas du lierre arborescent (*Hedera helix* L.) dans la foret alluvial. Acta Oecologia, Oecologia Plantarum. 9: 187-205.

Trumble, J.T. and L.T. Kok. 1982. Integrated pest management techniques in thistle suppression in pastures of North America. Weed Research. 22: 345-59.

Trumbo, J. 1994. Perennial pepperweed: a threat to wildland areas. California Exotic Pest Plant Council News. 2(3):4-5.

Tunison, J.T. 1992. Fountain grass control in Hawaii Volcanoes National Park: management considerations and strategies. In: Stone, C.P., C.W.

Smith, and J.T. Tunison (eds.). Alien Plant Invasions in Native Ecosystems of Hawaii: Management and Research, 1986. Hawaii Volcanoes National Park. Distributed by University of Hawaii Press, Honolulu, HI.

Tunison, J.T., N.G. Zimmer, M.R. Gates, and R.M. Mattos. 1994. Fountain grass control in Hawai'i Volcanoes National Park, 1985-1992. Cooperative National Park Resources Studies Unit, University of Hawai'i at Manoa. Technical Report 91. Manoa, HI.

Turner, S., P. Fay, E. Sharp, B. Sallee, and D. Sands. 1980. The susceptibility of Canada thistle (*Cirsium arvense*) ecotypes to a rust pathogen (*Puccinia obtegens*). Proceedings of the Western Society of Weed Science. 33:110-11.

Tutin, T.G., V.H. Heywood, N.A. Burges, D.H. Valentine, S.M. Walters, and D.A. Webb (eds.). 1964. Flora Europaea. Cambridge University Press, Cambridge, UK.

Tutin, T.G., V.H. Heywood, N.A. Burges, D.M. Moore, D.H. Valentine, S.M.Walters, and D.A. Webb (eds.). 1976. Flora Europaea, Volume 3. Cambridge University Press, Cambridge, UK.

Tutin, T.G., V.H. Heywood, N.A. Burges, D.M. Moore, D.H. Valentine, S.M. Walters, and D.A. Webb (eds.). 1976. Flora Europaea, Volume 4. Cambridge University Press, Cambridge, UK.

Tutin, T.G., V.H. Heywood, N.A. Burgess, D.M. Moore, D.H. Valentine, S.M. Walters, and D.A. Webb. 1980. Flora Europaea. Cambridge University Press, Cambridge, UK.

Tworkoski, T. 1992. Developmental and environmental effects on assimilate partitioning in Canada thistle (*Cirsium arvense*). Weed Science. 40:79-85.

US Air Force. 1996. Peacekeeper Rail Garrison and Small ICBM Mitigation Program, San Antonio Terrace, Vandenberg Air Force Base, California. Final report on the successful creation of wetlands and restoration of uplands. Unpublished report prepared by the Earth Technology Corporation, SAIC, and FLx for Department of the Air Force, Detachment 10, Space and Missile Systems Center, San Bernardino, CA.

US Congress, Office of Technology Assessment. 1993. Harmful Non-Indigenous Species in the United States, OTA-F-565, US Goverment Printing Office, Washington, DC.

US Department of Agriculture. 1970. Selected Weeds of the United States. Agriculture Handbook No. 366, USDA-ARS. Washington, DC.

US Department of Agriculture. 1984. Pesticide background statements. Vol. I. Herbicides. Agricultural Handbook No. 633, US Government Printing Office, Washington, DC.

US Department of Agriculture and US Department of Interior 1992. Guidelines for Coordinated Management of Noxious Weeds in the Greater Yellowstone Area. USDA, Washington, DC.

US Geological Survey. 1999. An assessment of exotic plant species of Rocky Mountain National Park.

Homepage of the Northern Prairie Wildlife Research Center.

Upadhyaya, M.K., R. Turkington, and D. McIlvride. 1986. The biology of Canadian weeds. 75. *Bromus tectorum*. Canadian J. of Plant Science. 66:689-709.

Uphof, J.C. 1942. Ecological relations of plants with ants and termites. Botanical Review. 8:563-98.

Uphof, J.C. 1968. Dictionary of Economic Plants. Verlag and J. Cramer, Lehre, Germany.

Ussery, J.G. and P.G. Krannitz. 1998. Control of Scot's broom (*Cytisus scoparius* (L.) Link.): the relative conservation merits of pulling versus cutting. Northwest Science. 72(4):268-73.

Valido, A. and M. Nogales. 1994. Frugivory and seed dispersal by the lizard *Gallotia galloti* (Lacertidae) in a xeric habitat of the Canary Islands. Oikos. 70:403-11.

van der Meijden, E. 1971. *Senecio* and *Tyria* (Callimorpha) in a Dutch dune area: a study on interaction between a monophagous consumer and a host plant. Proceedings of the Advanced Study Institute for Dynamic Numbers in Populations. Oosterbeck, The Netherlands.

van der Meijden, E. and R.E. van der Waals-kooi. 1979. The population ecology of *Senecio jacobaea* in the Netherlands sand dune system. I. Reproductive strategy and the biennial habit. J. of Ecology. 67: 131-54.

van der Walt, J.L. 1955. The camel-thorn bush. Farming in South Africa. 30:401-03.

Van der Westhuizen, F.G. and J.G. Joubert. 1983. The effect of cutting during anthesis on carbon dioxide absorption and carbohydrate contents of *Ehrharta calycina* and *Osteospermum sinuatum*. Handl. Weidingsveren. South Africa. 18:106-12.

Van Hook, S.S. 1983. A study of European beachgrass, *Ammophila arenaria* (L.) Link: Control methods and a management plan for the Lanphere-Christensen Dunes Preserve. Unpublished report. The Nature Conservancy, Lanphere-Christensen Dunes Preserve, Arcata, CA.

Van Staden, J. and P.H. Bennett, 1990. Effect of galling on assimilate partitioning in Crofton weed (*Ageratina adenophora*). S. Afr.Tydskr. Plantk. 57:2.

Van, T. and K. Steward. 1990. Longevity of monoecious hydrilla propagules. J. Aquatic Plant Management. 28:74-76.

Vartanian, V., The Nature Conservancy, Newport Beach, CA, personal communication, 1998.

Vermont Department of Environmental Conservation, Lakes and Ponds Unit, Eurasian Watermilfoil Control Program. 1997. A threat to Vermont's lakes: Eurasian watermilfoil, a non-native aquatic plant. Waterbury, VT.

Vila, M. and C.M. D'Antonio. 1998. Fruit choice and seed dispersal of invasive versus noninvasive *Carpobrotus edulis* (Aizoaceae) in coastal California. Ecology. 79:1053-60.

Villegas, B. and E. Coombs. 1999. Releases of the bull thistle gall fly, *Urophora stylata*, on bull thistle in

California. In: Woods, D.M. (ed.). Biological Control Program Annual Summary, 1998. California Department of Food and Agriculture, Plant Health and Pest Prevention Services, Sacramento, CA. P. 49.

Vines, R.A. 1960. Trees, Shrubs, and Woody Vines of the Southwest. University of Texas Press, Austin, TX.

Vitousek, P.M. 1986. Biological invasions and ecosystem properties: can species make a difference? In: H.A. Mooney and J.A. Drake (eds.). Ecology of Biological Invasions of North America and Hawaii, Springer-Verlag, NY. Pp. 163-76.

Vitousek, P.M., C.M. D'Antonio, L.L. Loope, and R. Westbrooks. 1996. Biological invasion as global environmental change. American Scientist. 84: 468-78.

Vitousek, P.M. and L.R. Walker. 1989. Biological invasion by Myrica faya in Hawaii: plant demography, nitrogen fixation, ecosystem effects. Ecological Monographs. 59:247-65.

Vitousek, P.M., L.R. Walker, L.D. Whiteaker, D. Mueller-Dumbois, and P.A. Matson. 1987. Biological invasion by Myrica faya alters ecosystem development in Hawaii. Science. 238:802-04.

Vivrette, N. and C.H. Muller, 1977. Mechanisms of invasion and domination of a coastal grassland by Mesembryanthemum crystallinum. Ecological Monographs. 47:301-18.

Vogel, W.G. 1981. A Guide for Revegetating Coal Mine Spoils in the Eastern United States. Gen. Tech. Rep. NE-68. US Dept. of Agriculture, Forest Service, Northeastern Forest Experiment Station, Broomall, PA.

Voigt, G.K. and F. Mergen. 1962. Seasonal variation in toxicity of ailanthus leaves to pine seedlings. Botanical Gazette. 123:262-66.

Wallander, R.T., B.E. Olson, P.K. Fay, and K. Olson-Rutz. 1991. The effects of intensive grazing on ox-eye daisy. Proceedings: Western Society of Weed Science. Pp. 91-94.

Wallen, B. 1980. Changes in structure and function of Ammophila during primary succession. Oikos 34: 227-38.

Warren, D.K. and R.M. Turner. 1975. Saltcedar (Tamarix chinensis) seed production, seedling establishment, and response to inundation. J. of the Arizona Academy of Science. 10:135-44.

Warren, L. 1996. Green grow the artichokes. San Diego Union-Tribune, April 21, 1996.

Washburn, J. and G. Frankie. 1985. Biological studies of iceplant scales Pulvinariella mesembryanthemi and P. delottoi: (Homoptera; Coccidae) in California. Hilgardia. 53:1-27.

Washington Water Quality Program. 1998. Washington State Department of Ecology's web site: www.wa.gov/ecology/wa/plants/index.html.

Watson, S. 1880. Geological Survey of California: Botany, Volume II. Welch Bigelow, Cambridge, MA.

Watson, H.K. 1977. Present weed control and projections for the year 2001. Unpublished manuscript on file at The Nature Conservancy, Western Regional Office, San Francisco, CA.

Wear, K.S. 1995. Hybrid lupine (Lupinus arboreus x L. littoralis) on the Samoa Peninsula. Unpublished document. The Nature Conservancy, Arcata, CA.

Wear, K.S. 1998. Hybridization between native and introduced Lupinus in Humboldt County. Master's Thesis, Humboldt State University, Arcata, CA.

Wear, K.S., The Nature Conservancy, Arcata, CA, personal communication, November 12, 1996.

Welsh, S.L., N.D. Atwood, S. Goodrich, and L.C. Higgins (eds.). 1987. A Utah Flora. Great Basin Naturalist Memoir No. 9. Brigham Young University, Provo, UT.

Westerdahl, H. and K. Getsinger (eds.). 1988. Aquatic Plant Identification and Herbicide Use Guide; Vol. II: Aquatic Plants and Susceptibility to Herbicides. Technical Report. A-88-9. US Army Engineers Waterways Experimental Station, Vicksburg, MS.

Westman, W. 1990. Park management of exotic plant species: problems and issues. Conservation Biology. 4(3):251-60.

Wheatley, W.M. and I.J. Collett. 1981. Winning the thistle war. Agricultural Gazette of New South Wales. 92:25-28.

Wheeler, D.J., S.W. Jacobs, and B.E. Norton. 1982. Grasses of New South Wales, University of New England, Armidale, Australia.

Wheelor, C.T., O.T. Helgorsen, D.A. Perry, and J.C. Gorde. 1988. Nitrogen fixation and biomass accumulation in plant communities dominated by Cytisus scoparius L. in Oregon and Scotland. J. of Appied Ecology. 24:231-37.

Whisenant, S.G. 1990. Changing fire frequencies on Idaho's Snake River Plains: ecological and management implications. In: McArthur, E.D., E.V. Romney, S.D. Smith, and P.T. Tueller (eds.). Proceedings: Symposium on Cheatgrass Invasion, Shrub Die-off, and Other Aspects of Shrub Biology and Management, Las Vegas, NV. USDA Forest Service Intermountain Research Station General Technical Report INT-276. Washington, DC. Pp. 4-10.

Whisenant, S.G. and D.W. Uresk. 1990. Spring burning Japanese brome in a western wheatgrass community. J. of Range Management. 43:205-08.

White, D.E. 1979. Physiological adaptations in two ecotypes of Canada thistle, Cirsium arvense (L.) Scop. Masters thesis, University of California, Davis, CA.

White, D.J., E. Haber, and C. Keddy. 1993. Invasive Plants of Natural Habitats in Canada. Canadian Wildlife Service, Environment Canada and Canadian Museum of Nature, Ottawa, Ontario.

White, R.E. 1990. Lasioderma haemorrhoidale now established in California, with biological data on Lasioderma species (Coleoptera: Anobiidae). The Coleopterist's Bulletin. 44(3):244-348.

Whitlow, T.H. 1980. Voucher specimen of *Spartina patens* collected in Benicia State Park, just east of Carquinez Straits, South Hampton Bay, Solano County. Davis Herbarium, University of California, Davis, CA.

Whitson, T.D., (ed.). 1990. Weeds of the West. Pioneer of Jackson Hole, Jackson, WY.

Whitson, T.D., L.C. Buril, S.A. Dewey, D.W. Cudney, B.E. Nelson, R.D. Lee, and R. Parker. 1991. Weeds of the West. Western Society of Weed Science, Jackson, WY.

Whitson, T.D., M.A. Ferrell, and S.D. Miller. 1987. Purple starthistle (*Centaurea calcitrapa*) control within perennial grass species. 1987. Research Progress Report, 40th Meeting of the Western Society of Weed Science Annual Meeting, Boise, Idaho, March 1987.

Whitson, T.D., R.D. Williams, R. Parker, D.G. Swan, and S. Dewey. 1985. Pacific North West Weed Control Handbook: January 1985. Extension Services, Oregon State University, Corvallis, OR.

Wiedemann, A.M. and A. Pickart. 1996. The *Ammophila* problem on the northwest coast of North America. Landscape and Urban Planning. 34:287-99.

Wijte, A.H. and J.L. Gallagher. 1992. The importance of dead and young live shoots of *Spartina alterniflora* (Poaceae) in a mid-latitude salt marsh for overwintering and recoverability of underground reserves. Botanical Gazette. 152:509-13.

Wilken, D.H. and E.L. Painter. 1993. *Bromus*. Hickman, J.C. (ed.). The Jepson Manual: Higher Plants of California. University of California Press, Berkeley, CA.

William, R.D., D. Ball, T.L. Miller, R. Parker, K. Al-Khatib, R.H. Callihan, C. Eberlein, and D.W. Morishita (eds.). 1996. Pacific Northwest 1996 Weed Control Handbook. Extension Service of Oregon State University, Washington State University, and University of Idaho. Oregon State University, Corvallis, OR.

Williams, D.G. and R.A. Black. 1993. Phenotypic variation in contrasting temperature environments: growth and photosynthesis in *Pennisetum setaceum* from different altitudes on Hawaii. Functional Ecology. 7:623-33.

Williams, D.G. and R.A. Black. 1994. Drought response of a native and introduced Hawaiian grass. Oecologia. 97:512-19.

Williams, D.G., R.N. Mack, and R.A. Black. 1995. Ecophysiology of introduced *Pennisetum setaceum* on Hawaii: the role of phenotypic plasticity. Ecology. 76:1569-80.

Williams, N. 1991. Why I killed trees on Earth Day. Boulder County Parks and Open Space Department, Boulder, CO.

Williams, R.D. and S.H. Hanks. 1976. Hardwood Nurseryman's Guide. Agricultural Handbook 473. US Department of Agriculture, Forest Service, Washington, DC.

Willis, J.C. 1973. A Dictionary of the Flowering Plants and Ferns, 8th ed.

Willoughby, J.W. and W. Davilla. 1984. Plant species composition and life form spectra of tidal streambanks and adjacent riparian woodlands along the lower Sacramento River. In: Warner, R.E. and K.M. Hendrix (eds.). California Riparian Systems: Ecology, Conservation, and Productive Management: Proceedings of a Conference, 1981 September 17-19, Davis, CA. University of California Press, Berkeley, CA.

Woodall, S.L. 1982. Herbicide tests for control of Brazilian-pepper and *Melaleuca* in Florida. USDA Forest Service Research Note SE 314. Southeastern Forest Experiment Station, Asheville, NC.

Woods, D.M. and V. Popescu. 1997. *Ascochyta* seedling disease of yellow starthistle, *Centaurea solstitialis*: inoculation techniques. In: Woods, D.M. (ed.). Biological Control Program Annual Summary, 1996. California Department of Food and Agriculture, Division of Plant Industry, Sacramento, CA.

Woods, D.M., D.B. Joley, M.J. Pitcairn, and D. Griffin. 1995. Field testing of alternate hosts of *Bangasternus orientalis* (Capiomont). In: Bezark, L.G. (ed.). Biological Control Program Annual Summary, 1994. California Department of Food and Agriculture, Division of Plant Industry, Sacramento, CA.

Wotring, S.O., D.E. Palmquist, and J.A. Young. 1997. Perennial pepperweed rooting characteristics. In: Management of Perennial Pepperweed (tall whitetop). Special Report 972. USDA, Agricultural Research Service and Agricultural Experimental Station, Oregon State University, Corvallis, OR.

Wyman, D. 1954. Vines for winter beauty. Plants and Gardens. 10(1):46-50.

Wyman, D. 1969. Shrubs and Vines for American Gardens. MacMillan, London, UK.

Young, F.L., D.R. Gealy, and L.A. Morrow. 1984. Effect of herbicides on germination and growth of four grass weeds. Weed Science. 32:489-93.

Young, J.A. 1992. Ecology and management of medusahead (*Taeniatherum caput-medusae* ssp. *asperum* [Simk.] Melderis). Great Basin Naturalist. 52(3):245-52.

Young, J.A. and R.A. Evans. 1978. Population dynamics after wildfires in sagebrush grasslands. J. Range Management. 31:283-89.

Young, J.A. and C.E. Turner. 1995. *Lepidium latifolium* L. in California. California Exotic Pest Plant Council News. 3(1):4-5.

Young, J.A., R.A. Evans, and R.E. Eckert, Jr. 1969. Population dynamics of downy brome. Weed Science. 17:20-26.

Young, J.A., R.A. Evans, R.E. Eckert, Jr., and B.L. Kay. 1987. Cheatgrass. Rangelands. 9:266-76.

Young, J.A., R.A. Evans, and B.L. Kay. 1971. Germination of caryopses of annual grasses in simulated litter. Agronomy J. 63:551-55.

Young, J.A., R.A. Evans, and J. Major. 1972. Alien plants in the Great Basin. J. Range Management. 25: 194-201.

Young, J.A., D.E. Palmquist, and R.R. Blank. 1997. Herbicidal control of perennial pepperweed (*Lepidium latifolium*) in Nevada. In: Management of Perennial Pepperweed (tall whitetop). Special Report 972. USDA, Agricultural Research Service and Agricultural Experimental Station, Oregon State University, Corvallis, OR.

Young, J.A., D.E. Palmquist, and R.R. Blank. In press. The ecology and control of perennial pepperweed (*Lepidium latifolium*). Weed Science.

Young, J.A., C.E. Turner, and L.F. James. 1995. Perennial pepperweed. Rangelands. 17(4):121-23.

Young, J.A., range scientist, US Department of Agriculture, Agricultural Research Service, Reno, NV, personal communication.

Zabkiewicz, J.A. 1976. The ecology of gorse and its relevance to New Zealand forestry. In: Chavasse, C.G.R. (ed.). The Use of Herbicides in Forestry in New Zealand. New Zealand Forestry Service, Forestry Research Institute Symposium. 18:63-70.

Zabkiewicz, J.A. and R.E. Gaskin. 1978. Effect of fire on gorse seeds. Proceedings of the 31st New Zealand Weed and Pest Control Conference. New Zealand. Pp. 47-52.

Zamora, D.L. 1993. Effect of herbicides on common mullein. Western Society of Weed Science Research Progress Report, I-63.

Zedler, P.H., C.R. Gautier, and G.S. McMaster. 1983. Vegetation change in response to extreme events: the effects of a short interval between fires in California chaparral and coastal scrub. Ecology. 64:809-18.

Zimmerman, P., San Francisco State University, unpublished data.

Zohary, M. 1966. Flora Palaestina. Part 1, Equisetaceae to Moringaceae, Jerusalem.

Zuniga, G.E., V.H. Argandona, H.M. Niemeyer, and L.J. Corcuera. 1983. Hydroxamic content in wild and cultivated Gramineae. Phytochemistry. 22(12): 2665-68.

Zwolfer, H. and P. Harris. 1984. Biology and host specificity of *Rhinocyllus conicus* (Froel.) (Col. Curculionidae), a successful agent for biocontrol of the thistle, *Carduus nutans* L. Zeitschrift fur Angewandte Entomologie. 97:36-62.

Resources

CalEPPC contact info on the web at http://www.caleppc.org

Team Arundo: Mark Norton, 909-785-5411.
Team Arundo del Norte: Tom Dudley, 510-643-3021.
Team Arundo can be contacted by e-mail at Team_Arundo@CERES.CA.GOV and visit the website at http://CERES.CA.GOV/tadn/

The Weed Wrench is available from New Tribe, 541-476-9492.

The Root Jack is available from Mike Giacomini, 415-454-0849.

If you have questions about any potential biocontrol agents, contact the CDFA Biological Control Program, 916-262-2048.

Detailed information on herbicides is available in the Weed Science Society of America's *Herbicide Handbook* (Ahrens 1994) and *Supplement* (Hatzios 1998). This publication gives information on nomenclature, chemical and physical properties, uses and modes of action, precautions, physiological and biochemical behavior, behavior in or on soils, and toxicological properties for several hundred chemicals (WSSA, Allen Press, P.O. Box 1849, 810 E. 10th St., Lawrence, KS 66044, or phone 800-627-0629).

Critical reviews of several common herbicides are available at a small charge from the Northwest Coalition for Alternatives to Pesticides (NCAP, P.O. Box 1393, Eugene, OR 97440, or phone 541-344-5044, or info@pesticide.org, http://www.pesticide.org./default.htm).